Processes of Life

Processes of Life

Essays in the Philosophy of Biology

John Dupré

OXFORD
UNIVERSITY PRESS

OXFORD
UNIVERSITY PRESS

Great Clarendon Street, Oxford OX2 6DP

Oxford University Press is a department of the University of Oxford.
It furthers the University's objective of excellence in research, scholarship,
and education by publishing worldwide in

Oxford New York

Auckland Cape Town Dar es Salaam Hong Kong Karachi
Kuala Lumpur Madrid Melbourne Mexico City Nairobi
New Delhi Shanghai Taipei Toronto

With offices in

Argentina Austria Brazil Chile Czech Republic France Greece
Guatemala Hungary Italy Japan Poland Portugal Singapore
South Korea Switzerland Thailand Turkey Ukraine Vietnam

Oxford is a registered trade mark of Oxford University Press
in the UK and in certain other countries

Published in the United States
by Oxford University Press Inc., New York

© in this volume John Dupré 2012

British Library Cataloguing in Publication Data
Data available

Library of Congress Cataloging in Publication Data
Library of Congress Control Number: 2011944132

Typeset by SPI Publisher Services, Pondicherry, India
Printed in Great Britain
on acid-free paper by
MPG Books Group, Bodmin and King's Lynn

ISBN 978-0-19-969198-2

1 3 5 7 9 10 8 6 4 2

Contents

IV. Humans

Preface

As with all academic work, my debts are more numerous than I can list. The essays in this collection were all written while I was Director of the ESRC Centre for Genomics in Society (Egenis), and I am deeply indebted to the ESRC (Economic and Social Research Council, UK) for this very generous support. I have also received major support from the Arts and Humanities Research Council (AHRC) and the Leverhulme Trust, both of which have made crucial contributions to making this work possible.

Directing a Research Centre has given me the great privilege of working day-to-day with a team of people with both shared interests and diverse expertise, something that I have found indispensable in trying to get to grips with the complexities of contemporary biology. A particular pleasure has been working with the original co-Directors of Egenis, Barry Barnes and Steve Hughes. Providing, respectively, the insights of a pioneer and leader in the sociology of science, and the insider knowledge of a scientist with a lifetime's experience at the forefront of genomics, these two colleagues and friends were essential to the intellectual environment and success of the centre. With Barry I also had the pleasure of co-authoring a book on genomics (Barnes and Dupré 2008), a project that taught me a lot about molecular biology, but also gave me a much better understanding of Barry's distinctive and important insights into science in general.

Another obvious debt is to Maureen O'Malley, with whom I co-authored three of the papers here reprinted. Among many talents, Maureen has a unique capacity for absorbing and synthesizing scientific ideas. She came to Egenis after three years in the laboratory of the leading microbiologist Ford Doolittle in Dalhousie (to whom I am also thereby indebted), and brought with her an extraordinary grasp of the state of several areas of current biology, including microbiology. Her collaboration made it possible for me to make something of a long-held but inchoate suspicion that attention to microbes could transform the philosophy of biology. I must also make special mention of Christine Hauskeller, a founder member of Egenis, now replacing Barry Barnes as co-Director. Among her many contributions, she has sometimes had occasion to remind me—in the face of ever more fascinating engagements with contemporary biology—that there is more to philosophy of science than science. Other present and past colleagues at Egenis from whom I have learned much include Jane Calvert, Paul Griffiths, Susan Kelly, Sabina Leonelli, Staffan Müller-Wille, and Paula Saukko.

Egenis has been fortunate to have had a stream of fine PhD students passing through and on to other, if not greater, things, and many of them have also left marks on my thinking. I would particularly mention Ann Barwich, Adam Bostanci, Jonathan Davies, Trijsje Franssen, Richard Holdsworth, Ingrid Holme, Pierre-Olivier Methot, Mila Petrova, Alex Powell, and Kai Wang. And I must apologize for not mentioning a number of other students with whom I did not work closely personally, but who contributed in many ways to the intellectual and social life of the centre.

Mentioning the staff and students of Egenis is the relatively easy part, though even here I fear that there may be egregious omissions. But we have also been fortunate to be able to host a good number of conferences and workshops, visitors over various periods of time, and an almost weekly series of seminar speakers. On top of this, I have had the opportunity to present parts of this work at conferences, seminars, and lecture series in many parts of the world. Consequently I have benefited from interactions with far more people than I could possibly recall to give proper thanks. I started to write a list, but after the first dozen or so names, I realized the task was hopeless; I have been in the academic world too long to count my debts or, more happily, to list my friendships. But thanks to all; I hope you know who you are. Some specific debts are mentioned in footnotes to individual chapters.

A major determinant of whether life is a pleasure or a burden as a Centre Director is the administrative staff, and here I have been very fortunate. Very special thanks go to Cheryl Sutton, the perfect research administrator. Cheryl effectively ran the Centre for seven years, shielding me from vast quantities of administrative routine and annoyance and thereby doing more than anyone to allow me to write. Her successor, Sue Harding, has an impossible act to follow, but has made a good start. Egenis and I have also been splendidly served by Annalisa Macnamara, Sarah Silverman, Saira Kidangan, and Laura Dobb; and by our communications officers, Ginny Russell (subsequently a PhD student and Research Fellow here) and Claire Packman.

Finally, as always, my greatest debt is to my partner, Regenia Gagnier. Not only has she had to put up with my occasional stress and cover for my domestic absences as various demands of research directing spilled into my home life, but she has remained my most reliable and insightful critic and commentator. That her work is not more cited in these essays reflects the depth of her involvement in it, which makes specific contributions impossible to identify. And special thanks to Alex Powell for his splendid work on the index. The book is dedicated to our collaboration on the processes of life, our sons Gabriel and Julian Gagnier Dupré.

Acknowledgements

Permission is gratefully acknowledged from the original publishers to reprint the following:

'The Miracle of Monism', from *Naturalism in Question*, ed. David MacArthur and Mario de Caro, Cambridge, MA: Harvard University Press, 2004, pp. 36–58.

'What's the Fuss about Social Constructivism?' *Episteme* 1, 2004: 73–85.

'The Inseparability of Science and Values', from *Value-Free Science: Ideals and Illusions*, ed. H. Kincaid, J. Dupré, and A. Wylie, New York: Oxford University Press, 2007, pp. 27–41.

The Constituents of Life, Amsterdam: Van Gorcum, 2008.

'Understanding Contemporary Genomics'. *Perspectives on Science* 12, 2004: 320–38.

'The Polygenomic Organism', from *Nature After the Genome*, ed. S. Parry and J. Dupré, Oxford: Blackwell, 2010, pp. 19–31.

'It is not Possible to Reduce Biological Explanations to Explanations in Chemistry and/or Physics', from *Contemporary Debates in Philosophy of Biology*, ed. R. Arp and F. J. Ayala, New York: John Wiley, 2010, pp. 32–47.

'Postgenomic Darwinism', from *Darwin*, ed. W. Brown and A. Fabian, Cambridge: Cambridge University Press, 2010, pp. 150–71.

'Size Doesn't Matter: Towards a More Inclusive Philosophy of Biology'. *Biology and Philosophy* 22, 2007: 155–91.

'Metagenomics and Biological Ontology'. *Studies in the History and Philosophy of the Biological and Biomedical Sciences* 38, 2007: 834–46.

'Varieties of Living Things: Life at the Intersection of Lineage and Metabolism'. *Philosophy and Theory in Biology* 1, 2009 (http://hdl.handle.net/2027/spo.6959004.0001.003).

'Emerging Sciences and New Conceptions of Disease: Or, Beyond the Monogenomic Differentiated Cell Lineage'. *European Journal for the Philosophy of Science* 1 (2011): 119–31.

'Against Maladaptationism: Or, What's Wrong with Evolutionary Psychology', from *Knowledge as Social Order: Rethinking the Sociology of Barry Barnes*, ed. M. Mazzotti, Aldershot: Ashgate, 2008, pp. 165–80.

'What Genes Are, and Why There Are No "Genes for Race"', from *Revisiting Race in a Genomic Age*, ed. Barbara A. Koenig, Sandra Soo-Jin Lee, and Sarah Richardson, New Brunswick, NJ: Rutgers University Press, 2008, pp. 39–55.

'Causality and Human Nature in the Social Sciences'. *Kölner Zeitschrift für Soziologie und Sozialpsychologie*, 50th Anniversary Special Edition: Controversies in Sociological Theory, 2010: 507–25.

Introduction

Our knowledge of the living world has surely grown further in the last fifty years than in the preceding millennia of human history. The generation of biological knowledge is increasing exponentially, diverging into multiple subdisciplines, and developing a bewildering array of specialized vocabularies. The Beijing Genomics Institute, currently the largest gene sequencing laboratory on Earth, has plans to sequence the gnomes of 1,000 plants and animals, and 10,000 microbes. Yet as we squirrel away these petabytes of genetic data, we are also adjusting to the realization that genetic data can tell us much less than we once imagined. As well as DNA sequence, scientists are now recording the 'epigenetic' modifications to the molecule that influence whether bits of sequence are transcribed into RNA. And they are also collecting libraries of RNA molecules, once thought of as merely an intermediary between DNA and proteins, the primary functional molecules in living cells, but now recognized as a whole new layer of cellular regulation. Far from the behaviour of a cell being 'programmed' in the DNA, it can now be seen to be jointly determined by a bewildering array of molecules and subcellular structures, many of these in turn being open to important influences from outside the cell. How to make sense of all these layers of complex interaction remains a vast challenge, so that while it is certainly true that we know far more than we once did, a good deal of that is knowledge of how little we know. As a philosopher of biology it is a challenge to achieve a degree of understanding of these almost bottomless molecular complexities sufficient even to talk sensibly enough about them to be taken seriously by practitioners. A combination of alertness, reliable informants, and good luck is needed to keep one sufficiently *au courant* to avoid quite basic mistakes and misunderstandings. However, genomics, which I construe as the successor science to genetics that takes account of these (and other) complexities, is a field of knowledge that the philosophy of biology can ignore only at its peril.

Of course, as a philosopher of biology one's primary goal is not to advance the science (though of course it's very nice if the occasional practitioner claims to find one's work useful); nor is it merely to explicate the science as a means to promoting public understanding. As a philosophical naturalist (see Chapter 2), one would like to draw— or at least support—philosophical conclusions from science. Life, after all, not least but not only human life, is a central object of philosophical interest and one that philosophers engaged with biology should aspire to have something to say about. And finally, philosophers should surely hope to deal with the discipline they study in a critical manner. We do not merely try to understand the science, but we stand ready to point out difficulties with the logic of scientific arguments, or ways in which different scientific findings appear to come into conflict, or uncritical acceptance of background assumptions that may import culturally local ideology into the supposedly objective claims of science.

These various philosophical desiderata are extremely difficult to achieve simultaneously, but the attempt has never been more important. The rapid changes in biology are not merely taking place unnoticed in the labs and conferences where scientists talk to one another, but they are of great interest to many other specialists and to a much wider public. Fundamental changes in biology are not only affecting our images of who we are, and how we came to be here, but also more practical questions of how we work and how our bodies sometimes malfunction, what changes are possible in the way we live or organize our societies, and what may be our future. To put the matter as banally as possible, if human life matters, then so does biology.

In this book I shall focus especially on three biological ideas that I think require assimilation and discussion far beyond the professional corridors of biological science. The first of these is an idea that has been more promulgated by philosophers than biologists; it is itself a reflection on biological ideas rather than an idea straight from the laboratory bench. This is what has come to be called Developmental Systems Theory (DST) or perhaps better, the developmental systems perspective (Oyama et al. 2001). DST, in my view, best encapsulates what is wrong with a common and naïve understanding of genetic inheritance, and equally important, of the evolutionary theories built upon that understanding. DST teaches—or reminds—us that there is much more than the passage of genes needed to transmit living form from one generation to the next. Even at the most microscopic level, parental organisms pass on entire cells, replete with a pharmacopeia of chemicals and minute but intricate structures to their offspring. Larger, more complex organisms pass on as well, and among much else, care, food, training, and knowledge. Not only does this simple observation show the inadequacy of the view that sees every feature of biological form somehow encoded in a single molecule, DNA, but it points us towards a fundamental fact about organisms: they are not properly understandable in terms of one set of properties, say those of the adult organism, but are ultimately processes. The life cycle is what is basic, the adult, or child, or egg is a particular time slice from this more basic reality. It is the robustness of the cyclical process that makes possible the recurrence of the various stages rather than the stability of the stages that makes possible the cycle. And crucially, evolution consists of sequences of overlapping and interacting life cycles rather than just a series of adult organisms.[1]

---

[1] A topic that I am aware is conspicuously absent from this book is evolutionary developmental biology, or evo-devo. There are influential theorists who believe that this is far more significant than DST and, moreover, has largely supplanted the latter (Gilbert 2003). My own view is that this misrepresents the relation

The second point I want is to stress the importance of a more esoteric topic, the science of epigenetics. One way of beginning to think about epigenetics is to realize that the genome, as much as the organism, is a process rather than a static thing. It is very common to represent genomes as strings of letters, As, Cs, Gs, and Ts, representing the chemicals, 'nucleotides', that are strung together to form DNA molecules.[2] It is often suggested, moreover, that this sequence provides the code, the blueprint, the recipe, or something of the sort, for the organism in which these molecules are found. This is entirely mistaken. The unique role of DNA in transmission of form has already been undermined by DST, but epigenetics points out further that the view of DNA that underlies these obsolete images is fundamentally wrong. A code, or a blueprint, must remain highly stable so that the message it encodes can be reliably extracted from it; and a sequence looks well suited to secure this stability. But while it is true and important that there is something for which sequence codes and for which, therefore, its stable sequence is important, this is at a far simpler level than the structure of a whole organism.[3] And genomes are much more than sequences of these canonical letters.

First, the appearance of stability is spuriously constructed by ignoring relatively transient changes in the nucleotides. Best known of these is the change of the nucleotide C, for Cytosine, to the related molecule 5-methyl-cytosine. This is a process called methylation, and it is a very important means by which chemical changes to the chromosome, often induced in functional ways by the cellular environment, alter the behaviour of the DNA molecule. In particular, methylation tends to suppress the expression of the sequence in which the methylated molecule occurs. In fact there are quite a number of other, less familiar, nucleotides that can occur in wild DNA, though none has been attributed comparable importance to 5-methyl-cytosine.

And second, genomes do not merely have sequence, they have shape. The DNA in a human genome would stretch to about 2m, whereas the diameter of the human cell is typically around 20μm; evidently this requires some careful folding. DNA is intricately spooled on special proteins called histones, but the tightness of this spooling, or 'condensation', can vary a great deal, and again in ways that are affected by chemical interactions with the histones. And as the DNA becomes more condensed it becomes difficult or impossible for it to be transcribed into RNAs, the process that initiates its activity in the cell. So in summary, the genomes in cells are highly dynamic entities,

between the two projects. DST, in my view, remains the perspective best suited to presenting philosophical issues in evolutionary theory. Evo-devo has some wonderful biological insights to its credit (see e.g. Carroll 2005), which I am sorry not to have found occasion to discuss. But I cannot see it as a threat to the perspective provided by DST.

[2] This representation of the genome, its uses, and its limitations are discussed in detail in Barnes and Dupré (2008).

[3] Specifically, DNA can code for the sequence of elements (amino acids) in proteins, but that is all (see Godfrey-Smith 2000).

constantly changing both in their physical conformation and in their chemical constitution. These changes affect their activity, and they are effected by interactions with the
wider environment. The genome is not a mere repository of biological form, but a full
participant in biological process. The nature of an organism, finally, is not inscribed in
its DNA but constantly recreated by a cyclical process of which the DNA is part.

The third and final biological point that informs these essays is an emphasis on the
fundamental importance of microbes and the distortions that have been induced in
biological thinking by their widespread neglect. In the words of Stephen Jay Gould,
'We live now in the Age of Bacteria. Our planet has always been in the Age of bacteria.
Bacteria are—and always have been—the dominant forms of life on Earth.'[4] It is not
merely that the vast majority of living things are microbes, though they are, or that
microbes are able to occupy a far greater diversity of environments than are complex
multicellular organisms, though they are. It is rather that the picture of microbes as
something primitive and distinct from the 'higher' organisms that have evolved in
more recent times is confused. Higher organisms did not evolve as separate from the
pre-existing microbes, but as parts of complex symbiotic systems that always included
microbes, usually in vast numbers. Indeed, the boundaries between ourselves and
microbes, I argue in later chapters, are not as clear as we might wish. Viewing the
biological world from the perspective of these dominant organisms can transform our
view of life.

I mentioned at the beginning of this introduction that one of the main challenges for
philosophy of biology is the need to bring distinct areas of biological knowledge into
contact. Microbiology provides a striking example of just such a challenge. The claim,
explored in detail in the chapters in part III of the book but anticipated in earlier
chapters, is that quite revolutionary findings from this domain have not been properly
assimilated in other central parts of biology and philosophy of biology; and that their
assimilation would transform aspects of fields as diverse as evolution, systematics,
ecology, and medicine. Taking an appropriately microbe-centred view of life,
I argue, poses problems for traditional views of the biological individual; it largely
undermines standard ideas about species and the Tree of Life; and it threatens
many traditional ideas about evolution and evolutionary relations between kinds of
organisms.

★ ★ ★

[4] See http://www.stephenjaygould.org/library/gould_bacteria.html. Nowadays it is common to distinguish unicellular organisms into three kinds: Bacteria and Archaea (or Eubacteria and Archaebacteria), which
until a few decades ago were both thought of as Bacteria, and the many and diverse single-celled organisms in
the domain Eukarya. Bacteria, Archaea, and Eukarya are the three most fundamental divisions in contemporary taxonomies of terrestrial life. Eukarya, distinguished by generally larger and more complex cell
architecture, and a separate nucleus in which the genome resides, are the class to which fungi, plants, and
animals also belong. The term 'microbe' is sometimes used to refer just to the first two groups, but may also
be construed to include the many unicellular Eukaryotes and even the non-cellular viruses. These distinctions
are not important here, but will be discussed in detail in later chapters.

These and other developments in the science of biology have radically changed our beliefs about life: what it is, how it came about, how it is divided into individuals and kinds, and much else. But the essays in this book are not just about the content of science, they are also about science itself, about which I also have some heterodox views though not, I like to think, as heterodox as they were when I first began to articulate them in the 1980s and 1990s. The first section of the book contains three chapters that lay out some of the broader views of science that inform the more narrowly focused essays that follow. And these chapters also provide some links between the more recent work presented here and the earlier views that were articulated in my 1993 monograph, *The Disorder of Things*, and my 2002 collection of essays, *Humans and Other Animals*.

Chapter 1, 'The Miracle of Monism', most clearly establishes this continuity in my thinking. The central thesis of *The Disorder of Things* was the defence of pluralism against a certain kind of monism, the latter often inferred from a certain interpretation of physicalism. In briefest summary, while I agree with most contemporary philosophers that there is no stuff but physical stuff, I take it to be equally important not to let this agreement conceal the fundamental diversity of the kinds of things which are composed of that stuff. This metaphysical pluralism is closely connected in my thinking with an epistemological or methodological pluralism: there is no unique method for investigating all the many different kinds of things there are in the world. The first half of this chapter summarizes some of my reasons for espousing this pluralistic view of science: the various projects of enquiry that fall under the general rubric of science share neither a methodology nor a subject matter. Ontologically, I claim, sciences need have nothing in common beyond an antipathy to the supernatural. There is nothing of interest to science beyond the physical world; but being part of the physical world gives us little general guidance about how we can most successfully investigate a range of phenomena.

On the epistemic side I do defend one overarching virtue: empiricism. By this I mean only that scientific knowledge must ultimately be to some extent answerable to some aspect of our experience. In terms of Francis Bacon's memorable epistemological entomology, we should shun the ways of the spider, the arch-rationalist whose theories are spun entirely from its own substance. The ant, a too radical empiricist, who merely collects material (empirical data) and stores it, may be little more useful; but it is to the bee, who both collects and processes material from the world, that we should look for inspiration. But one moral of empiricism, I claim, is the rejection of monism. Prima facie science is as diverse as the world it studies; and rejection of this prima facie diversity in favour of a spurious aspiration to unity is grounded in a priori assumption, not experience.

I should perhaps add here a point that was perhaps insufficiently emphasized in my earlier work, that by disunity I by no means want to imply disconnectedness. Indeed, a major task of this book is to point out the relevance of particular developments in some parts of biology to others (microbiology to evolution, for instance) or even their relevance to quite distant issues in other sciences or philosophy. I see the various areas of scientific enquiry as connected by a dense web of mutual information; but such connections in no way threaten the autonomy of these different areas. My objection to the 'tree' of knowledge, a tree rooted in physics, and with implications running only in one direction towards the sciences of the increasingly complex, is structurally analogous to the critique of the Tree of Life that emerges in the third part of the book, on microbiology. In each case a set of relations traditionally understood as branching in a unidirectional, always divergent, pattern from a single origin needs to be reconceived as a web of relations with no uniquely privileged origin or direction.

Chapter 1 also anticipates some themes from the following chapters by exploring possible motivations for monism, or functions that it serves for some of its adherents. One suggestion is that it may serve to distribute authority from epistemically confident parts of science to far more dubious projects that claim, unlike their rivals, to be part of the One Unified Science. More disturbing is the possibility that reductive versions of monism may encourage certain kinds of interventions over others, for example in cases of serious mismatch between human behaviour and the social environment, favouring pharmacological interventions into the functioning of people's brains over changes to the environment. I do not, I hasten to add, accuse philosophical defenders of monism of malicious intent. My point is rather that there are reasons to consider the implications and uses of philosophical doctrines as well as the arguments that may be offered for their truth or falsity.

Recalling Francis Bacon's entomology, it is appropriate that the two insects that employ some kind of proper empiricism, the ant and the bee, are social insects. For a main theme of Chapter 2 is that science is an irredeemably social activity. Uncontroversially, science involves the cooperation of many people, not only in its day-to-day practice, but as that practice is grounded on the results of many others. Much more controversially, socially embedded beliefs, assumptions and values are inscribed in the way science is done. I suggested above that reductive views of science might motivate particular interventions, for example treating people with psychotropic drugs rather than attempting to ameliorate their environments—hence, perhaps, the millions of children being treated for attention deficit hyperactivity disorder (ADHD) with fairly dangerous stimulants such as Ritalin. But it should also be clear that the extension of the concept of ADHD reflects social values concerning, for example, appropriate behaviour for children. What is the correct amount of attention, and how far short of this must one fall to suffer from a disorder? Inability to concentrate in classes is deplorable, no doubt, but may be a very typical characteristic of children, especially as some classes are (or at least used to be) quite dull. Outside fairly extreme forms of brain damage, it is difficult to imagine a characterization of mental health free of normative assumptions about how people should behave.

It is this kind of manifestation of the sociality of science that has led to the controversy over social constructivism, a doctrine sometimes violently attacked by philosophers of science. In a modest way, in this chapter I defend social constructivism.

I do so by rehearsing a view that I have been promoting for several decades, that classifications of natural objects are not purely discovered, but are designed to serve specific goals. This is not to say that they are simply made up; differences between natural objects are discovered and these enable us to formulate models, or theories, that may serve our epistemic or practical goals. There are real distinctions in nature, but there is no unique set of distinctions; different classifications, serving different purposes, may overlap and cross-classify. This is the doctrine I first referred to thirty years ago as promiscuous realism, and I am still convinced it is correct. In this chapter I show how cases that I have considered in detail, species and genes, illustrate the doctrine and the way it inevitably introduces a social aspect in the construction of scientific theory.

In Chapter 3 I pursue the most contentious aspect of the sociality of science, the idea that in much of science there are elements of embedded value. As with the previous chapter, the argument is grounded in pervasive features of language, in this case the impossibility, in many areas of language, of separating the factual from the evaluative connotations. There are countless examples. Murder or fraud both specify actions of a certain kind, and imply that they are to be condemned; killing in self-defence or in war do not count as murder because they are generally considered permissible. Slightly more controversially, health and illness both refer to states of the body and evaluate them. There are also more subtle ways in which values become embedded in scientific concepts. An example discussed in this chapter is the concept of inflation. As is well known to economists but not always appreciated by lay consumers of economics, the measure of inflation depends entirely on which commodity prices are used in constructing a particular index. Experienced inflation varies greatly from one person to another with the different commodities they consume. Meanwhile, publicly quoted values are used to determine decisions from macroeconomic policy to the pay rises of workers. Constructing such an index is a highly political matter.

★ ★ ★

Science, then, is not the one truth about the world, but truth from a particular perspective, answering a particular set of questions, and often serving a particular set of interests. Nonetheless, I insist, when done well it often gives us truth, or all the truth that is to be had. In the second section of the book I turn explicitly to biology, a science, or really a cluster of sciences, many parts of which have, I believe, been very well done in recent decades. With regard to the science the next few chapters, therefore, are generally interpretive rather than critical. The section begins with two chapters, 4 and 5, based on lectures entitled 'The Constituents of Life' I delivered in 2006 at the University of Amsterdam, which give an overview of the most important philosophical themes that I see as having emerged from the last half century or so of biological enquiry. Subsequent chapters in this section explore the topics introduced in these lectures in greater detail.

A central goal of 'The Constituents' was to understand why it had proved so difficult to pin down with any sort of precise definition the central concepts through which we understand living things. Two such concepts that have received serious attention from philosophers of biology are the species and the gene, and these provide the main focus of Chapter 4. The species is generally thought of as the fundamental unit of classification of organisms, though it has also been popular in recent years to consider it an individual thing, specifically a particular genealogically connected chunk of the Tree of Life. I briefly rehearse the difficulties that have been found in providing a unified account of these concepts, and that have led many philosophers to advocate a pluralistic view: these concepts refer to a variety of different kinds or entities in a variety of contexts. Two explanations of this conclusion are also canvassed. The first, a topic that has already been stressed in this introduction, is that our concepts and patterns of thought struggle with the profoundly processual nature of living things. From the countless processes taking place at the level of molecular biology, through the processes of ontogeny that span the various scales of different life cycles, to the sometimes aeon-scaled processes of evolution, biology is about change. In fact, living things sustain themselves only by constantly changing. It is arguable that our thinking is for deep reasons anchored to conceptions of objects describable in static terms; certainly many biological concepts are described in such static terms. But such concepts can only capture particular frozen time slices through the more fundamentally processual biological reality. And this points to an obvious reason why our choice of such concepts may be seriously underdetermined by the reality that they purport to capture, and may thus be different according to the purposes for which the concepts are wrought.

A second general problem I identify is that, contrary to a still very widely endorsed reductionist conception, biological entities cannot be understood adequately in terms of properties of and relations between their structural constituents. What they are—not only their role in a wider biological economy, but even their causal properties—depends simultaneously on their relation to that context. This argument is sketched for a range of cases in the chapters reprinted from 'The Constituents'.

Chapter 5, introducing a theme that will be taken up in several later chapters, extends the pluralistic perspective to the crucial case of the organism. The argument here requires for the first time some more detailed discussion of the theme of the neglect of microbes. Serious attention to microbes brings into sharp focus a phenomenon already noted, the ubiquity of symbiosis. Here I make for the first time the argument that the omnipresence of symbiosis should be seen as undermining the project of dividing living systems unequivocally into unique organisms, a conclusion I refer to in later chapters as 'promiscuous individualism', in parallel with my doctrine of promiscuous realism about species or, more generally, natural kinds. This chapter concludes with some reflections on possible implications for the practice of biology, including the currently fashionable, and potentially very exciting, project of systems biology.

Chapter 6 addresses in more detail an area of biology that has been central to my approach to the topics of this book, genomics, the study of genomes. This is most familiar to the casual observer of biological science through the excitement that surrounded the Human Genome Project. Much of the publicity around this project suggested that sequencing the three billion nucleotides, or base pairs, that constitute the human genome would provide us with a complete map of the human organism, a blueprint of human life. In fact, though I have no wish to depreciate the significance of the achievement it represents, the Human Genome Project has done nothing of the kind. Rather, it has made increasingly clear that, while the DNA that makes up the genome is a fundamentally important constituent of human life, it is far from uniquely so. The remarkable success in understanding the chemical basis of genetics has, indeed, undermined the simplistic views of inheritance that preceded it (Barnes and Dupré 2008). Unfortunately, these earlier ideas still remain dominant in much public and media discussion of genetics, and even in parts of science not directly involved with these molecular investigations.

Chapter 7 resumes the topic of the individual organism, and connects this with some of the points made about genomes in the previous chapter. It is often supposed that something fundamentally important that the parts (cells) of a complex, multicellular organism share is the genome sequence found in all their cells. But in fact it is not true that all the cells in a multicellular organism share the same genome sequence. There are a number of processes that produce genomic mosaicism or chimerism in an organism in ways that hardly threaten its individuality. Even more importantly, and anticipating the topic of the next section of the book, functioning biological individuals are typically symbiotic wholes involving many organisms of radically different kinds. The idea mentioned earlier, that much of the ambiguity in defining biological entities derives from our difficulty in dealing with process is not explicitly discussed in this chapter, but I hope its relevance is obvious. For example, the phenomenon of epigenetics that is briefly discussed in the chapter is conceptually problematic in raising the question whether chemical modifications of the genome that do not affect its sequence should count as generating different genomes. This concern reflects the importance that has often been attached to genome sequence; but nucleotide sequence is merely one description of a constantly changing entity—a process—that is treated as a biological fixed point through a particular chemical abstraction. This is a very useful abstraction, without doubt, but an abstraction nonetheless.[5]

Chapter 8 considers an issue that has been mentioned several times in preceding chapters, reductionism. This piece was originally written as a debate with the historian and philosopher of biology Evelyn Fox Keller (Keller 2010), but I think it stands quite well on its own. One reason for this is that despite the format, the differences between myself and Keller are not very great; the sense in which she supports reductionism is a

---

[5] I do not say much in these essays about the positive virtues of genome sequence. Above all, it is a wonderful device for comparison: of the parts of a thing as for instance in criminal forensics; or of different things in constructing phylogenies. Determining the behaviour of the entity in which the sequence is found is quite another matter.

fairly rarefied one. Opposition to reductionism is certainly a majority opinion among philosophers of biology, a state of affairs that has, perhaps surprisingly to some, developed in parallel with the growing success of molecular biology. In this chapter I try to spell out in some detail my own reasons for opposing reductionism, and also to highlight some philosophical corollaries of this opposition. The chapter will say something about emergence, the often obscure but nonetheless important idea that complex objects possess properties that are in some sense autonomous from the properties of their constituent parts, and about what I take to be the closely related thesis of downward causation. The central point that in my view explains both these phenomena is that what a thing is and what its properties are depend simultaneously on its structure and its context.[6] A simple, if somewhat surprising, example is the property of pathogenicity of microbes. As I have said, humans live in symbiotic relations with a multitude of microbes. However, many of these when introduced into the wrong part of the body, for example through a wound, can become seriously harmful. The microbe is neither pathogenic nor beneficial intrinsically; these are properties that emerge in its different possible relations to the remainder of the complex system. The property depends on context. I argue that this is typical of biological properties. The chapter also includes some further remarks on the developments in biological methodology involved in the emergence of systems biology.

The final chapter in this section, Chapter 9, turns to what has been the dominant topic in philosophy of biology for several decades, evolution. (It is an implicit thesis of this book that this emphasis has been excessive, and has diverted attention from many equally important topics. However I have no wish to deny that evolution is a matter of great importance.) The central point of this chapter is that recent developments in molecular biology, in microbiology, and in evolutionary theory itself have presented major difficulties for a canonical and widely popularized understanding of evolution, the so-called New Synthesis that emerged in the 1920s and 1930s, and the core of which has survived various subsequent refinements of that basic picture. A number of central features of that view of evolution are increasingly open to question. One assumption that has seldom been questioned is the idea that it is possible to arrange the kinds of living things in a branching but never converging pattern, the Tree of Life. This assumption is increasingly seen as untenable for microbial life, however, and therefore for the great majority of evolutionary history in which there was no other kind of life, and it is even open to question for more complex recent organisms. A broader point, to be developed further in Chapter 12, is that the traditional view emphasizes competition almost to the exclusion of cooperation, but in fact both are very important. Cooperation is embodied in a wide range of biological interactions of which the extreme form is complete merger between independent evolving lineages. Some criticisms of the New Synthesis, emphasizing much greater diversity of inheri-

---

[6] This idea is developed further and in greater detail in a paper co-authored with Alexander Powell (Powell and Dupré 2009).

tance mechanisms, notably epigenetic and cultural inheritance, have even questioned the ultimate taboo in evolutionary theory and suggested that there may be a significant place for Lamarckian processes, or the inheritance of acquired characteristics (Jablonka and Lamb 1995). This chapter is a speculative one, but the conclusion that evolution is a theory in flux, in opposition to the somewhat rigidified and dogmatic endorsement of the New Synthesis that remains standard, is one which I do not think is in serious doubt.

★ ★ ★

The third section of this book finally addresses in detail the topic of the nature and importance of microbial life. This is also the topic emphasized in this book that is furthest from the central agenda of the philosophy of biology. The philosophy of microbiology is a project that has been central to my research for several years, in collaboration with my colleague Maureen O'Malley, and the first three chapters in this section were originally co-authored with Maureen. Chapter 10 is something of a manifesto for the inclusion of microbes in our general thinking about life and biology. We begin by proposing that, since the vast majority of organisms are generally referred to as microbes, it is remarkable, and indicative of the imbalance in our attention between microbes and non-microbes, that we have no general word for the latter. We advocate the obvious candidate, 'macrobe'.

Microbes dominate terrestrial life. They were the only life forms for 80 per cent of life's history, and are by far the most numerous today.[7] They have colonized a vast number of environments, from high-altitude clouds to deep in the Earth's crust, at temperatures from over 100°C to −20°C, many of which are quite intolerable for more familiar plants and animals. Equally important, it turns out that all or almost all macrobes actually live in obligatory symbiotic relations with vast numbers of microbes. So even someone who insisted that macrobes were sufficiently more interesting than microbes to justify an exclusive philosophical focus on the former would be unable to carry through that policy coherently.

Chapter 10 also provides a very brief history of microbiology, up to and including the development of metagenomics, the methods that have been developed for in-ventorying the genetic resources of whole microbial communities. We then introduce one of the most crucial points in coming to terms with microbes, that far from being typically isolated individual cells, microbes are typically found as parts of complex multicellular communities. (This suggests, incidentally, that the microbe/macrobe distinction may not ultimately be tenable; most microbes are parts of macrobe-like entities.) We also consider recent ideas about the evolution of microbial life, and about its taxonomy and diversity. Crucial to all these topics is a phenomenon we think has yet to be adequately assimilated by philosophers of biology, and indeed many biologists,

---

[7] It is estimated that if extracellular plant material is excluded, microbes make up more than half the living mass on Earth.

lateral gene transfer. The evolutionary models that have mainly concerned philosophers and biologists have tracked the evolution of traits through a series of parent/
offspring transmissions. Among many microbes, and indeed probably many macrobes,
transmission of genes, and traits to which these genes give rise, are also sometimes
transmitted horizontally, between sometimes very distantly related individuals, and the
implications of this lateral transmission are currently hotly debated. Seen another way,
we encounter here again the question whether life can indeed be arranged in a tree, or
whether we must rather see it as a web or net.

A theme introduced in this chapter, but developed in much more detail in the next
two, is once again the topic of the nature of the biological individual. The largest part
of Chapter 11 is devoted to the topic of metagenomics, a very important scientific
development that deserves much more attention from philosophers. In traditional
microbiology an entity could not be subject to serious study unless it could be cultured
on its own *in vitro*. Recently it has become clear that this excluded 99 per cent of
bacteria from scientific scrutiny, in part because most bacteria can exist only in relations
with other organisms. Metagenomics uses state-of-the-art sequencing methods to
investigate the entire genetic resources of microbial communities, a perspective on
microbial life that is providing radically new understanding of this fundamental realm
of biological processes. It is possible, for example, to investigate metabolic processes
that are not confined within single lineages of cells.

The question naturally arises whether there is a robust ontological reading of the
findings of metagenomics: whether, that is, we should recognize a transorganismic
entity, the metagenome, and even an individual to which this belongs, the metaorganism. We argue that this is indeed the way to go. In the end, if metagenomic
perspectives give a richer understanding of what microbes are and what they do than
traditional conceptions of the organism, then this must be taken as strong evidence that
this perspective captures a way in which living material is genuinely organized. An
implication of this conclusion is that we must also rethink traditional neo-Darwinian
conceptions of cooperation and competition, according to which cooperation is on the
whole only to be expected between genetically similar organisms or cells. These last
points, on metaorganisms and cooperation, are pursued in greater depth in the
following chapter.

Chapter 12, 'Varieties of Living Things', is one of the most philosophically ambitious in the collection: it proposes a general account of what constitutes an entity as
living. At the same time it suggests that the narrow neo-Darwinian conception of
cooperation be subsumed under a much broader concept that we call 'collaboration',
and which characterizes a range of ways in which different entities may participate in
the complex processes that make up living things. One point we are particularly
concerned to stress in this chapter is the diversity of biological entities. We discuss a
range of these from prions and plasmids to the familiar multicellular organisms. Of
particular significance towards the lower end of the scale are viruses. While generally
assumed not to be living things, because they depend on other biological entities for

metabolism and reproduction, viruses are enormously widespread and surely very significant parts of the living world. There are estimated to be around 10^{31} viral particles on Earth; it is estimated that placed end to end they would stretch for about 100 million light years. Many viruses have significant effects on their cellular host, ranging from death and disease to protection against other pathogenic bacteria or viruses.

The central argument of the chapter, alluded to in its subtitle, is that the paradigmatic living thing contains parts of many distinct biological lineages in more or less cooperative relations. Excessive focus on evolutionary theory, we argue, has led to the assumption that a living individual must be part of a lineage, an evolving sequence of ancestors and descendants. But to function properly such segments of lineage almost invariably need to be in complex relations with parts of other lineages (as humans are with their many bacterial symbionts). We don't want to insist that all these relations are cooperative in the sense of conferring net benefits to all parties. Rather, we suggest a broader concept of collaboration that can include the various kinds of symbiosis (mutualism, commensalism, parasitism) that connect the various more or less obligate members of the complex systems characteristic of living things. The most radical implication of this view, perhaps, is that there is no unique way, suited to all purposes, of dividing the living world into individuals. Evolutionary models may indeed need to focus on the individuals that form parts of lineages; but for most parts of functional biology, and even evolution in so far as it considers the actual processes and units of selection, the more complex entities we describe will be more appropriate. This, again, is a central instance of the difficulties resultant from the division of biological process into more or less statically described individuals.

Chapter 13 explores one set of consequences of this vision of living things in general and humans in particular, the implications for our understanding of medicine and disease. It might equally well have been placed in the next section of the book, but at any rate provides a bridge between this section and the next. I'm sorry to say that I cannot offer a radical new solution to the problem of the burden of disease. I do suggest, however, that we might learn to see the issues a little differently. If O'Malley and I are right, and we are actually complex symbiotic systems, it would be astonishing if this did not have some implications for medicine. One area in which there are certainly such implications is for the understanding of the immune system. The entire interface between the human (traditionally conceived) and the outside world is populated by microbial communities. This therefore is the first port of call for potentially harmful microbes to insinuate themselves into the system. Moreover, certain diseases of the skin and the gut, for instance, are increasingly acknowledged to be malfunctions of the relevant microbial systems. And very importantly, there is no reason to think that the health implications of microbial symbionts will apply only to the surface of the body. There is known to be much chemical communication and interaction between microbial and 'human' cells, and there is growing evidence that disorders of the gut microbiome, in particular, may have medical impact on distant parts of the body. Since we are all inclined to agree that what we eat can have profound and diverse effects on our health, none of this should be surprising. Here I touch on one very important area in which it may matter a good deal to update our conception of the living human organism. More far-reaching implications will emerge in the concluding chapters.

★ ★ ★

The final section of the book will explore implications of the biological ideas discussed in earlier sections for our conception of ourselves, and particularly for the ever problematic concept of human nature. These chapters continue the development of a thesis I have been defending for over twenty years, that the complexity of human development is such as to prevent biology by itself, whether evolutionary theory or molecular genetics, from providing much insight into the details of human behaviour. Chapter 14 renews the criticism of a target I have been engaging with for many years, Evolutionary Psychology,[8] and does so starting with one of its most characteristic theses, that human psychology is adapted primarily to the life of Stone Age humans and therefore presumably is poorly adapted to the very different conditions of twenty-first-century urban humanity. At the core of this argument is the belief that evolution happens too slowly for conditions in the last few thousand years to have had much effect on human nature. Drawing on (and recapitulating) many of the biological ideas elaborated in earlier chapters, I argue that evolution is in fact capable of moving much faster than the models assumed by Evolutionary Psychology suppose, and hence that there is no reason to doubt the intuitively plausible view that we are in fact pretty well adapted to modern life.

Chapter 15 addresses a set of questions about human races. The central issue is whether races constitute real divisions among people or merely socially imposed categories serving political or other interests of the imposers. Social scientific opinion has tended very strongly in recent years to the latter view, but the realist alternative has been revived in recent years by claims that sophisticated genetic techniques enable us to distinguish human races with considerable accuracy. The chapter reviews some of the problems with the concept of a gene introduced earlier in the book, and attempts a fairly detailed taxonomy of the ways this word is used. This is then deployed to elaborate the claim that in no sense of the word 'gene' do genes help to give biological legitimacy to the concept of a human race. This, in turn, makes possible a more critical analysis of these recent claims to be able to distinguish races through genetic testing. The central point is just that self-identified race is no doubt in many cases strongly correlated with geographic origin (for example self-identified African-Americans and

---

[8] Here, I follow an emerging convention—first proposed by David Buller—of using the capitalized title to refer to the influential school of evolutionary studies of human behaviour associated especially with Leda Cosmides and John Tooby at the University of California, Santa Barbara. Needless to say, I do not doubt that human psychology evolved or object to all attempts to study the process by which it did so, others of which will be referred to as evolutionary psychology (lower case).

relatively recent West African origin), and geographical origin will inevitably be correlated with many genetic differences (alleles). But geographic origin in the relevant sense is something very different from any socially significant concept of race; and most of the alleles characteristic of particular geographic origins have nothing to do with the phenotypic differences generally associated with particular races. The chapter concludes with some reflections on the difference between racial and sexual differences—the latter, unlike the former, are biologically real if often badly misunderstood; and on the risks and benefits of engaging in debates about the biological status of these socially problematic categories. I suggest that despite some real risks, it is generally worthwhile to aim for clarity on these matters.

I had originally planned to include a chapter on the biology of sex and gender; however, on further reflection I decided not to do so. This is a topic that I have treated at length in the past (see Dupré 2000, 2001, 2003), and I have seen no reason to revise the views, specifically a strong dissent from the approach to the issue pursued by Evolutionary Psychologists, presented in that earlier work. Moreover, what I have written more recently (e.g. Dupré 2010) goes over a lot of ground covered in other chapters in the present book and thus threatens a quite different kind of redundancy. I do think there are important emerging issues potentially connecting sex and gender to current biology. For example, the ideas about development that figure prominently in this book should surely open up new avenues for investigation of the development of gender and sexuality; and the concerns about individuality raised in part III suggest new ways of thinking about the relation of mother and foetus. But these are topics that must await future work.

The final chapter in the book treats the philosophical issue of human nature in a more general way, asking what role, if any, the concept has in a proper social science, orienting the discussion also to the causal status of the concept. This chapter ties together a number of earlier themes, moving from sociobiology and Evolutionary Psychology to a more adequate Developmental Systems perspective, by way of epigenetics, niche construction, and cultural evolution. The conclusion of all this is that human nature as something fixed and constant throughout the human species is an illusion. This is not, *pace* a certain line of attack by Evolutionary Psychologists, because the human mind is a blank slate on which culture can write whatever it likes, but because of the complexity of the developmental interactions between a wide variety of internal and external factors. I try to say briefly what role this does leave for attributions of genetic causality to human behaviour. The chapter and the book conclude with some reflections on a perennial philosophical issue, the freedom of the will. Denial of genetic determinism is a long way short of defending human free will. However, restating a line of thought that I have tried to develop over a number of years, I claim that there is an important sense in which a radical rejection of determinism does in fact make possible a robust thesis of human autonomy. In a brief afterword inspired by a talk I heard recently by John Perry, I record a reconception of my position, which I have described in the past as a version of voluntarism, as instead an indeterminist compatibilism. Compatibilism here does not, of course, mean compatibility with determinism but merely with the normal causal order.

★   ★   ★

As I explained at the beginning of this introduction, there are themes that recur with some regularity in the following chapters, concepts which, in my view, have fundamentally transformed our biological understanding. Central among these are Developmental Systems Theory, both as a corollary of the rejection of genetic determinism and also as representing evolution as a process that cannot be reduced to a sequence of time slices; epigenetics, especially as pointing to the deep impossibility of separating 'nature' from 'nurture' in the explanation of development; and the centrality of microbes to a proper understanding of biology.

I make no apology for carrying on about these ideas; indeed a core part of my message is that they have implications across diverse areas of biological thought. Less happily, the fact that the chapters of the book were written with various goals and with no intention at the time that they should end up in the present form has inevitably generated some unnecessary and perhaps annoying repetition. Most obviously, several chapters review, in slightly different ways, the recent scientific insights that have so deeply problematized earlier, and still widely assumed, conceptions of genes. And had I had the foresight to envisage the reproduction of these papers in book form, I would have gone to more trouble to use different examples to illustrate key concepts. I regret, for instance, that in most or all of my references to epigenetics I have used the same, admittedly canonical, example of maternal care in rats. This might give the impression that a lot of weight was being rested on a single example in support of what is admittedly still a partly controversial thesis (see Miller 2010 for an expression of scepticism about some of the morals often drawn from research on epigenetics). However, there are other examples. I could, for instance, have illustrated my discussions with the conclusion of Kurcharski et al. (2008), that eating royal jelly, which determines that bee larvae will turn into queens rather than workers, operates through epigenetic suppression of a particular gene (Dnmt3). Or I could have mentioned the detailed neurochemical work that has established the role of histone methylation induced by ingestion of cocaine in producing addiction (Maze et al. 2010). I hope that any readers who read the book sequentially from start to finish will forgive these infelicities of production history. I could, of course, have cut and altered the chapters to increase the smoothness of a continuous reading, but as a source of reference, there is something to be said for providing access to essentially the same text as that found in the sometimes difficult to access papers here reprinted. And I suppose many readers will read in a less linear and perhaps less complete fashion. In the event, at any rate, I have made only trivial stylistic changes and very minor corrections.

I said at the beginning of this introduction that one of the things that made the study of contemporary biology exciting was its pace of change. Inevitably this means that in some respects a book such as this will be out of date as soon as it is written—and some of the chapters were written a few years ago. Moreover, there are areas of biology closely connected to themes of the book on which I have hardly touched. I have said little about systems biology and nothing about synthetic biology; I haven't written about the burgeoning field of transcriptomics, in which new categories of RNA molecule are being described practically on a weekly basis, and a whole unexpected level of cellular control external to the genome is beginning to appear. (This, incidentally, has pretty much disposed of the once-popular concept of 'Junk DNA'.) And as mentioned in footnote 1, above, I haven't touched on evo-devo.

However, these essays were not intended as a survey of contemporary biology. I do hope, though, that they will provide at least a sketch of a plausible general view of living processes that is emerging from our best science, point out the flaws in some still popular misconceptions about life, and assist the less expert reader in reaching more judicious assessments of some of the marginally scientific claims with which we are all constantly confronted. These are ambitious goals enough; indeed, for anyone like myself who takes it as obvious that we are ourselves fall within the subject matter of biology, but also that this obvious truth requires that we develop a conception of biology expansive enough to include creatures as strange and complex as ourselves, there can be few goals more ambitious or inspiring.

PART I

Science

1

The Miracle of Monism

As Barry Stroud nicely displays in his APA Presidential Address 'The Charm of Naturalism', naturalism is far from an unequivocal idea (Stroud 1996). Versions of it are as old as philosophy, and many ideas frequently associated with naturalism are either too vague to make much sense or too banal to have much interest. As Stroud notes, one substantive naturalist theme is anti-supernaturalism. This has been philosophical orthodoxy for at least a century, if never without its dissenters, so it fails to capture the reasons why many philosophers appear to think that naturalism is a fairly recent philosophical movement. But perhaps anti-supernaturalism may nevertheless provide a good way into some of the ideas associated with contemporary naturalism. One thing I want to suggest in this chapter is that among the many philosophical defects of monism is the fact that it involves more than a whiff of supernaturalism. So by my title I mean not merely that it is miraculous that so many otherwise sensible philosophers subscribe to the doctrine of monism, but also that the doctrine to which they subscribe is itself a doctrine with a miraculous dimension.

By anti-supernaturalism I mean something like the denial that there are entities that lie outside the normal course of nature. It is easier to point to some of the things that are agreed to lie outside the normal course of nature than it is to characterize the normal course of nature. Central cases of such outliers are immaterial minds or souls, vital fluids, angels, and deities. My aim in this chapter is to investigate some of the reasons that contemporary naturalism has often come to be associated with a doctrine that seems to me wholly incredible: monism. One part of the answer, I suggest, is that monism is a gross exaggeration of what is indeed plausible about anti-supernaturalism. The grounds for the denial of the existence of souls and vital fluids have been taken as grounds for the denial of minds and, in a sense, even bodies. But in fact the argument should go exactly the opposite way. The arguments against souls and suchlike are really the kinds of arguments that should lead us to reject monism. Or so I shall claim.

What is wrong, according to most philosophers, with souls and vital fluids? One of their central defects is their being immaterial. There are perfectly respectably immaterial entities—concepts, numbers, or hypotheses, for example—but souls and suchlike are not the right kinds of things to be immaterial. Part of the reason for this is that they are taken to be subjects of causal agency, and immaterial causes are seen as contravening naturalism. This is not an entirely straightforward matter. One might well claim that the concept of evolution, or of class struggle, had changed the world. But I think it is fairly clear that we don't want to treat this as the exercise of a causal power by an entity, or at any rate we don't if we are any kind of naturalists. I'm not exactly sure why we feel this way. Perhaps the answer is that knowledge of causality does, as a number of influential philosophers have argued, derive ultimately from our material transactions with things. Even for material things that are much too small for us to interact with, there is considerable force in Ian Hacking's much-cited remark about electrons: if you can spray them they are real (Hacking 1983). Perhaps there is a cruder picture here, that nature is ultimately composed of material things pushing and pulling at one another. Pushes and pulls from outside this material universe are just the sorts of supernatural interventions that naturalists rule out. At any rate, I propose to take for granted that there is something importantly fundamental to our ontology about material things, material here merely in the Cartesian sense of things that occupy space, albeit some-times, like electrons, not very much, or, like gases, not very fully. Being in the same space as we are is, of course, a minimal condition of being in a position to engage with us in pushings, pullings, and sprayings.

Important threats to this ontological primacy of space occupiers have come from physics. Physics tells us about fields that, while they have particular strengths at particular locations, don't appear to occupy the locations where they occur. And worse, quantum mechanics tells us that what might have seemed like the most fundamental space occupiers, physical particles, are actually not, from some perspec-tives, particles at all, but waves, which are apparently not space occupiers at all. These claims are particularly important to the contemporary naturalistic philosophy that I am here considering, because most contemporary naturalists include among the central commitments of naturalism not materialism, but physicalism. The move from materi-alism to physicalism, in large part motivated by such developments in physics, aimed to avoid being committed to a long-superseded, broadly Newtonian physics. Physicalists instead committed themselves to take whatever physicists ended up saying about the nature of reality as an ultimate ontological criterion for reality. If the ultimate con-stituents of reality couldn't make up their minds whether to be particles or waves, or turned out instead to be ten-dimensional bits of string, so be it.

I have, naturally, no objection to allowing physicists whatever authority there is to be had about the ultimate structure of matter. What might be a worry, however, is whether physicalism, as just described, is in the same line of business as the materialism it has replaced. Materialism, as I introduced it, was part of the expression of anti-supernaturalism. If you can't kick it, or at least spray it, you should treat it with some suspicion. We can kick things or spray them because they live in the same space as we do, and there is nothing and nowhere else. Materialism is, of course, also contrasted with dualism, a specific form of supernaturalism, and a doctrine that explicitly insists on the existence of things that are not in the space of electrons, stones, and ourselves, and in most versions insists on the causal efficacy of such things.

Materialism, as just described, has no obvious connection to any particular scientific doctrine, though no doubt it was substantially motivated by the success of certain scientific projects. More specifically, and here is a central naturalistic argument, science has proved increasingly successful at solving all kinds of explanatory problems in basically materialistic ways. To subscribe to anything supernatural, an explanatory *deus ex machina*, is mere pessimism, and profoundly unwarranted pessimism in the light of the successes of science. It is in connection with this sort of argument that monism insinuates itself into the naturalistic viewpoint. Reflection on 'the successes of science' may well motivate an interest in what exactly science is, and an answer to that question will potentially limit severely what kinds of entities are amenable to the investigation through science that naturalism mandates. And the deliberately vague phrase 'in basically materialistic ways' that I used to qualify scientific explanatory projects may invite much sharper specification in terms of the actual conceptual resources of the physical sciences. Very summarily, naturalism is explained in terms of anti-supernaturalism, which is in turn cashed out in terms of materialism; the developments of radically new conceptions of matter by the physical sciences leads from materialism to physicalism; and physicalism is often understood as entailing monism.

I don't propose to explore in great detail these routes from naturalism to monism. Rather, I want to emphasize a quite different philosophical commitment often associated with naturalism: empiricism. And what I want to claim is that the move to monism violates this commitment. Monism, far from being a view of reality answering to experience, is a myth. And myths are just the sort of thing that naturalism, in its core commitment to anti-supernaturalism, should reject.

The myth of the unity of science

While my ultimate target in this chapter, monism, is a metaphysical thesis, as I have just indicated a main bridge from naturalism to monism is through a commitment to the explanatory reach of science. If this is combined with the idea that science is a largely continuous and homogeneous activity, and even more specifically that its explanatory resources depend on its sole concern with the material structure of things, then we are well on the way to naturalistic monism. But monism, I claim, is a myth. And it is a myth that derives what credibility it has from its connection to another myth, the unity of science. So I shall now explain in some detail why this latter doctrine, the unity of science, is indeed a myth.

There are at least two minimal implications of calling something a myth, and in so referring to the unity of science I intend both of them. First, myths are literally false. No doubt there are some complications here. Perhaps it is some kind of a solecism to refer to a myth as either true or false, since the conveying of factual information is not what they are for. Still, myths do include statements with literal meanings, even if these are not the relevant meanings for sympathetic understanding of a myth. As philosophers are wont to say, 'Athene emerged fully armed from the head of Zeus' is true if and only if Athene emerged fully armed from the head of Zeus. Most of us, doubting even the historical existence of Athene or Zeus, and even more sceptical of this medical marvel, are confident that nobody emerged fully armed from anyone's head, and hence that the statement 'Athene emerged fully armed from the head of Zeus' is false.

I pretend to be no authority as to what segment of the population of ancient Greece believed in the literal truth of Greek mythology. Certainly many contemporary myths are widely believed. I have heard that more Americans currently believe in alien abductions than believe in the theory of evolution; and myth or fact, I know that quite a few Americans believe in evolution. And very many people now believe in the virgin birth or the resurrection from the dead of Jesus Christ. From a contemporary perspective one person's fact is another's myth. This brings me to the second implication of calling something a myth. Those of us who think that virgin births, resurrections, or alien abductions are myths generally hold that they are nevertheless something more than merely unsuccessful attempts at stating the facts. Religious myths, it is often said, provide wider meanings to people's lives, or offer consolation for the unavoidable tragedies of disease and death. Even those who believe in the literal truth of religious claims will often acknowledge such functions; and they will almost certainly acknowledge the significance of such functions to the epistemically benighted adherents of other, false religions. So myths are, minimally, false stories that serve often central and important functions in the lives of their adherents distinct from stating how things are. This is the sense in which the unity of science, I hold, is a myth. So I want now to say why I take the doctrine of the unity of science to be false, and to say something about the non-truth-dependent functions I think it serves.

An obvious prerequisite for thinking seriously about whether science is unified is an account of the extent of this subject of unification. There is a threat of vacuity in the offing. Would the discovery that, say, microeconomics failed, on some account of unity, to be unified with the rest of science, refute the account of unity in question, or show that microeconomics was not a science? But recalling that the object of the present enquiry is to explore the consequences of the idea that science should provide the right tools for investigating whatever is real, it is clear that the relevant brand of unified science must potentially provide us with an account of the world that is complete and exhaustive. There are some obvious worries about this idea—where does it leave common sense or history, for example?—and I shall develop some of these concerns later in this chapter. For now I note only that unless unified science is, if not the only way of finding out about the world, at least unequivocally the best, then it will be of no relevance to any argument for establishing the truth of monism.

In view of this point, it seems that science must be taken to include any projects of enquiry that have in fact produced worthwhile empirical results. And the discovery that some empirical enquiry has produced valuable results, but cannot be unified with the rest of science, should be taken as a refutation of the conception of scientific unity in question. But this presents some immediate difficulties. Many practical activities, for example violin-making, have achieved considerable success and empirical knowledge. To be sure, the goal of violin-making is not to discover how to make the best violins, but just to make the best violins. But to exclude from science projects that are practically directed would remove a great deal of what we take as clearly scientific (most of medical research, or the investigation of nuclear fusion, for instance). 'Applied science' is surely not an oxymoron. And then consider history, sometimes classed with the social sciences, sometimes with the humanities, and surely a repository of more empirical knowledge than many prima facie sciences. We would like some more principled ground for its classification.

This raises more generally the question of the social sciences. Should the word 'sciences' here be taken seriously? This is a point at which the issues surrounding the unity of science really make a difference. Much work done in the social sciences, especially work that is quantitative or based on some kind of mathematical modelling, quite explicitly aims to be scientific. On the other hand, other social scientists, or at any rate other people with the same departmental and disciplinary allegiances as the first group, oppose these methods as inappropriate to the study of humans. They claim, for various reasons, that such quantitative approaches are inadequate to the subtleties of a human culture and insist on the necessity of something like semantic interpretation. These disputes are often bitter. Anthropology departments can sometimes only resolve their disputes over such questions by separating into distinct disciplines of cultural anthropology and (aptly named) physical anthropology.

It is not my intention here to adjudicate this dispute. My point is rather that the thesis of the unity of science threatens to resolve it from outside. If science is the one and the true path to knowledge of the world, then we should certainly back those projects that meet the basic conditions for integration into science. Such an appeal is clearly part of the rhetoric of advocates of 'hard' social sciences—mathematical economists and neo-sociobiologists, for instance, deriding the fuzziness of cultural anthropologists or interpretive sociologists.

So we cannot assume at the outset which disciplines are to be joined together in unified science. The natural strategy is to start with a minimal conception of science. I take it that a unified science that didn't include at least physics, chemistry, and biology would be of little interest. Should we establish a unified account of at least these sciences, we could then ask what other disciplines could be included in this conception and decide whether the excluded disciplines should be seen as something less than scientific, or as counterexamples to the conception in question. If, on the other hand, an account of unified science cannot be sustained even for this minimal extension, we should have no difficulty in rejecting the unifying project.

Several features might be held to provide science with unity. The most important division among such factors is that between those that propose a unity of content to science and those that provide for a unity of method. Paradoxically, while unity of scientific method is intuitively a far more plausible thesis than is unity of content, contemporary philosophical defenders of science generally defend the latter rather than the former. So let me begin by mentioning some reasons why the idea of unity of scientific method has gone into decline.

I suggested that the idea of a unique scientific method has some plausibility. At least, there is still a good deal of talk, not all of it ignorant, about the 'Scientific Method'. A little reflection, on the other hand, soon suggests that the idea of a single scientific method is problematic. If one thinks of the daily practice of a theoretical physicist, a field taxonomist, a biochemist, or a neurophysiologist, it is hard to believe that there is anything fundamentally common to their activities that constitutes them all as practitioners of the Scientific Method, though all are engaged in activities that fall within the minimal extension of science just proposed.

Note that an account of the Scientific Method must also serve as a criterion of demarcation, a criterion, that is, that can answer the question of whether any practice should count as scientific. If there is just one Scientific Method, then a practice is scientific if and only if it follows that method. It would be very nice to have a criterion of demarcation, but considerations that have already begun to emerge suggest that there may not be one to be had. My own view, to which I shall return at the conclusion of the chapter, is that the best we can do is to draw up a list of epistemic virtues and apportion our enthusiasm for knowledge-claiming practices to the extent that they meet as many as possible of such criteria. Such epistemic virtues will include certainly coherence with empirical data and with other things we take ourselves to know, and these virtues will be subject to detailed elaboration. They will surely include other things: perhaps aesthetic virtues such as elegance and simplicity, perhaps even moral virtues. There will no doubt be an unavoidable element of boot-strapping in this project. To some extent our enthusiasm for epistemic virtues will derive from their conspicuous role in providing us with what we take to be outstanding examples of scientific knowledge. This is not viciously circular to the extent that, in the end, we can ground this enthusiasm in genuinely empirical support.

Returning to more ambitious theses, probably the last account of a unique scientific method to be widely accepted was Karl Popper's theory of falsificationism. According to this well-known view, an investigation was scientific to the extent that it attempted to falsify hypotheses within its domain. And Popper also deployed this with some gusto as a criterion of demarcation against some practices, notably Marxism and psychoanalysis, that he considered pseudoscientific. Popper's ideas had a great deal of influence with scientists and surely had a significant effect on the kind of scientific work that was carried out. It is my impression that many scientists still consider Popper's the last word on scientific method; and no doubt this is especially true among those scientists employing quantitative or experimental methods in fields also explored by more qualitative and discursive approaches. But although there are still a few able defenders, among philosophers of science Popper's view of science has been very largely rejected.[1]

---

[1] The locus classicus for the general rejection of any account of scientific method is Paul Feyerabend's *Against Method* (1975).

There are some serious conceptual problems that have contributed to this, most centrally a persisting worry about the great difficulty of falsifying hypotheses: given a recalcitrant observation, how does one decide whether the observation was inaccurate, some unknown factor has interfered, some unquestioned background assumption is erroneous, or finally, that a hypothesis under test is false? It seems that this variety of options always leaves it open to a scientist to rescue a hypothesis. And the work of Thomas Kuhn and others has even made it plausible that this is almost always the right thing for a scientist to do.

But I don't want to pursue this kind of objection in detail here. For present purposes I would rather point simply to the prima facie inadequacy of Popper's thesis to the huge variety of activities that form parts of the practice of science. Consideration of this diversity should, moreover, make us suspicious of any unitary account of scientific method. So although the official target of this discussion is falsificationism, I hope it will also indicate the enormous difficulty that would be faced by any alternative attempt to provide a uniform account of scientific method. I shall briefly compare four admittedly schematic examples of relatively indisputably scientific work, and consider how useful falsificationism is in understanding what is happening:

1. *The attempt by physicists to detect a new particle.* This activity has been quite widely described by historians and other students of science (Galison 1987; Traweek 1988). Particle physicists divide themselves into theoreticians and experimentalists. The former, unsurprisingly, devise theories that the latter attempt to test. These theories have often involved predictions that under certain conditions, for example high-energy collisions in an accelerator, specific particles should be produced. The experimentalists then try to create the relevant conditions and observe the particles. For a moment this might look like good Popperian methodology. We have hypotheses, and even a specialized caste of hypothesis generators, predictions, and experiments. The problem, however, is that no one would imagine for a moment that the failure to observe the sought-after particle would refute the theory under test. In accord with the line of argument sketched above, there are far too many alternative explanations of the failure of the experiment for it to make sense to reject the hypothesis at any particular stage of the experimental process. Whether or not one has observed a subatomic particle is hardly a trivial matter to decide. And in fact historical accounts of this kind of work make it clear that it is typically a long and difficult process to make such experiments work and convince the community of physicists that they have worked.

2. *The attempt by molecular biologists to find the genetic basis of cancer.* First, a vast amount of background to this project is not a candidate for falsification. I suppose that a persistent enough failure to progress with the project might eventually contribute to the rejection of the view that there is a genetic basis to cancer, but testing this view is not part of the work of contemporary molecular geneticists.[2] One might suppose that

---

[2] [Note added to this edition] The exact role of genetic changes in cancer has become more problematic in recent years, especially with advances in epigenetics. See Chapter 13 for discussion.

the real science here involved testing hypotheses, such as that gene X is implicated in the development of pancreatic cancer; and no doubt something like this could be said to happen in this kind of research. The trouble is that of at least equal, and perhaps much greater, significance are the processes by which such hypotheses are produced. A fundamentalist Popperian, I suppose, would have the geneticist randomly select sequences of DNA and test the hypotheses that these sequences constituted genes involved in the development of a particular cancer. This would be a slow-moving science. What predominantly determines progress in this area is the search procedure by which likely candidates for relevantly interesting genes are selected. These search procedures will include synthetic elaboration of current theoretical ideas about the matters in question. Why should the development of such procedures be somehow less fundamentally scientific than the subsequent empirical testing of hypothesized effects on the organism?[3]

3. *The classification of the beetles of a hitherto unexplored terrain.* The background the coleopterist brings to this project is an extensive knowledge of the classificatory scheme applied to the already discovered beetles. (Not, of course, a detailed knowledge of every recognized species, for there are hundreds of thousands of these, but a knowledge of the outlines of the hierarchy into which they are to be arranged.) The task will then be to collect as many specimen beetles as possible, and try to assign them either to a particular already-known species, or to a new species suitably related to the existing hierarchy. The scientist might recognize that a novel specimen belonged to the family Silphidae or the particular genus *Necrophorus*, and invent an appropriate name for it. Perhaps these attributions could be conceived as hypotheses. If the coleopterist is a cladist, she might claim, for example, that this is a sister species to *Necrophorus vestigator*, and this would be a fairly precisely defined thesis. But there is no process of testing this hypothesis that routinely follows its formulation. It is really better seen as a judgement based on the accumulated knowledge and expertise of the investigator. It may subsequently be challenged on a variety of grounds by other scientists, but more commonly it is not challenged, but simply accepted as a part of our overall taxonomic system put in place by a competent contributor. In this case as in the last, it is the method by which the piece of scientific knowledge is acquired, in this case through the judgement of a properly trained practitioner, that matters more than any subsequent testing to which the hypothesis may be subjected.

4. *The statistical investigation of a sociological hypothesis.* Here, I think, is where we see the most superficially Popperian domains of scientific practice. Indeed, when investigating a hypothesis such as, for instance, being female causes one to be paid less, it is

---

[3] [Note added to this edition] If I were writing this paper today I would pay more attention to the increasingly studied topic of data-driven science, the idea that in many parts of science the task today is to extract a meaningful signal from massive quantities of data generated by technologies such as gene sequencing. This topic is addressed in a forthcoming issue of *Studies in History and Philosophy of the Biological and Biomedical Sciences* edited by Sabina Leonelli, See her introduction to the special issue: 'Editorial: Making sense of data-driven research in the biological and biomedical sciences'.

natural to start with an investigation of the extent to which pay is correlated with gender. The fact that women are, on average, paid less than men provides prima facie, though naturally not conclusive, evidence for the hypothesis. Unfortunately, just the reasons that prevent this from being conclusive evidence for the hypothesis are equally reasons why the lack of such a correlation could not refute the hypothesis. Women could be paid less because they are qualified for less skilled and hence less well-paying jobs, or they could be paid no more than men despite being better qualified and hence doing generally better-paid jobs. So we cannot see the statistical investigation as a Popperian attempt to falsify the hypothesis. We can strengthen the evidence for or against the hypothesis by considering more and more potential causal factors. But the hypothesis cannot thereby be irrevocably proved or disproved. Now in fact it is common to express results of this kind in an extremely Popperian-looking way, specifically by noting that the null hypothesis, the hypothesis that the correlation between being a woman and being poorly paid comes about by chance, can be rejected. (Or, more accurately, that the probability of this correlation having come about by chance is very low.) The origins of this kind of talk are complicated, and I don't claim to know how much, if anything, their popularity has to do with Popper. It does, at least, illustrate nicely the way in which negative results provide limited information. All the falsification of the null hypothesis accomplishes is the demonstration that there is some kind of causal chain relating the factors under investigation. No amount of falsification can replace the hard work of building up a positive case for a specific causal claim by a suitable sophisticated and varied set of empirical investigations.

I have focused on the inadequacy of Popper's falsificationism to illuminate the ways that various kinds of scientific work contribute to the growth of scientific knowledge. However, my point is not simply or even mainly to criticize Popper,[4] but rather to reinforce the suggestion that the variety of scientific practices makes any uniform account of scientific method unlikely. Methodologies have developed in wholly different ways in response to different kinds of problems, and the methodologies we have accumulated are as diverse as those questions. I therefore turn to what is currently a more popular conception of scientific unity: unity of content.

How could the sciences be, in any sense, all about the same thing? The simple answer, and one that is still quite widely accepted, is that the only science, ultimately, is physics, so that all the sciences are really about whatever physics is about. The classical version of this doctrine is the doctrine of physicalistic reductionism. According to this theory, the sciences should be thought of as arranged in a hierarchy, with particle physics at the base, then (perhaps) chemistry, molecular biology, and organismic

---

[4] Indeed, enthusiasts for Popper's work will no doubt object, quite rightly, that I have presented no more than a caricature of Popper's most sophisticated development of his views. My response is that the kinds of ways that crude falsificationism misrepresents these various practices are highly diverse, and attempts to conform them to sophisticated falsificationism would take the theory in diverse directions that would not leave anything like a unified account of scientific method.

biology, and at the top ecology and the social sciences. All sciences other than elementary particle physics were to be reduced to the next lowest level in the hierarchy by characterizing the entities with which they dealt in terms of their structural make-up, and deriving the behaviour of those entities from the laws governing the structural parts of which they were made. Ultimately, therefore, everything was to be understood as an arrangement of elementary particles and the behaviour of everything was to be derived from the laws governing the behaviour of such particles.

As an account of the real workings of science, this picture has been widely, though by no means universally, abandoned. But its spirit has by no means been abandoned, and indeed continues to govern thinking in central areas of philosophy. It also, and no doubt more importantly, continues to have a profound effect on the way many scientific problems are approached. I shall next point to some of the reasons for the abandonment of the classical version of reductionism, and then discuss the somewhat weaker doctrines in which its spirit continues to live on.

A pivotal arena in which conceptions of reductionism have been tested has been genetics. Detailed studies of the inheritance of the characteristics of organisms had occurred for half a century before the famous chemical analysis of DNA in 1953. A good deal of information was collected in this period about patterns of inheritance, and it was natural to suppose that this information could subsequently be translated into the successor language of molecular genetics. This appears not to have been possible, however. The simplest explanation of this failure is in terms of the impossibility of correlating organismic traits, of the kind the inheritance of which was the subject matter of earlier transmission genetics, with anything describable in the terminology of molecular genetics. This, in turn, is because typically many genes are involved in the production of any trait, and any gene is involved in the production of many traits. This suggests a more general perspective on the problem: attempts to understand phenomena at a particular organizational level determine schemes of classification at that level. Such schemes of classification need not, and typically will not, correlate in any manageable way with schemes of classification at lower levels. The force of this point is still controversial, and there are other kinds of objections to reductionism. There has, however, been extensive discussion of the failures of reductionism as a practical project, including my own book-length treatment (Dupré 1993), which includes a variety of arguments against the possibility of various putative reductions. So in this chapter I shall take this failure for granted and look at some of the main reactions to it. The most widespread reactions attempt to maintain the basic metaphysical assumptions underlying scientific unity. I want to argue that such responses are not the most plausible, and certainly not the responses that should appeal to a committed empiricist.

The first such move is suggested immediately by my brief remarks about genetics. If classical genetics cannot be correlated with contemporary molecular genetics, so much the worse for classical genetics. Ultimately it is genes that determine the transmission of traits, and so we must aim to produce a theory of trait transmission based directly on the real molecular causes of the phenomena. The concepts and theories of traditional transmission genetics will very probably not appear at all in this successor science. A little reflection on the hopelessness of this reductionist project will also indicate the kinds of difficulties typical of many or most such enterprises.

Doubts about the present case start with worries about the assumption that 'genes determine the transmission of traits'. This, I think, has become a commonplace, and the fact that it has become a commonplace reflects the grip of this kind of reductionist thinking. Nevertheless, it is a highly problematic assumption. What does it mean? Certainly not that any aspect of the genome is by itself capable of guaranteeing that, say, I will have brown eyes. Clearly a vast array of causal conditions throughout a good part of my early development are also necessary conditions for the development of this trait. More promising might be the idea that the factors that differentiate me and others with brown eyes from those with eyes of a different colour are genetic. But this is still by no means obviously true. We should note in passing that even in the case of eye colour a genetic characterization of the class of people with a particular eye colour will be much more complicated than is often supposed, and will involve several or even many genes. But for most traits no such characterization is possible in principle. Many complex constellations of genetic and environmental factors can lead to someone being six feet tall, say. And when we move to standbys of present investigations of inheritance such as intelligence, the variety and complexity of the possible causal backgrounds would be hard to exaggerate. The temptation to say that only the genetic parts of these constellations are inheritable should quickly be resisted. Wealth, education, and so on, themselves of course diverse and complex factors, are certainly transmitted from one generation to another, and certainly affect the inheritance of height, intelligence, and so on. Once we have given up the idea that genes determine the inheritance of traits, we can see that the replacement of classical genetics by molecular genetics can only be an abandonment of the subject matter. There is no translation between the language of transmission genetics and the language of molecular genetics. And quite typically there is much more to the causal basis of the phenomena to be reduced (in this case trait transmission) than merely the subject matter of the putative reducing science (in this case [molecular] genes).[5]

To mention one more very prominent example, consider the prima facie absurd proposal of a number of contemporary philosophers that we should replace talk of the mind with talk of bits of the brain. This remains absurd on more careful reflection. The

[5] An interesting speculation here is that the term 'transmission genetics' already carries the seeds of error. The development of genetics as a science involving hypothetical factors capable of transmitting traits, and the emphasis on experiments such as Mendel's and later work on *Drosophila* involving traits that did indeed vary in systematic response to different genes, led to the assumption that inheritance was to be explained in terms of genes. When the molecular structure of genes was subsequently elucidated it became natural to think of this as the discovery of the molecular basis of inheritance. But while there is no question that genes play a fundamental role in inheritance, it is essential to realize that this is only an essential contribution to a very complex process.

problem is simply that to replace mind talk with brain talk requires that the latter can serve the purposes of the former. But it is exceedingly unlikely that this is so. Even if, in some sense, we are talking about the brain when we refer to features of our mental lives, there is not the slightest reason to believe that, say, my belief that the US stock market will crash soon can be identified with some well-defined part of my brain; still less that the same part of my brain will consistently correspond to just this belief; and least of all that everyone has a structurally identical part of their brain if, and only if, they believe that the US stock market will crash soon. And it seems that it is this last that would be needed if there were to be some piece of brain talk with which, in principle, one could replace this bit of belief talk. (I suggest, indeed, that this is a place where the supernatural qualities of monism appear clearly. Magical powers are being attributed to brain cells on the basis of no empirical evidence, merely from metaphysical commitment.)

Finally, even if one or other of these reductionist projects could somehow go through, it should not encourage us to see the strategy of replacement as generally applicable. The kinds of obstacles to reduction I have mentioned seem likely to occur in most or all cases, forcing us to resort at every stage in the hierarchy to some kind of replacement. Thus we will not have reached physicalistically impeccable categories until we have gone all the way down to fundamental physics. But the suggestion that we could replace talk about beetles, people, or ecosystems with talk about physical particles seems to have lost all contact with the realities of scientific work.

Despite the absurdity of taking this replacement talk seriously, it is clear what is the intuition underlying it, and this is an intuition that seems far from absurd to most philosophers. This is the idea that whatever we may say about beliefs, intentions, and the like explaining or even causing our behaviour, there is also a set of physical causes that must simultaneously fully explain our physical movements. This brings us, finally, to the doctrine of the completeness of physics. A remarkable amount of contemporary work in the philosophy of mind, in the philosophy of biology, and in the philosophy of the social sciences is concerned with the attempt to reconcile the assumed completeness of physics with the perceived failure of attempts to reduce higher-level science in the direction of physics. One solution is the eliminativism I have just been discussing, the view that the march of science will eventually sweep away the vocabularies of biology, folk psychology, or sociology. A more modest solution is the instrumentalism defended by those who admit that we cannot, perhaps for reasons of principle, get by without biology, psychology, or the social sciences, but add that since the entities of which these sciences speak are incommensurable with those of physics, the former cannot be recognized as ultimately real.[6]

A much better solution, it seems to me, is to abandon the dogma of the completeness of physics, together with the doctrine of the unity of science that it underpins. Let me

---

[6] A good example of this view of biology can be found in Rosenberg (1994). In the philosophy of mind such a view has led, among other oddities, to the bizarre but widely discussed panpsychist dualism of Chalmers (1996).

mention a few reasons why we should be happy to abandon this dogma. Foremost among these is the commitment to empiricism insisted on earlier in this essay, and the observation that there is essentially no evidence for the completeness of physics. We can begin to see this by noting the extent to which the failure of reductionism in various crucial areas of science undercuts the plausibility of the completeness of physics. There are, of course, no actual accounts of the behaviour of mice or men, or even bacteria, as flowing from the physical properties of their smallest physical parts; and no such accounts appear to be in the offing as reductionist science develops. The belief that such an account must exist in principle, or in the mind of God, is at best an inference from what we do know about the behaviour of systems simpler by many orders of magnitude, and hence an inference that goes beyond any decent empiricist strictures. It is, in short, a supernaturalist belief.

It will be said that these simpler systems give us knowledge of laws of nature, and these laws can then (in principle) be applied to far more complex systems. But what could possibly license this extension in the admitted absence of any direct evidence for the applicability of the laws to these more complex systems? As Nancy Cartwright (1983) pointed out many years ago, even as apparently simple and robust a law as the law of universal gravitation only applies to situations in which no other forces (electromagnetic, for example) are acting. Naturally, we have good procedures for dealing with some situations in which there are forces of different kinds acting. But it is still the case that the principles by which we combine different forces are principles distinct from and additional to those laws describing the sources of specific forces. When we move from systems of a few particles under the influence of one or two fully understood forces to systems involving trillions of complexly organized particles subject to many, perhaps very many, different kinds of forces, we are embracing a rather more ambitious augmentation of the basic physical laws.

These objections to the completeness of physics are strongly reinforced by even the most cursory consideration of the actual methods by which physical laws are discovered and confirmed. Accounts of work in high-energy physics laboratories, the central testing ground for contemporary work in fundamental physics, show that conducting experiments in this arena is a very difficult matter. A great deal of time and energy is required to get such experiments to work (Galison 1987). The point is not that this should cast doubt on the results of these experiments, but only that we should reflect on what makes this work so difficult. And the answer, plainly enough, is that it is the process of separating an effect of interest from all the interferences, unwanted effects of the apparatus, and so on, that is required before any meaningful result can be attained. Hence the results we do finally attain are results that apply to quite simple, effectively isolated systems. Their extrapolation even to interactions with other simple factors is not guaranteed. Their extrapolation to every system in the universe, however complex, is easily resistible. Some people seem to be very impressed by the very high precision of predictions from quantum mechanisms that are sometimes experimentally confirmed. This seems to me a red herring. Precise confirmation of predictions of course adds to the impressiveness with which the application of laws to the systems in which they are being tested is supported. It is also a testament to the skill with which the experimenters have succeeded in isolating the effect from unwanted interferences. But it has little or no effect on the plausibility of extrapolating to the applicability of those laws to quite different kinds of systems.

My conclusion is that there is very little reason to believe in any kind of unity of science. The idea of a unity grounded in method fails to survive a cursory scrutiny of the variety of methods employed in science. The idea of a unity of content, grounded in the completeness of physics, seems to lack any convincing rationale and, most importantly, any empirical support. And since it is a powerful and, I think, counterintuitive view, I am inclined to think its hold on us is best explained either in terms of a degenerating historical tradition or, more interestingly, in terms of non-epistemic functions it serves. It is to this possibility I now turn.

The functions of the myth

The foregoing arguments have, it seems to me, been thoroughly naturalistic in spirit. They assume that our knowledge of the world should be derived from our experience of it and our interactions with it. First and foremost among these will be the projects— often remarkably successful, explicitly aimed at gaining knowledge about the world— that comprise the sciences. Naturalism suggests, however, that we not think about these in terms of Procrustean a priori epistemology, but rather as natural objects apt for empirical scrutiny. Such scrutiny reveals a diverse set of practices using methods some of which are common to a variety of investigations but others of which are peculiar to particular areas. Though it certainly reveals local attempts at reductions, it reveals few if any fully elaborated reductions of one level of phenomena to a lower level of structural complexity or replacements of accounts of one domain by a lower-level domain. I think one may go one step further. The universe-wide microphysical machine, the integrated realm of microscopic particles that forms the substance of reductionist fantasies, is not a product of naturalistic enquiry, but a supernatural construct of the scientific dreamer. Naturalists should reject the image not just because it lacks proper naturalistic credentials, but because it violates the most basic naturalistic commitment to the rejection of the supernatural. In this concluding section I offer some speculations about the appeal of reductionism and unity to many who see themselves as card-carrying naturalists. This should also suggest some of the likely and desirable consequences of abandoning this relic of supernatural thinking.

The first function that unity of science theses serve is as a ground of demarcation. Science carries an epistemic authority that generally greatly exceeds that of non-scientific practices of knowledge production. And I do not wish to question the assumption that our most compelling examples of epistemic excellence come from the sciences. It would presumably be a threat to the credentials of more dubious areas of science if scientific practices were evaluated piecemeal for epistemic worth. The impression that macroeconomic theory, say, or Evolutionary Psychology, has serious empirical credentials surely owes much to the idea that they are part of science, and that science, surely, has significant empirical warrant.[7] Unity provides solidarity and protects the weaker brethren. This would obviously be the case if science had a unity of method as then, presumably, we could suppose that that method was generally a reliable one. But a similar conclusion would at least be suggested by any kind of unity.

Unity, in short, distributes epistemic warrant. The claim to be scientific is not an important one for solid-state physicists or organic chemists; it is one they take for granted. But on the more controversial margins of science such claims are all-important. Economists claim to be scientific in ways that their more interpretive rivals among the social sciences cannot aspire to, and Evolutionary Psychologists claim to be uniting the study of humanity with science in ways that must spell the end for more traditional exceptionalist accounts of our species. The importance of such rhetorical moves, it seems to me, depends on some version of the thesis of the unity of science.[8] If science is no more than an overlapping collection of practices for investigating the world, with diverse assumptions and methods, and often incommensurable results, then it is unclear why anything much hangs on the claim to scientificity. The status of 'science' might, on such a view, much better be used as an honorific to be bestowed on investigative practices when they have provided convincing evidence of success in their investigations. (Of course it is also possible that it is just such an honorific that economists and others intend to bestow upon themselves.) On the other hand, if there is just one system of interconnected truths that constitutes science, a science moreover that ultimately, at least in principle, exhausts the truth about the world, then everything depends on establishing the claim of one's practice to belong to this totality. And if such could be done on general grounds that do not require the demonstration of actual empirical successes, the relevance of such claims will obviously be greater still. Here I suggest we see Science as a whole in its supernatural guise. Just as membership of the True Church guarantees redemption, so membership of the One True Science guarantees credibility. In both cases the details of how this trick is pulled off remain obscure. And neither strategy should appeal to a naturalist.

The consequences of the ideology of scientific unity are not limited to matters merely theoretical. Reductionist models of scientific unity have a particular and potentially damaging effect on the practice of science. The ultimate goal of articulating unified science in its full glory leads naturally to a preference for seeing phenomena as depending on the internal structure of the entities that produce them rather than emphasizing the influences of the environment. Probably the most serious practical

---

[7] This point is developed at length in my book *Human Nature and the Limits of Science* (Dupré 2001). There I try to show both the defects of some questionable scientific projects, most especially Evolutionary Psychology, and also how the motivation for much bad science draws on the mythology of unity.

[8] It is no coincidence that John Tooby and Leda Cosmides begin what has become a classic defence of Evolutionary Psychology with a discussion of the unity of science (and the important place of the former in the latter) (Tooby and Cosmides 1992).

consequences of this tendency are in the human sciences, and most especially in the medical sciences. Consider, for instance, the several million American children (mostly boys) recently discovered to be suffering from attention deficit hyperactivity disorder (ADHD) but, happily, being treated with apparent success with the drug Ritalin. It is somewhat surprising that such a widespread disorder should have been unknown a few decades ago. But this doesn't mean that there were not numerous sufferers. Albert Einstein is sometimes mentioned as a prominent victim.

No doubt among these millions are some seriously sick children. But I do not find it a bit surprising that many children now, and in the past, have had difficulty paying attention in schools. I do doubt whether this proves that there is something wrong with these children's heads that is appropriately treated with psychotropic (and, apparently, addictive) drugs. Schools are, after all, often boring. The fact that powerful drugs can alleviate the manifestations of the syndrome shows very little. Threats of violence may be equally effective at concentrating the minds of recalcitrant students, but this would not prove that they were suffering from corporal punishment deficiency syndrome. There are many ways of influencing behaviour. It is evident that there is some kind of mismatch between the dispositions of the problem child and the social context in which that individual is placed. Such a mismatch could, on the face of it, be addressed by changes to the child, to the environment, or both. I do not deny that changes to the child brought about by the ingestion of psychotropic substances may, in the end, be the best solution in many cases, though it is disturbing that anyone distributing such substances at the school gate rather than in the doctor's office would be risking decades of incarceration. My worry is that the reductionist perspective on science makes this sort of response look natural, if not inevitable. Millions of drugged children—or, to take a different case, slashed, burned, and poisoned patients in unrelievedly carcino-genic environments—are, arguably, the price we pay for action on the basis of this myth. The suggestion that this sort of thing is explicitly part of the function of the myth is perhaps uncomfortably close to conspiracy theory or even serious paranoia. It is of course true that drug companies make many billions of dollars from their expertise in adjusting people's minds to the demands of the environment, and it is surely also true that it is much easier and generally cheaper for governments to point to the defects of individuals than to attempt to make positive changes in the environments to which people appear maladapted. Whether there are connections between the prevalence of reductionist thinking and its obvious advantages for some of the most rich and powerful is not, at any rate, something about which I pretend to have any evidence.

Conclusion

The main point of this chapter has been to emphasize that monism, and the unity of science doctrine on which contemporary monism rests, far from being inevitable concomitants of a properly naturalized philosophical perspective, are elements of a new mythology. Monism is surely not grounded on empiricism. For one thing, if it were, there would be no need of the vast amounts of work expended in the elaboration of eliminativist, instrumentalist, and supervenientist theses designed to explain the empirical failures of monism. More simply, our empirical experience of nature is, on its face, an experience of a huge diversity of kinds of things with an even huger diversity of properties and causal capacities. Some of these properties are open to casual inspection; others require careful, even inspired, scientific investigation. Neither casual experience nor detailed investigation suggests that all these properties are best understood through attention to the physical stuff of which things are made. The advance of science does indeed lend credence to the view that we do not need to appeal to supernatural things in explaining phenomena. One variety of supernatural things are those that are made out of non-physical stuff, like angels or Cartesian minds. So we may allow that naturalism commits us to the monism that insists that all stuff is material, even physical, stuff. The corollary that insight into the properties of stuff holds the key to understanding the properties and behaviour of all those diverse things that are made of that stuff is another matter altogether. And this indeed is the kind of doctrine that suggests the attribution of supernatural powers to physical stuff in a way wholly inimical to naturalism.

Casual and not so casual inspection of the world suggests that it contains a great diversity of kinds of things. Similarly, inspection of science, our best ways of finding out about the world, suggests that there is a great diversity of projects that can be directed towards the investigation of the natural world. Does this leave us without resources for distinguishing the credible (scientific) projects from the worthless (unscientific) projects? Not quite. There is no a priori criterion for distinguishing successful projects of enquiry, but that, I think, is what we should expect as naturalists. A posteriori we can distinguish success well enough in terms of familiar epistemic virtues such as understanding, explanation, prediction, and control. Many parts of the physical and biological sciences are unmistakable successes on the basis of their production of such goods. A more abstract level of reflection allows us to see particular methodological virtues that have contributed to the success of these projects. Many investigative projects that have been championed at different times lack both the products of success and the virtues that have been seen to lead to success. It is of course a difficult matter of judgement when we should conclude that such a project is irredeemably flawed, but it is surely a judgement we must sometimes be willing to make. However, Science as a unified reification and as a fig leaf behind which broken-backed investigations can hide is nothing but an obstacle to the process of deciding which projects belong to science, in the merely honorific sense.

There are further benefits to be gained from a pluralistic naturalism. C. P. Snow's (1959) famous jeremiad about the Two Cultures, the opposed forces of science and humanism, is outdated in one sense. No cabal of Oxbridge dons now effectively opposes the application of science to such problems as world hunger for which science has essential contributions to make. Even in Britain, science has gained the upper hand over its humanist enemies. On the other hand, the opposition between the two camps sometimes seems as strong as ever. In the contemporary Science Wars,[9] in which angry scientists rail against the humanists alleged to have discussed science with insufficient respect and expertise, it is still clear that science is assumed to form a hegemonic unity, and there is a strong tendency for the opposition to be constructed as an equally unified Luddite whole. Enemies, real or imagined, are a classic device for reinforcing solidarity or unity.

In the hope that this may tend to reduce these hostilities, I suggest that the rejection of the unity of science is the path to what there is of value in a genuine unity of knowledge. The sciences offer us a diverse group of practices with a diverse set of partially overlapping virtues. Some encompass large ranges of empirical facts; others enable us to predict or control natural events; others again provide theoretical equipment for thinking more clearly about complex phenomena; and so on. But some of these virtues also characterize the traditional non-sciences. History is an essential repository and organizer of empirical fact; studies of the Arts provide us with insight into aspects of human life that are still far removed from the more mechanical approaches of psychology or the social sciences; even philosophy provides us tools for thinking about the natural world; and so on. What is most valuable about this picture of diverse and overlapping projects of enquiry is that it makes unsurprising what seems empirically to be the case, that complex phenomena are far more likely to be understood if a variety of distinct but complementary approaches are brought to bear on them.[10]

But having insisted on a variety of practices of knowledge production with diverse and overlapping epistemic virtues, I can now allow that there are characteristics that distinguish much of science from much academic work outside of science. For example, science tends to aim at the expansion of empirical knowledge, whereas humanistic disciplines are often more theoretical and more critical. And this leads to perhaps the most essential form of interaction between the two. One of the contributions that has made Kuhn's *The Structure of Scientific Revolutions* (1976) so fundamental to current understanding of science is the insight that science does not, on the whole, promote criticism. Of course, scientists have very heated disputes about the details of their empirical or theoretical claims, but these take place within a context that is not, on the whole, called into question. The contemporary studies by sociologists of the details of how scientific work is really done; by historians of how it has been done in the past and how the doing of it has led to the consensuses that exist in various contemporary scientific fields; the attempts by philosophers to expose and analyse those assumptions that are not called into question in the doing of science; and much else that goes under the very general rubric of science studies, not least in the

---

[9] [Note added to this edition] Contemporary at the time of first writing. At the time of reprinting this essay in the present volume, they seem thankfully to have receded.

[10] The broadest theme of my (2001) is that human nature is a topic that can only be adequately addressed from such a plurality of perspectives.

philosophy of science, has surely improved our understanding of science and even, in the end, may encourage the doing of better science. The division between C. P. Snow's Two Cultures can only be deepened by those who, for whatever reasons, insist that there are indeed two, and that those on one side have no right to say anything about what goes on on the other side. But there are not two grand cultures, but many small and overlapping subcultures. And knowledge will advance to the extent that the members of different subcultures make whatever efforts they can, even if sometimes critical efforts, to inform the work of those in others.

And this leads finally to a point at which pluralistic naturalism does really diverge from more standard recent versions of naturalism. Quine, famously, held that philosophy was continuous with science. Given that science is perceived as a unity, and as a unity that will contain all the genuine insight into the nature of the world that we manage to acquire, the only possible positions for philosophy appear to be as part of science or in the dustbin of history. In light of that choice, Quine's decision is understandable. No such predicament faces a pluralist. Indeed, for a pluralist the thesis that philosophy is continuous with science faces the obvious question, Which science? Philosophy, on the contrary, emphasizes rather different epistemic virtues from most sciences and has typically different goals. Its characteristic epistemic virtues are, perhaps, analytic rigour and clarity of argument, though certainly it is subject to others, such as sensitivity to empirical fact, that are central to most sciences. Its goals are typically more abstract, theoretical, and critical. There are, surely, fundamentally important classes of questions for which the methods of philosophy are particularly suited and even essential, though I certainly do not propose here to make any claims on the controversial matter of what those questions are. However, the view of philosophers as a kind of laboratorially challenged scientist seems to me to have been a total failure. Pluralistic naturalism offers a way out of this backwater, but one that does not consign the philosopher to even more unpromising undertakings in the noumenal world or Plato's heaven.

2

What's the Fuss about Social Constructivism?

The topic of this chapter is social constructivist doctrines about the nature of scientific knowledge. I don't propose to review all the many accounts that have either claimed this designation or had it ascribed to them. Rather I shall try to consider in a very general way what sense should be made of the underlying idea, and then illustrate some of the main points with two central examples from biology. The first thing to say is that, on the face of it, some doctrine of the social construction of science must self-evidently be true. The notion of science as progressing through the efforts of solitary geniuses may have had some plausibility in the seventeenth century, but it has none today. Science is a massively cooperative, social, enterprise. And surely it is constructed. Scientific knowledge doesn't grow on trees; it is produced through hard work by human agents. Putting these two banal points together we conclude that science is socially constructed.

The second thing to say, on the other hand, is that a great many philosophers appear to think that social constructivism is not only false, but positively pernicious. Given the first point, this presents something of a paradox. Presumably more is taken as implied by this apparently harmless phrase than appears from my banal gloss. One such implication emerges from a less naïve interpretation of 'social': the question is not just whether science is cooperative, but whether the values or beliefs of a particular society influence scientific opinion. I shall return to this point shortly. Some anti-constructivist philosophers would also object at once that my first paragraph has begged the question by talking about scientific *knowledge*. For one interpretation of constructivism is that it somehow implies that what scientists say about the things they study is not true, or that the things they purport to describe do not exist. It is associated, in other words, with various forms of anti-realism. No doubt it is the case that some who claim to be social constructivists also subscribe to some brand of anti-realism, though, as I shall insist, it need not imply any such thing. Many things are, after they have been constructed, real. The problem, of course, is that the notion of construction may suggest that science is, in some sense, made up. Traditionalist philosophers want to insist that scientific truths are discovered, not in any sense made up.

What I want to suggest in this chapter is that there is solid middle ground between the extremes of science as discovered and science as made up. Science, I take it, is produced by people in interaction with nature. Nature doesn't determine what science we produce, because there are indefinitely many sets of truths we could articulate about nature. Which we choose to articulate will depend on the interests of a particular investigating community. We want, as Philip Kitcher (2001) has recently put it, to distinguish the *significant* truths. And though truths are only significant for someone or some society, they may, nonetheless, be *truths*. I shall attempt to illustrate such a position by focusing on the determination of schemes of classification. While nature does not, I think, determine how we should classify phenomena, something that is in fact importantly guided by our interests, schemes of classification can be more or less effective vehicles for genuine illumination of nature. First I shall say a bit more in defence and explanation of this general view.

Consider the suggestion that if our scientific stories are constructed, made up even, they may be made up for reasons quite distinct from the traditionally assumed goal of simply describing how things are. Actual critics of actual constructivists do often assume something like this, and accuse constructivists of holding that scientific claims are devised largely to serve personal, social, or political goals of their authors. And it is certainly true that constructivists have tended to pay more attention to determinants of scientific belief of these kinds than have more traditional philosophers of science. When it is further supposed that these secondary goals are sufficient to explain scientific belief on their own, this view is both implausible, and implausibly attributed to most serious thinkers who have embraced the term 'constructivist'. Few deny that scientific belief has some important dependence on interactions with the world. On the other hand, scientists are people, and undoubtedly they do have personal, social, and political goals. If, as many philosophers have insisted, appropriate interactions with nature are insufficient to determine scientific belief, it is hardly implausible that these other goals may have some role to play.

This, I think, gets us close to the genuine philosophical issue that underlies much of the controversy over social constructivism. Those most strongly opposed to constructivism are those who believe that interactions with nature completely determine scientific belief and, in connection with proper scientific methodology, determine true beliefs about the world. Of course there is a good deal of nuance possible to such views. No one supposes that all beliefs about nature ever held by honest scientists are true; it is acknowledged that science may be very difficult, and take many attempts. And it is generally acknowledged that there are, in principle, indefinitely many true facts about the world, so some account is needed of the process by which a finite set of these is selected as being worth including in the corpus of science. However, such optimists will at least assume that science approaches the truth, and they are likely to believe that in some areas it has even reached the truth. They will also want to explain failures to achieve truth in terms of some kind of failure of scientific method—perhaps failure unavoidable at that point in history, perhaps excusable in many other ways, but nevertheless in some way a deviation from the ideal interaction with nature that will, in the end, lead to the true story of how things are. Thus one concrete issue that tends to divide constructivists from their critics is the universality of scientific knowledge. If nature will eventually tell us the true story, then all sufficiently diligent enquirers should converge on the same story. If knowledge involves also something that we contribute, then different communities of enquirers may contribute different things, and come out with different knowledge.

Another approach to genuine disagreement here is through a classic statement of social constructivism, the symmetry principle famously associated with the Edinburgh School of the sociology of scientific knowledge. This principle states that there is no difference in principle between the explanation of scientific beliefs that we take to be true and those that we take to be false (Barnes 1974; Barnes and Bloor 1982). Though apparently quite innocuous, the principle in fact seems to block the anti-constructivist project of explaining the failures of scientific process to produce scientific truth.

There is surely something importantly right about the symmetry principle. What causes scientists to believe what they believe is at one level a psychological question. And it would be absurd to suppose that there is a systematic psychological difference between those scientists who got it right (as we now think) and those who got it wrong. Or, indeed, that there is some different psychological process at work when the same scientist gets it right or wrong—as if Newton must have been in a different psychological state when he asserted the existence of absolute space and when he proposed the laws of motion. This point is quite independent of whether we are inclined to believe that such explanations are more likely to be found in his relations with his mother during early infancy or in rational deliberation on the deliverances of experience.

Nonetheless, it will be said, there is one kind of explanation that is available for true beliefs but not for false beliefs, namely the explanation of the belief that p by appeal to the fact that p. Thus we may be tempted to say that the belief that there once were dinosaurs is most fundamentally explained by the fact that there once were dinosaurs. But of course this kind of explanation always requires a further step. Facts only explain beliefs via evidence or reasons. The belief that there were dinosaurs is explained by such things as observation and analysis of fragments of bone, the results of carbon 14 tests, and so on. These facts are no doubt explained in turn by the erstwhile existence of dinosaurs. But the dinosaurs themselves cannot explain the difference between the beliefs of the palaeontologist and the person who believes that the world was created in 4004 BC. Neither has direct contact with the dinosaurs. One, sensibly, takes the bones as evidence for the existence of long extinct creatures, the other, less sensibly, as evidence for God's ingenuity in testing our faith. Even were I to travel back in time to the Jurassic and observe real live dinosaurs it is my observation of them rather than the dinosaurs themselves that explain my much firmer belief in their existence. I do not mean here to introduce a strange mental object, merely to note that it is in interaction with the world that beliefs about it are produced, and this interaction is more than purely passive reception of pre-interpreted data.

Hence both special creationist and evolutionary accounts of such things as the provenance of dinosaur bones are socially constructed in the stronger sense introduced at the beginning of this chapter. We can very well ask which are reasonable, defensible constructs. And we can give general accounts of the way science works as part of the grounding for the distinction between reasonable and unreasonable epistemological constructs. It is then important to note that this distinction by no means maps exactly on to the constructs we now take to embody true knowledge and those we do not. Newton's mechanics was a brilliant and reasonable construct, even if we now take it to have been wrong in some very important respects. Democritus was perhaps right in certain respects in his atomic theory of matter, but it is open to question whether his beliefs were properly founded. Social constructivism is often taken to deny that there can be any distinction of epistemological merit between different epistemological social constructs (theories, let us say). What I have pointed out is that there is nothing in the idea of social constructivism that mandates this conclusion. And, as a matter of fact, I doubt whether many of those thinkers prominently associated with social constructivism hold any such opinion.

Before considering some examples in more detail, I need to say a bit more about the senses in which science is a 'social' activity. As already noted, it is largely uncontroversial that science is a highly cooperative, that is social, activity. However, social constructivism invokes different connotations of this term, in accordance with which it is supposed that various social and political agenda affect the outcome of science. In fact there is a huge range of actual and possible positions here. At one extreme it is possible to note the collaborative nature of much contemporary science with equanimity and conclude only that science has become more complicated and difficult. What could once be done by a solitary investigator now requires the contributions of several or many.

A growing number of philosophers, without going so far as to embrace the designation 'constructivist', have recently taken the social dimension of science much more seriously, however. There is a growing body of work under the rubric of 'social epistemology' specifically addressed to exploring the consequences for our understanding of knowledge of a social perspective (see, for example, Solomon 2001). One particularly interesting project is Helen Longino's (1990, 2002) argument that objectivity can only be understood as arising from a dialogue between different interested perspectives, and that science needs to be organized socially in ways that will best promote such dialogue. In a somewhat similar vein, Philip Kitcher (2001) has recently argued for the importance of 'well-ordered science' science that is directed optimally at the production of the knowledge that will genuinely benefit (and, importantly, not harm) the members of a society. Well-ordered science, naturally, will tend to generate (socially) significant truth. At the most epistemologically radical, and explicitly constructivist, end of the spectrum, we might instance, for example, Bruno Latour's (1987) examination of the process by which a scientific claim comes to be generally accepted, a process Latour describes in terms of the recruitment and regimentation of allies.

Though all of these philosophical projects consider science as much more socially involved than simply asking nature to answer our questions, none of them comes close to denying that there can be any truth to what is believed. Latour, no doubt the closest to such an outcome, may indeed be sceptical of scientific truth for a variety of reasons. But as far as the project just summarized, we need only note that among the 'allies' successful scientists recruit are such things as biological organisms or even inanimate objects. And the ability to control such allies surely depends on having some real insight into what they are and how they work.

On the other hand, even the less radical positions delineated view science as a social enterprise in which there is space for the negotiation of scientific agenda and, to some extent, outcomes. Processes are distinguished that allow plenty of scope for social and political forces to have an impact on the progress of science. I don't have space here to examine these processes in detail. In the examples that follow, I shall rather attempt to illustrate a somewhat less controversial aspect of the construction of scientific knowledge through which, it may be argued, social agenda can have an impact on scientific belief. This will leave open the importance of such an influence. The aim will rather be to define an important part of the playing-field in which serious debates between constructivists and their opponents take place.

More specifically, I plan to show that in biology, at least, the conceptualization of phenomena, as displayed in schemes of classification, is not determined by nature (as a still important philosophical movement would have it) but is developed, or constructed, in the light of human goals and interests. As indicated above, however, I do not take this to be incompatible with a generally, if selectively and critically, realist attitude to the phenomena they describe. The broader thesis of which this is part, a thesis which will be unwelcome to both extreme positions in the Science Wars, but will seem commonplace to a growing number of moderate intermediate positions, is that science is a process of interaction between social agenda and often growing insight into the natural world. Since I can make no sense of the idea of a final complete insight into nature, I assume that science is likely to remain a domain of partial and partially interested insights into nature.

The scientific construction of kinds

So much for global views of social constructivism. My conclusion so far is that there is no issue here that should be very controversial. Social constructivism is, if modestly construed, largely self-evident. Less modest anti-realist or sceptical positions sometimes associated with constructivism should be treated separately on their merits or lack of them. I turn now to a view which I wish to defend, and a view which is congenial to certain aspects of constructivism though certainly not to anti-realism or scepticism. This view is a thesis about the classificatory systems which are applied to domains of phenomena for purposes of scientific investigation, and the kinds that are distinguished by such classifications. The claim is that rather than being discovered in the course of investigation, delivered by nature to the properly diligent investigator, classificatory schemes are constructed. Though such construction is hardly oblivious to interactions with nature, it is also likely to be motivated in part by factors that may be described as social. Let me begin with two general points about this issue.

First, scepticism about the ability of nature to provide us with the classification of her products has nothing to do with any kind of anti-realism or scepticism about the things classified. I might, not very promisingly, begin a scientific investigation of the category of things that I particularly like. This might generate a miscellaneous assortment of people, kinds of food, pieces of music, and so on, which would not strike many as a category straightforwardly discovered in nature. But this casts no doubt on the existence of these things. There is, however, a much more elusive question about the reality, or perhaps the objectivity, of the category. In one perfectly legitimate sense a category is real if anyone bothers to distinguish it, provided only that it offers a more or less reliable criterion for inclusion within it. On the other hand, there is a strong inclination to attribute a much more robust existence to, say, the category of acids than to the one defined by my peculiar tastes. One reason for this is that whereas there are both a range of typical characteristics of acids and a general explanation of these common characteristics, in the case of the things I like we might well doubt whether there was anything to be discovered about this class of things beyond the initial defining characteristic, my positive feelings about them. The greater reality of the former category has something to do with its having more robust and interesting properties. This also suggests that the latter category would be poorly chosen for scientific purposes: it is doubtful whether there are any general claims to be made about its members, let alone claims that might amount to scientific laws. And the reason that many philosophers hope that nature herself will provide us with categories is that they believe that only categories with this provenance are likely to figure in scientific laws.

The second point is that the question of the social component in categorization is likely to be a local and variable one. There is no reason why nature should provide us with categories for social science even if she is generous enough to do so for the purposes of particle physics, for example. It is very widely supposed, in fact, that the basic kinds of physics and chemistry are purely discovered and that the kinds of particles recognized by physicists and the atoms distinguished by chemists are the uniquely correct kinds in terms of which physics and chemistry must be developed. I shall say nothing here for or against this view. On the other hand, it is almost as widely believed that the classifications used in social science are chosen to serve particular needs and interests and are, for that reason, contingent. This idea is taken a stage further by Ian Hacking (1999) in his development of the idea that the choice of human kinds, at least, may have profound effects on the humans who are the members of those kinds, an idea captured in his conception of 'looping' kinds. What seems indisputable is that the social and human sciences will deploy a variety of classificatory schemes, and that these will cross-cut one another. A particular individual might be, for instance, a manual worker, a member of the Labour Party, homosexual, an atheist, and so on, classifications which he would share with distinct further sets of people. Any of these classifications may be relevant for some social scientific project.

Biology, generally seen as lying somewhere between the physical sciences and the social sciences on many relevant dimensions, is therefore a particularly salient dimension in which to consider the respective roles of natural and social factors in the determination of classificatory schemes. Before turning to this, I shall illustrate some of the central themes I want to develop with an example from the social sciences.

Consider the human category of criminals. It is hard to imagine that social science could do without this category, and indeed there is a whole discipline, criminology, based on it. Yet it is quite obvious that the category is in important respects constructed by society. Who counts as a criminal is determined by the laws currently enforced in a particular society, and as legislation is introduced and withdrawn the extension of the kind 'criminal' changes. Clearly the category must be interpreted relative to a specific society: a twenty-year-old drinker, for example, is a law-abiding member of most European societies, but a criminal in the USA. The staggering incarceration rates in Russia and the USA (about 1 person in 200) do not reflect an unusual prevalence of a particular natural kind, but social decisions about what counts as serious criminality. The complexity of the processes of criminal formation and of the social motives underlying these processes has been famously and brilliantly explored by Michel Foucault (1977).

Despite its central category being socially constructed and relative to a particular society, criminology has some true things to say. There are established links between, say, criminality and poverty, and statistically true psychological generalizations may well be discovered about criminals. I say 'statistically true' because there can be little doubt that in every respect apart from their criminality, criminals will surely be a heterogeneous group. Think only of the stereotypical images of pickpockets, pimps, Mafia hitmen, corrupt executives of multinational corporations, or politicians engaged in launching illegal wars. Without endorsing the stereotypes, there is surely not much in common between these groups beyond the necessary contempt for social norms. The diversity of cross-cutting kinds will, I suppose, inevitably put quite modest limits on the possibilities for generalizations about kinds of humans. The question for biology, then, is whether we should see biological kinds as more like human kinds or chemical kinds or, perhaps most plausibly, somewhere in between.

Let me now turn to two biological examples which show quite different patterns of interaction between elements of social construction and elements of the naturally given. The first of these is a topic with which biologists and philosophers of biology have been concerned for many years, the classification of organisms into species. The second is a more recent concern, but one that is increasingly central to theoretical debates in biology, the classification of parts of the genome into genes.

Species

The so-called species problem has been a concern of biologists and philosophers at least since the general acceptance of Darwin's theory of evolution, and in different forms, at least since the Greeks. The modern version centres on the attempt to reconcile a conception of biological kinds generally, and species in particular, with the theory of evolution. A natural point of entry is with the thought that if evolution implies that one species can, over time, turn into another, there can be no sharp divisions between species of the kind that appears to exist between, say chemical kinds. Of course, it is now possible to transmute one element into another, but this is a discrete process and does not entail the existence of a range of kinds intermediate between the two elements.

The assumption that species are the entities which are the primary subject of the theory of evolution has led many philosophers and some biologists to conclude that species are not kinds at all, but concrete, if dispersed, individuals. I shall not be concerned with this issue in this chapter. This is because whether or not it may sometimes be necessary to treat species as individuals, it is also necessary for many purposes to classify organisms (see Dupré 2002: ch. 4). This will be true, for instance, for many purposes in ecology or for the measurement of biodiversity. The question that will concern me is whether such a classification is something provided by nature or, rather, something in part constructed for particular human purposes. The former answer would be consistent with the position that in classifying individuals we were in fact distinguishing parts of individuals somehow constituted as such by nature.

There is a notoriously large number of opinions on what distinguishes organisms as members (or parts) of particular species. By far the most popular, however, are those based on a criterion of reproductive isolation or reproductive coherence, and those based on genealogy. The trouble with a criterion of reproductive isolation is its limited applicability. The context in which it has greatest intuitive plausibility is for a species in which a range of genetic and behavioural constraints limit reproductive relations to a well-defined set of relevantly similar organisms. Reproductive isolation can also be seen as potentially serving the important function of maintaining the integrity of a coherent and successful genotype. But whereas this seems to apply very well to many species of birds and mammals, for instance, in many parts of the biological world it is much less successful. To begin with, many species are asexual. The attempt to apply this conception of the species to asexual organisms would imply that every distinguishable clone of such a species, being reproductively isolated, was a distinct species. This would lead to a massive proliferation of species and, ironically, imply that the vast majority of species were local clones the application to which of this criterion would seem lacking in intuitive plausibility. At any rate, for a large proportion of the history of life all species were asexual and, incidentally, appeared to show very little in the way of reproductive isolation. Increasing realization of the extent of lateral genetic transfer between apparently quite distantly related organisms in prokaryotes is beginning to problematize even the assumption of a specific genome reliably characteristic of a branch of the Tree of Life. (The chapters in Part 3 of this book aim to cast doubt on the whole idea of a Tree of Life.)

It is very widely the case that the criterion of reproductive isolation fails to deliver intuitively satisfactory categories. A classic example is a group of American oaks between which there is considerable gene flow, but which have retained morphological distinction over very long periods of time (Van Valen 1976). The theoretical conclusion, given the conception of the species in terms of reproductive isolation, is that this has been discovered to be a single and variable species. But for the forester or ecologist who has reason to distinguish these different types, the criterion of reproductive isolation fails to provide the required kinds. At the other extreme, there are morphologically homogeneous species that exist in a number of geographically distinct populations. These may be reproductively isolated one from another for long periods of time, and therefore would appear to qualify as distinct species. But such a conclusion is driven solely by theory, and serves no useful practical purpose.

The second, and currently most popular, conception of the species ties species directly to evolutionary history by treating the species as a distinct part of the genealogical nexus. Cladism, the dominant version of this idea, requires that species be monophyletic, which is to say that they should include all and only the descendants of a particular ancestral grouping. An immediate difficulty for this approach is that monophyletic groups are likely to occur at many different scales, and it is unclear how to decide which of these is a species. Isolated populations dying out in marginal habitats will form monophyletic groups as will a whole series of increasingly inclusive groups. This difficulty has led a number of theorists to conclude, plausibly enough, that the extent of the species is a matter of convenience and hence, in a certain sense is determined by scientists rather than by nature. But the requirement of monophyly itself presents serious conflicts with intuitive judgements about what organisms should be classified together. This is most familiar and striking at the level of larger groupings. Since birds are believed to have descended from a species of dinosaur any monophyletic group that includes all the dinosaurs will also include the birds, showing, apparently, that birds are a kind of dinosaur. It is often reported that science has discovered that birds are dinosaurs, and if the monophyletic criterion of classification is accepted, then this is a correct statement. On the other hand, given that birds have diverged rather significantly from their dinosaur ancestors another plausible conclusion is that the requirement of monophyly is either misguided or shows that scientific classifications are of limited use for the intuitive project of classifying like with like.

A final difficulty, alluded to above, is that inheritance is not limited to the transfer of genetic material from parent to offspring. It is increasingly clear that in many groups of organisms there is substantial lateral transfer of genetic material through a variety of mechanisms. This phenomenon raises a doubt as to whether the criterion of monophyly is even intelligibly applicable in a general way, since a group of organisms may have a degree of descent from a variety of perhaps distantly related ancestors. The general strategy of classification into monophyletic groups assumes the traditional view of evolution as generating an always divergent tree. Lateral genetic transfer suggests that a better representation may be a densely connected net of hereditary relations. This seems to be increasingly the case for prokaryotes and to a lesser extent even for eukaryotes. This move threatens to undermine the entire project of phylogenetic classification.

I don't propose to examine in detail here the various pros and cons of various strategies for biological classification some of which I have not yet even mentioned (for example morphological and ecological classifications). The important point, which I have argued in much more detail elsewhere, is that there is no criterion discoverable through biological investigation that provides us with a unique and privileged system for organizing biological diversity. Particular criteria break down in many cases and, more relevantly to the present topic, they present us with classifications often poorly suited to the applications to which a variety of scientific and non-scientific users may wish to put them. If there were, nonetheless, some unequivocally given classificatory system one might look at this failure of usefulness as merely regrettable. But given the controversy over which system is the best, and given the deep flaws that each of them has in some areas, no such conclusion is reasonable. A more reasonable response is to see particular classifications as selected in the light of particular goals. And this, of course, is to say that classifications are constructed by people to serve their interests.

The observation that even within properly scientific biology principles of classification must be selected to serve particular theoretical or practical ends removes any serious reason for doubting that even thoroughly non-scientific activities may legitimately develop their own classifications of biological entities. Elsewhere I have suggested that such practices as gastronomy, forestry, herbal medicine, and so on may all legitimately divide organisms in ways that do not coincide with those found most useful for scientific purposes (Dupré 2002: chs. 1, 2). There is also no doubt that such interests have left their mark on canonical scientific classifications. At the very least, degree of interest has a major effect on fineness of classification. So, for instance, the various closely related rosaceous fruits (apple, pear, quince, medlar) would surely have been assigned to the same genus but for their economic salience (Walters 1961).

All these possibilities arise from the fact that nature appears to have underdetermined the taxonomy of her products, an outcome that is easy to understand when one reflects on the process, evolution, that we take to have generated those products. However, it is important to stress that the claim is not that there are no natural divisions to be found between kinds or organisms. Rather, there are too many. We have to choose which to focus on, and such choice will inevitably and appropriately be constrained by the theoretical ends that our taxonomies are designed to serve. In this respect the situation is parallel to the social case. There are countless divisions we could emphasize between humans, and many have actually been emphasized. Which are emphasized, investigated, and perhaps thereby deepened, will depend on our interests and goals. These interests drive the production—or construction—of the great range of biological categories we distinguish and in turn make possible the various kinds of knowledge that different groups of us develop about the biological world. This range of categories, however, represents only an infinitesimal fraction of the distinctions that could, in principle, be made.

Genes

The project of identifying genes within the organismal genome is a very different one from the project of identifying and classifying organisms. The best point of entry to the former is historical. Whereas organisms have been classified probably since the dawn of language, genes have been with us for about a century. The word 'gene' is generally attributed to Wilhelm Johannsen, in 1909. The basic idea for which it has most generally come to stand is that of particulate units of inheritance, and is standardly traced to the experiments of Gregor Mendel, an Austrian monk, in the 1860s. Mendel's famous experiments on peas disclosed numerical ratios in crosses between strains with different characteristics that suggested the existence of factors responsible for the characteristics that were inherited intact, without blending or dilution, to subsequent generations. These factors were what later came to be called genes, and were the subject of an intensive research programme in the first half of the nineteenth century, most famously the experiments on the fruitfly, *Drosophila melanogaster*, by Thomas Hunt Morgan, Hermann Müller, and others. Genes were identified in this period entirely by the observable trait to which they gave rise. The theory developed that these units of inheritance were arranged along chromosomes and measures of the frequency with which particular traits were inherited together enabled inferences to be drawn about the order of this arrangement. The first genetic map, including half a dozen genes for eye colour, wing shape, and suchlike, was published by Alfred Sturtevant in 1913.

There was a natural hope that the physical basis of these particulate causes of inheritance would be found, and with the famous discovery of the structure of DNA in 1953, it was widely supposed that just this had been achieved. Whereas work in the first half of the century had involved inference down from properties of the phenotype to properties and relations of genes, it now became possible to work from the bottom up, attempting to move from chemically characterized bits of DNA to the phenotypic properties that they were held to produce. Unfortunately, however, these top-down and bottom-up projects have failed to meet in the middle. Long before 1953 and its sequel, it was well known that the relation between genes and phenotypes was generally a complex one, and that genes interacted with one another to produce varying effects. One gene might be correlated with a wide variety of phenotypic traits and, on the other hand, most traits required a variety of genes for their manifestation. However, as the mode of action of DNA became clear, namely as providing a template for the production of protein sequences, it seemed possible that genes could be unequivocally identified in terms of the protein sequence the production of which they directed. But even this hope has proved to be far too optimistic.

A first point is that very little of the genome has turned out to be composed of even prima facie candidates for being genes. Most of the genome appears to code for nothing and has even been thought to be lacking any function at all (though it is increasingly clear that this merely reflects unwarranted presuppositions about the kind of function being sought). Of the parts that are known to be functional, a large proportion does not code for protein production, but serves a variety of regulatory and other functions. But more importantly, even those parts of the genome that do code for proteins, typically bear no simple relation to the products they are involved in constructing. Coding sequences may be read in different ways, depending on where their transcription begins, and may be part of overlapping transcribed sections. Some sequence may even be read in both directions. Hence there may be several possible immediate products from a particular bit of sequence. I say 'immediate product' because the RNA which is first transcribed from a DNA sequence is only the beginning of an often complex sequence of interactions that will finally produce one or several functional proteins. The RNA may be edited, cut, or spliced onto other fragments before it is translated into amino acid sequence which, in turn may undergo similar modifications before final functional proteins are produced. (For more detailed exposition of these complexities and their philosophical significance see Moss 2003 and several of the essays in Beurton 2000.)

The processes leading from DNA to phenotype, then, are extraordinarily complex from the very start. To expect in general that identifiable bits of the genome will have privileged relations to particular traits of the phenotype, given that they do not typically even have unique relations to particular functional proteins, would be hopelessly unrealistic. The notion of the genome as composed of a series of genes 'for' particular phenotypic traits has gone the way of phlogiston.

Given this situation, how are we to understand the apparent successes of Mendelian genetics and the continuing relevance of something similar in medical genetics? Actually, one answer to this question is relatively straightforward. Although it makes no sense to identify bits of the genome with bits of the phenotype, changes to the genome will often have quite predictable effects on the cascade of developmental processes, and lead to predictable changes in developmental outcome. Most often this involves a dysfunctional effect on a protein deriving in part from the altered protein. A classic example is what is often treated as a paradigm of human Mendelian genetics, blue and brown eyes. Blue eyes result not from a blue pigment, but from the absence of brown pigment. Various mutations may result in the failure to produce this pigment and will result in blue eyes (ignoring, for the sake of simplicity, various other genes that affect eye colour). Since in high latitudes there is no significant loss of fitness to blue eyes, such defects have tended to accumulate in high latitude populations. Thus there is no gene for blue eyes, and not even a specific localized gene for brown eyes. Nonetheless, the mutations that disrupt the pigment producing processes will exhibit classical Mendelian patterns of inheritance. For example, if neither parent has the capacity to produce brown eye pigment they will reliably produce blue-eyed children.

Essentially the same story can be told for most or all cases of interest to medical genetics. What most typically exhibits Mendelian inheritance patterns is a harmful mutation or one of a set of harmful mutations. It is curious if common to refer to such a mutation as a gene for a disease, and just wrong to refer to the undamaged sequence as a gene for the absence of the disease. Familiar Mendelian phenomena outside the realm of pathology—double-jointedness, and suchlike—are simply alterations in the genome with functionally harmless effects on the developmental process leading to the phenotype.

Where does this leave the concept of the gene, and the question to what extent this concept is socially constructed? One way of telling the story would be to trace the divergence of two histories of genetics. On the one hand, there is a popular development of genetics, which has led to the central cultural role of genes as naturalizing inheritance and, perhaps more importantly, grounding various accounts of the inflexibility of developmental outcomes. On the other hand, there is the story I have briefly summarized, leading from genetics to genomics, one plausible conclusion from which would be that genes have turned out not to exist at all. In the process of discovering and describing the genome we have failed to find any place within it for the genes that originally led us to it.

A natural reflection on the relation of these two stories would be to draw a strongly anti-constructivist moral. While society has embraced the gene concept and reconfigured earlier concepts of inheritance in a new naturalistic way by appeal to the gene concept, science has gradually deconstructed the very same concept. Of course, scientists have constructed contemporary genomics in the banal sense noted at the beginning of this essay, but the direction this construction has taken has been substantially driven by quite unexpected findings about the processes investigated. Science ultimately corrects the errors that derive from social influence. This thought should, however, be qualified by the reflection that the situation described is at a provisional and very possibly unstable stage. The insights into cellular function that I have briefly described are very recent and their effects both on science and on public reception of science remain to be seen. Science has provided powerful resources for destabilizing social understandings of inheritance, but we should not assume that it will quickly or easily do so. Scientific undermining of essentialist views of species has been available for much longer than have genomically based critiques of traditional genetics, but it is doubtful whether essentialism has declined much outside very narrow and specialized discourses.

But there are also rather different ways of thinking of the story. One might think of the molecular genetics leading from 1953 to contemporary molecular biology as a spin-off from, rather than a continuation of, genetics. In favour of this view it might be noted that Mendelian genetics has continued to this day with empirical inheritance studies, in theoretical population genetics, in medicine, and no doubt other areas. Consistency with molecular genomics requires that its central concept receive a somewhat more instrumental and less realistic treatment than might have been supposed, but it has also turned out that this is not a fatal flaw in the project. Why has genetics continued apparently so oblivious to the undermining of its foundations? Presumably because its concepts, genes for biological properties that matter to us, have been constructed to serve a huge range of our interests in things biological. Contemporary genomics may show us limitations to its utility in addressing these interests, but in the absence of some more useful substitute we are likely to persevere with classical genetics as the best thing going.

Does contemporary genomics provide a substitute for classical genetics? Evidently not, precisely because it refuses to offer concepts connected closely with the phenotypic properties we care about. Of course genomics has notoriously been advertised as exactly a successor to genetics, but a more powerful one that will lead us to cures for the diseases classical genetics traces, and enable us to produce the organisms, and perhaps the babies, we have sought much more slowly with traditional breeding and eugenics.

But the truth is that genomics has so far done little to realize these goals, and perhaps is not well equipped to do so. It is (though this is not something that anyone will want to advertise in the present utilitarian and short-termist climate) currently very close to a project of pure enquiry. The classificatory division of the genome within genomics proper, therefore, is one driven very much by theoretical considerations, and is little affected by social factors in the interesting sense of 'social'. If genomics eventually gives us a good understanding of development, then we might expect to derive real abilities to control developmental outcomes, human and otherwise. But given the demonstrable complexity of development and of its joint dependence on internal and environmental factors, the task is a daunting one. It is entirely possible that traditional genetics methods will remain more effective for traditionally understood goals for the foreseeable future. So whether the story will develop as a demonstration of how nature can eventually dictate the shape of a science, rather than as an illustration of a strongly constructivist thesis, remains to be seen.

It may perhaps be good to state explicitly that the historical narratives I have offered are grossly oversimplified, rationally reconstructed even, with the goal of presenting some extreme positions on the issue of constructivism. A more realistic and nuanced history of how these understandings were gained would unquestionably emphasize much more of the contingency of the process and of the personal interests and goals that motivated the contributors to it. As I have indicated, I take the reality of science to be a variable interaction between social construction for human goals, and a partly recalcitrant nature. If my narrative suggests, for instance, that genes are social constructs whereas genomes are real things, this should only indicate the caricatured nature of the historical narrative.

Conclusion

I have suggested in this chapter that social constructivism is in some ways a fairly banal doctrine, and that the controversy that has surrounded it derives from further claims that are wrongly alleged to follow from it. Indeed the reminder that science is a social process, something done by human beings for a variety of reasons, is a thoroughly salutary one. Nonetheless we can accept this reminder, and embrace such reasonable consequences of it as the symmetry principle, without any threat to the possibility of adopting a robustly realist attitude to some parts of science.

The greatest danger of social constructivism, perhaps, is one that it shares with most sweeping isms about science, that it should harden into a general and restrictive set of claims about science in general. As I have argued extensively elsewhere, I can see no reason to suppose there is any true such set of claims. Science is an extraordinarily diverse set of activities. The ways these activities are shaped by social forces are diverse, and the plausibility of adopting a realist attitude to the claims made within these activities is highly variable.

I have tried to give a sense of this diversity by looking at one of the central ways that science is shaped, whether by nature or society, the decision on how to classify the objects within the domain of the science. Unsurprisingly, social forces dominate this process in cases of highly politically charged social categorizations, and internal scientific processes have a much greater role for the case of quite technical concepts in quite technical sciences. The terminology of genetics is a particularly fascinating case because it is at the same time a terminology that has been developed in a highly technical scientific domain, and one that has from the start been absorbed into highly contentious public discussions. As should be clear, I remain undecided about how best to conceive this general area of enquiry. Lying uneasily between these technical and political contexts, genetics/genomics is a paradigm field for exploration of the interaction between broadly social and technically scientific forces.

3

The Inseparability of Science and Values[1]

There is a view of science, as stereotyped in the hands of its critics as in those of its advocates, that goes as follows. Science deals only in facts. Values come in only when decisions are made as to how the facts of science are to be applied. Often it is added that this second stage is no special concern of scientists, though this is an optional addition. My main aim in this chapter is to see what sense can be made of the first part of this story: that science deals only in facts.

The expression 'deals in' is intentionally vague. Two ways of dealing fairly obviously need to be considered. First is the question of the nature of the products of science. These are certainly to be facts. But there might also be a question about inputs. In generating a fact, dinosaurs are extinct, say, one needs to feed some facts in. (These are dinosaur bones. Our best tests suggest they are 80 million years old. No dinosaurs have been observed recently. And so on.) So these inputs had better be facts too.

There are some immediate worries. One might reasonably object to the suggestion that the only products of science are facts with the observation that science often produces things. Polio vaccines, mobile phones, laser guided missiles, and suchlike are often thought of as very much what science is in the business of producing. Recalling the stereotypic view with which I began, it may be replied that science produces laws and suchlike on the basis of which it is possible to create polio vaccines, mobile phones, and so on. And the trouble with this is that it seems so grossly to misrepresent how science actually works. A group of scientists trying to develop a vaccine do not try first to formulate general rules of vaccine development, and then hand these over to technicians who will produce the actual vaccines. No doubt they will benefit from the past experience, recorded in texts of various kinds, of past vaccine-makers. And perhaps, if they are successful, they will themselves add to the body of advice for future vaccine-makers. But it seems beyond dispute that the primary objective here is an effective vaccine, not any bit of fact or theory.

Let us ignore this concern for the time being, however, and concentrate on the question whether, in so far as science produces what we might think of as bits of

[1] I am grateful to Francesco Guala for helpful comments on an earlier draft of this chapter.

discourse, these bits of discourse are strictly factual, never evaluative. So we need to ask, What is the criterion for a bit of discourse being merely factual?

It is not hard to find some paradigm cases. 'Electrons have negative charge' is pretty clearly factual, whereas 'Torturing children is a bad thing to do' is pretty clearly evaluative (though we might note at the outset that the clarity of this judgement strongly invites the suggestion that it is also a fact). The existence of these and many other possible paradigms may tempt one to apply the criterion made famous for the case of pornography by Supreme Court Justice Potter Stewart, 'I know one when I see one.' But it is just as easy to find cases that are much less clear. Consider for instance, 'The USA is a violent country.' On the one hand, we can easily imagine a sociologist devising an objective measure of social violence—number of murders per capita, number of reported cases of domestic violence, and so on—and announcing that the USA ranked higher than most comparable countries in terms of this measure. But on the other hand, we can imagine someone describing this conclusion as a negative judgement on the country.

Of course there is a familiar response here. We have the fact and then the judgement. The fact is that there are certain statistics about acts of violence. The value judgement is that these statistics constitute a bad thing about the place where they were gathered. In support of this distinction we can point out that it is always possible to accept the fact and reject the value judgement. Some people approve of violent countries (they reveal the rugged independence of the populace, perhaps) and perhaps there are even people who think torturing children is a good thing. But this defence is beside the present point. That point was just that the statement 'The USA is a violent country' cannot be obviously assigned to either of the categories factual or evaluative. In case this is not clear compare the statement 'Sam is a violent little boy.' In any normal parlance this does not mean just that Sam is disposed to occasional violent acts—that is, after all, true of virtually all little boys—still less that his rate of violent act production reaches a certain level on a standard scale approved by the American Psychological Association. It is a criticism of Sam, and probably of his parents too. Anyone who doubts this should visit their nearest day-care centre and try out this comment on the parents collecting their precious charges there.

Suppose, as I have imagined with the case of social violence, that there is indeed a standard measure of violence for little boys. On this scale a violent child is defined as one who emits more than five acts of aggression per hour. Now when I, as an expert child psychologist, announce that Sam is a violent child my remark is entirely factual. Should his parents find the remark objectionable I shall point out that this is no more than a factual observation and it is entirely a subjective opinion, and one that I as a scientist shall certainly refrain from entertaining, whether it is a bad thing to be a violent child.

A possible conclusion at this point would be something like this. 'The USA is a violent country' and 'Sam is a violent little boy' are both potentially ambiguous. While both may often be used evaluatively, especially by regular folk, scientists will only use them after careful definition (operationalization) of their meanings. Thus when used by responsible scientists these statements will turn out to be merely and wholly factual. The statements under consideration are thus seriously ambiguous.

So perhaps scientists would do better to avoid these normatively loaded terms and stick to an explicitly technical language. To say that Sam scored 84 on the Smith–Jones physical assertiveness scale is much less threatening (even if this practically off the scale, the sort of score only achieved by the most appallingly violent children). And it is certainly true that psychologists or psychiatrists, to pursue the present example, are often more inclined to invoke technical diagnostic language, backed up by detailed technical definitions in standard nosological manuals, than to say, for instance, that someone is mad.

There is however, an overwhelming advantage to ordinary evaluative language: it provides reasons for action. To say that the USA is a violent country is a reason for politicians to act to reduce violence or mitigate its effects (for example by controlling the availability of dangerous weapons). It is, other things being equal, a reason not to live there. And so on. It is of no interest just to be given a number and told this is the violence index for a country or a city; we want to know whether it is high or low or indeed whether it is good or bad. Similarly, though here we tread on shakier ground, it might be valuable to know that someone is mad. It might be expedient to restrain them, or at least not put them in charge of security at the local nuclear power station.

There is a more general point. Once we move away from the rarefied environments of cosmology or particle physics, we are interested in scientific investigations that have consequences for action. And this undoubtedly is why, while often paying lip-service to operationalized or technical concepts, scientific language often gets expressed in everyday evaluative language.

The situation so far seems to me to be this. Many terms of ordinary language are both descriptive and evaluative. The reason is obvious. Evaluative language expresses our interests which, unsurprisingly, are things we are interested in expressing. When we describe things it is often, perhaps usually, in terms that relate to the relevance of things for satisfying our interests. Sometimes we try to lay down rather precise criteria for applying interest-relative terminology to things. These range from the relatively banal—the standards that must be met to count as a class 1 potato, for instance—to the much more portentous—the standards that an act must meet to count as a murder. In such cases we might be tempted to say that the precision of the criteria converts an evaluative term to a descriptive one. It is important to notice, however, that the precision is given point by the interest in evaluation. The same is often the case for operationalized terms in science. More often in everyday life, the terms are a much more indeterminate mix of the evaluative and the descriptive: crisp, soggy; fresh, stale or rotten; vivacious, lethargic, idle, stupid, or intelligent; or, recalling J. L. Austin's memorable proposal for revitalizing aesthetics, dainty and dumpy (Austin 1961: 131).

This, I think, is the language that we use to talk about the things that matter to us, and to understand such language requires that we understand both the descriptive criteria and the normative significance of the concepts involved. It seems to follow that there is no possibility of drawing a sharp fact/value distinction. Science may reasonably eschew some of these familiar terms on the ground that they are vague and imprecise, and may try to substitute them with more precisely defined alternatives. But, first, the use of these alternatives will ultimately depend on their capturing the evaluative force of the vaguer terms they replace. And, second, science does not, and almost certainly cannot, entirely dispense with the hybrid language of description and evaluation. This fact makes the assumption of a sharp fact/value distinction not only untenable, but often harmful.

So much for the general background of scepticism about the fact/value distinction. For the rest of this chapter I shall be concerned with more detailed specific examples. Two such examples will be used to illustrate more concretely how normativity finds its way into scientific work, and how its denial can potentially be dangerous.

Before continuing, though, let me make one more very general comment. The examples that I shall discuss will both be drawn from parts of science directly connected to human concerns. I have often heard the view expressed that though it is interesting and important that the human sciences should be contaminated with values, it is not altogether surprising. But what would really concern the advocate of the value-neutrality thesis with which this chapter began would be an indication that physics or chemistry or mathematics was value-laden. So, on such a view, I am dodging the really important task.

In reply to this let me first say that I do not propose to deny that many of the results of these sciences may well be value-free. The sense in which I am questioning the legitimacy of the fact/value distinction is not one that implies that there are no areas that human values do not infiltrate. It is rather that there are large areas, including the domain of much of science, in which the attempt to separate the factual from the normative is a futile one. What I want to say about physics is that if most or all of physics is value-free, it is not because physics is science, but because most of physics simply doesn't matter to us. Whether electrons have a positive or a negative charge, or whether there is a black hole in the middle of our galaxy are questions of absolutely no immediate importance to us. The only human interests they touch (and these they may indeed touch deeply) are cognitive ones, and so the only values that they implicate are cognitive values. The statement, 'Electrons have negative charge', is thus value-free in a quite banal sense: it has no bearing on anything we care about.

I said that these were matters of no immediate importance, and the word 'immediate' is crucial. It is often pointed out that physics also tells us how to build nuclear power stations or hydrogen bombs. Here, we are, to say the least, in the realm of values. There is no unique nuclear power station that physics tells us how to build, nor could there be a general theory that applied to the building of any possible power station. Physics assists us in building particular kinds of power stations, and particular kinds of power stations are more or less safe, efficient, ugly, and so on. I doubt whether anyone could seriously suppose that there was a value-free theory of nuclear power station building, let alone hydrogen bomb construction. The argument that physics is value-laden beyond the merely cognitive values mentioned in the last paragraph seems most plausibly to depend on some such claim as that physics really is, contrary to appearances or propaganda, the science of bomb-building. Examinations of the extent to which physics is a project funded by the military lends some credibility to such a view, but it is not one on which I shall offer any judgement. My point here is just that the value-freedom of physics, if such there be, has no tendency to show that science is in general value-free.

Rape

My first example is not a pleasant one. It is the Evolutionary Psychological hypothesis about rape.[2] The basic story goes something like this. In the Stone Age, when the central features of human nature are said to have evolved, females were attracted to mates who had command of resources that could be expended on rearing children. Perhaps they were also attracted to males with good genes—and perhaps these were simply genes for being, in the virtuously circular sense characteristic of sexual selection, attractive. Perhaps finally these ancestral females were smart enough to deploy some deception on the resource-rich males, and get their resources from the 'dads' and their genes from the more attractive 'cads'. At any rate, there would very probably have been males with neither competitive-looking genes nor resources, and they, like everyone else, would be looking for a sexual strategy. Since they have no chance of persuading any females to engage in consensual sex with them, this strategy can only be rape. As is generally the way with Evolutionary Psychology, once a form of behaviour has been proposed as a good idea in the Stone Age, it is inferred that a module for producing it must have evolved. So men, it appears, have a rape module, activated when they find their ability to attract females by any acceptable method falls to a low enough level.

Evolutionary Psychologists presenting such theories generally insist also on a quite naïve version of the fact/value distinction. Their claimed discoveries about rape are merely facts about human behaviour, certainly not facts with any sort of evaluative consequences. We can at least agree, contrary to what Evolutionary Psychologists sometimes accuse their critics of maintaining, that showing that rape is, in the sense just described, natural, doesn't mean it is good. Earthquakes and the AIDS virus are, discounting some paranoid speculations, natural, but not, thereby, good. But such theories certainly do have consequences for what would be appropriate policy responses to the incidence of rape. Even this indisputable fact is enough to refute the occasional claim that such theories have no evaluative consequences. They have at least the consequences that certain policies would be good or bad. The most obvious such policy response to the theory in question would be the elimination of poverty, since

---

[2] A standard reference is Thornhill and Thornhill (1992). The ideas were popularized by Thornhill and Palmer (2000). For detailed rebuttal see various essays in Travis (2003).

the hypothesis is that it is poor men who are rapists (because they lack the resources to attract women). Though certainly a good idea, this goal has unfortunately proved a difficult one to achieve. On some plausible Marxist analyses it is a goal that could not be achieved without the elimination of capitalism—an equally tricky proposition—since, on these analyses, poverty is not an intrinsic property of people, but a relation between people, and a relation that is fundamental to capitalism. And it is interesting that such an analysis appears relevant to the sociobiological stories: it is not the intrinsic worthlessness of the failed caveman that doomed him to sterility or sexual violence, but his relative lack of worth compared to his more fortunate rivals.

But all of this is of course somewhat beside the point. Those who have thought seriously about contemporary sexual violence as opposed to the hypothetical reproductive strategies of imagined ancestors have observed that rape is not exclusively, or even mainly, a crime of resourceless reproductive predators lurking in dark alleyways, but has much more to do with misogyny, and more to do with violence than sex, let alone reproduction. Its causes appear therefore to be at the level of ideology rather than economics.

The existence of such divergent theories, and the fact that they do indeed have implications for policy, indicate that the stakes are high in theorizing about matters of this moment. But these issues do not get to the heart of my present argument. So far I have spoken as if there is no problem whatever in deciding what, in the context of this theoretical enquiry, we are talking about. Indeed to make research simpler, sociobiologists often begin their investigation of rape with observations of flies or ducks. If we have a good understanding of why sexually frustrated mallards leap out from behind bushes and have their way with unwilling, happily partnered, passing ducks, then the essential nature of rape is revealed, and we can start applying these insights to humans. And of course one thing that this blatantly ignores is the fact that human rape (and I doubt whether there is any other kind) is about as thoroughly normative a concept as one could possibly find. Someone who supposed they were investigating the causes of rape but, since they were good scientists, were doing so with no preconceptions as to whether it was a good or a bad thing, is deeply confused: they lack a grasp of what it is that they are purporting to investigate.

All this is perfectly obvious when one looks at real issues rather than pseudoscience. A more serious perspective on rape is that it involves a profound violation of the rights of its victims. When, not long ago, it was conceptually impossible for a married man to rape his wife, this reflected a widespread moral assumption that, vis-à-vis her husband, a woman had no rights. Indeed the husband was supposed to have a right, perhaps divinely guaranteed, to whatever kinds of (generally legal) sexual relations he desired with his wife. Nowadays more complex debates surround the concept of date rape, and the exact tones of voice in which 'No' means 'Yes' and so on. Less controversially, it has long been understood that sexual intercourse with young children is a form of rape, since the relation between adults and small children does not permit meaningful consent. But the age at which consent becomes possible varies greatly from culture to culture and is often subject to renegotiation. As Ian Hacking (1995) powerfully illustrates, evaluating intergenerational sex in different historical eras is anything but simple.

The point of this is not to argue that there is no place for science in relation to such a topic. On the contrary, there are quantitative and qualitative sociological questions, psychological questions, criminological questions, and no doubt others, that are of obvious importance. The point is just that if one supposes one is investigating a natural kind with a timeless essence, an essence that may be discovered in ducks and flies as much as in humans, one is unlikely to come up with any meaningful results. Though this is an extreme example, in that the value-ladenness in this case is so obvious that only the most extreme scientism can conceal it, I think it is atypical only in that obviousness. As I argued in the opening section of this chapter, fact and value are typically inextricably linked in the matters that concern us. And we are most often concerned with matters that concern us.

Economics

My second example is a quite different one. Nowhere is the tradition of dividing the factual from the evaluative more deeply ingrained than in economics. In recognition of the fact that issues about the production and distribution of the goods on which human life depends do have a normative component there is, indeed, a branch of economics called normative, or welfare, economics. But this is sharply divided from the properly factual investigations of so-called positive economics, and it is hardly a matter of debate that it is the latter that is the more prestigious branch of the discipline. In common with traditional positivism and contemporary scientism, the underlying assumption of this distinction is that there is a set of economic facts and laws that economists are employed to discover, and that what to do with these is largely a matter for politicians or voters to decide.[3]

And in fact normative economics has itself tended to reinforce this perspective, and has therefore tried to limit itself to the question whether there are economic actions that are indisputably beneficial. This concern is expressed in the focus of attention on the criterion of Pareto optimality: an economic allocation is said to be Pareto optimal if there is no possible transfer of goods that would improve the lot of some agent or agents while harming no one. It may be that failures to achieve Pareto optimality should be addressed where possible (though even this may be called in question by some accounts of distributive justice). But the 'optimality' in 'Pareto optimality' is a dubious one. If, for example, I possess everything in the world and I derive pleasure from the knowledge that I own everything in the world, this distribution of goods constitutes a Pareto optimum. If some crust of my bread were diverted to a starving

---

[3] A classic paper by Friedman (1953) provides a well-known statement of this position.

child, I would no longer have the satisfaction of owning everything in the world, and similarly with any other possible transfer. So one person, myself, would be less well off. But this would be an unconvincing argument that this distribution was optimal, or even good. There are of course countless Pareto optima, which by itself suggests something anomalous in the use of the term 'optimum'.

The problem is perfectly obvious. While we can all agree that Pareto optimality is a good thing if we can get it, the issue of interest is which of the many Pareto optima we should prefer. Pareto optimality is really about efficiency, whereas we are interested in properly normative economics in matters such as justice. We should recall here the general assumption that science in general, and economics in particular, should aim simply to describe the mechanisms of economic activity and leave it to others to decide what to do with it. Not only is this assumption at work in positive economics, but it is even more starkly visible in much of the practice of normative economics, which is concerned not with how economies ought to be organized, but with efficiency.

I believe that this is a highly undesirable, and very probably incoherent, conception of the business of economists. One way to see that it is undesirable is to note that when we consult supposedly expert economists about what might be good economic policy we might naïvely suppose that they would have useful advice to offer us. But on the conception under review it turns out that, apart perhaps from pointing to the occasional departure from Pareto optimality, they have no relevant expertise whatever. They are, after all, experts in efficiency not policy. But since economists often seem willing to offer such advice, it seems disingenuous that they should deny that normative questions are part of their discipline. And if they do insist on this denial, they will presumably be of much less use to us than we had thought, and we could perhaps get by with rather fewer of them.

More worryingly, it is quite clear that there is an implicit normative agenda to the vast majority of economic thinking. Because economists believe they have something to say about economic efficiency, they are naturally inclined to think of this as a good thing. And as the clearest measure of efficiency is the ability to produce more stuff with the same resources, economists are often inclined to think the goal of economic activity is to produce as much stuff as possible. Even if this account of the aetiology of this goal is disputable, it is hard to dispute that many economists do assume such a goal; and assuming a goal is a good way of avoiding the vital intellectual labour of considering what the goals of economic activity really should be. Returning to the economists who offer advice on matters of public policy it is clear that very frequently they assume that what they are required to do is to advocate those policies that they believe, rightly or wrongly, will promote the production of as much stuff as possible.

It is, in fact, an enormously difficult question, even if we agree that something should be maximized by economic activity, what that something should be. Not infrequently positive economics assumes that the real question is about maximizing wealth measured in monetary terms, and tragically many politicians seem willing to accept this facile view. An obviously preferable goal would be something like standard of living, except that that would be little more than a marker for the difficult question of what constitutes standard of living. The work particularly of Amartya Sen[4] has made it clear that any satisfactory analysis of this concept will be only marginally related either to any standard account of utility, or to the accumulation of wealth. It is also clear that even if we knew what constituted standard of living we would still have to face the task of deciding how this should be distributed. Surely the utility of increases in standard of living declines as one reaches more comfortable levels, so greater good can be gained by distributing standards of living more equally. And there is also the question of who should be among the beneficiaries of a distribution. Should we care about the standards of living of foreigners for instance? Do the as yet unborn have any claim on a decent standard of living? Or must we consider the well-being of non-human animals, or the effects of economic activity on the environment?

Once again, however, the issue I want to emphasize here is the inescapably value-laden nature of the terms in which we talk about ourselves and our social existence. Consider a central idea in macroeconomics the measurement of which has had profound implications on economic policies throughout the world, inflation. Like earthquakes or AIDS, inflation is generally seen to be a bad thing. But also like earthquakes and AIDS it is seen as the sort of thing that can be described and theorized without regard to its goodness or badness.

The problem here is somewhat different from that for rape. The normative judgement is fundamental to the meaning of rape, and therefore fundamental to negotiations about what should and should not count as rape. With inflation, normativity comes in a little later. The primary problem, as has long been familiar to economists though it appears often to be surprising to others, is that there is no unequivocal way of measuring this economic property. It would be easy enough if everything changed in price by identical percentages, but of course that does not happen. How should we balance a rise in the price of staple foods, say, against a fall in the price of air travel? The immediately obvious reply is that we should weight different items in proportion to the amount spent on them. The problem then is that not all goods are equally consumed by all people, or even all groups of people. It is quite commonly the case that luxury goods fall in price while basic necessities rise. It might be that these cancel out under the suggested weighting, so that there is no measured inflation. But for those too poor to afford luxury goods there has manifestly been an increase in the price level.

How then does one decide how such an index should be constructed? The unavoidable answer, it seems to me, is that it depends on the purposes for which it is to be constructed. There are many very practical such purposes. People on pensions, for instance, may have their incomes adjusted to account for changes in the level of inflation. For such purposes the goal might reasonably be to maintain the value of the pension, in which case the ideal would be to enable typical pensioners to continue to

---

[4] A number of insightful discussions of the issue can be found in Nussbaum and Sen (1993).

afford the goods that they had previously consumed. Of course no pensioner is absolutely typical, but a case might be made for addressing particularly the case of pensioners dependent solely on the pension. Clearly for such ends it would be desirable to have specific indices designed for specific groups. But the goals might be quite different, calling for different measures. For example, and perhaps more plausibly, one such goal might be to save the taxpayer money.

Perhaps the central goal nowadays of inflation measurement is as an input into the decision procedures of central banks in determining interest rates. In Britain (I'm not sure how widespread the practice is) this leads to the rather bizarre habit of regularly announcing something called the 'underlying rate of inflation'. This is a measure of inflation that ignores changes in mortgage payments consequent on changes in interest rates. A plausible rationale for this measure might be that the article of faith on which much macroeconomic policy depends (or depended not long ago) is that inflation rate is inversely related to interest rate. Since increasing interest rates cause an immediate and large increase in the prices confronted by consumers, especially in debt-laden home-buying cultures, this central dogma would be constantly refuted if mortgage costs were included in the measure of inflation. Hence the importance of the underlying rate as a way of allowing the theory to be maintained, at least as true in some sufficiently long term. (I suppose this aspect of the matter is of more obvious concern to students of the theory-laden than of the value-laden.)

Another aspect of all this is that the assumption that inflation is objectively bad is by no means self-evident. In common with most middle-class Americans, I have spent substantial parts of my life owing large sums of money borrowed at fixed interest rates. From a personal point of view, therefore, I have always seen inflation as something to be enthusiastically welcomed. Since moving to the UK, and a regime of variable or short-term fixed interest rates, the situation is more complex as rises in interest rate have an immediate impact on my cost of living and are, moreover, reliably induced by increases in price level. At any rate, the deep horror with which inflation is now so widely perceived should lend support to those who believe that the world is mainly controlled by bankers.[5]

Some quite different aspects of value-ladenness could be introduced by considering another central macroeconomic concept, employment. Having work is widely perceived in many contemporary cultures as a necessary condition for any social status and even for self-respect. But what counts as work is a complicated and contentious issue and one that has profound implications for all kinds of economic policies. It is still frequently the case, for instance, that work is equated with the receipt of financial reward, with the consequence that domestic work from raising children to the domestic production of food counts, from an economic perspective, as a form of

---

[5] [Note added to this edition] I'm not sure how obvious this was when I first wrote it in the late 1990s. I fear that in the aftermath of the 2008 economic crisis largely induced by the surreal ventures in gambling by the investment banking community, it will seem merely banal.

unemployment. A quite different concept can be found in Adam Smith (and an earlier Adam who was required to make his living 'in the sweat of thy face') in which work is generally unpleasant—toil and trouble—and understood by its contrast to leisure or ease (see Smith 1994: 33). Quite different again is the idea most conspicuous developed by Karl Marx that work provides the possibility of human self-fulfilment. Both these conceptions are evidently value-laden, and the notion that there can be a purified economic conception of work somehow divorced from any of these varied normative connotations seems both misguided and potentially dangerous.[6] No doubt it is widely taken as obvious and hardly requiring argument that work is a good thing, and the promotion of high levels of employment is a good thing. But whether this is true because people must have food and shelter and work is the necessary, if otherwise undesirable, condition of providing these things, or because satisfying work is at the core of a valuable human life, makes enormous differences to the implementation of an employment policy.

There are, in sum, many ways in which values figure in the construction and use of many of our concepts and scientific concepts are no exceptions. For much of language the notion of separating the one from the other is altogether infeasible[7] and the pretence that this can be done can have very serious and damaging consequences on policy and action.

Conclusion

As I indicated earlier in this chapter, I am not claiming that there is no distinction between the factual and the normative. What I do claim is that this is not a distinction that can be read off from a mere inspection of the words in a sentence, or a distinction on one side or the other of which every concept can be unequivocally placed. For large tracts of language, centrally the language we use to describe ourselves and our societies, the factual and the normative are thoroughly interconnected. Where matters of importance to our lives are at stake, the language we use has more or less profound consequences, and our evaluation of those consequences is deeply embedded in the construction of our concepts. The fundamental distinction at work, here, is that between what matters to us and what doesn't. There are plenty of more or less wholly value-free statements, but they achieve that status by restricting themselves to things that are of merely academic interest to us. This is one reason why physics has been a sometimes disastrous model for the rest of science. We hardly want to limit science to the investigation of things that don't matter much to us one way or the other. The application of assumptions appropriate only to things that don't matter to those that do is potentially a disastrous one.

---

[6] These different meanings of work are discussed in more detail in Dupré (2001: 138–46) and Gagnier and Dupré (1995).

[7] For more detailed accounts of important aspects of value-ladenness in economics see Starmer (2000) and Guala (2000).

PART II
Biology

4

The Constituents of Life 1: Species, Microbes, and Genes[1]

The title of this chapter and the next, 'The Constituents of Life', refers to the things that are the subject matter of biology: organisms, the systems, organs, cells, and molecules to be found within them, and the larger systems, such as species or ecosystems which they, in turn, compose. It might not be obvious that there is much for a philosopher to say on this subject. We are all familiar enough with these things at a common-sense level, but it is surely for biologists to provide us with more sophisticated insight into what they do and how they do it. Yet attempting to provide philosophically adequate accounts of these various categories has proved extremely difficult, and such difficulties have been a major topic for my own academic specialty, the philosophy of biology. In these chapters I shall consider some of these kinds of things and the philosophical difficulties they present. A wider aim will be to try to locate some fundamental problems in our conception of life and its constituents, problems that more generally explain the difficulties in understanding central biological categories.

It is natural and traditional to think of life in terms of a structural hierarchy. We analyse an organism into a set of interacting organs and systems—livers, hearts, brains, circulatory systems, immune systems, and so on—and these in turn into smaller structural components, most notably cells. Cells, in turn, are understood as enormously complex ensembles of interacting molecules. And this picture extends in both directions. Molecules are complex structures of atoms; organisms are components of species, ecological systems, or social groups. And so on.

This vision has undeniably been fundamental to the extraordinary success the sciences have achieved in advancing our understanding of the natural world. This success has often been taken to lend support to a more general reductionist scientific

[1] This chapter and the next are based on the Spinoza Lectures delivered at the University of Amsterdam in May and June 2006. I am very grateful to the Philosophy Department of the University of Amsterdam for their hospitality, especially to Professor Frans Jacobs, Head of the Department, and to Ria Beentjes and Willy van Wier. My enjoyment of the visit, as well as the content of these chapters, were much enhanced by conversations with Michiel van Lambalgen, Martin Stokhof, Beate Roessler, Gerard de Vries, Veit Bader, and Tine Wilde. Wolfram Hinzen, and the students in the postgraduate seminar on Philosophy of Biology I taught with him, provided an excellent opportunity to discuss some of the issues with an appropriately sceptical audience. These two chapters were originally published as *The Constituents of Life*, Amsterdam: Van Gorcum, 2008.

methodology. Reductionism, in its classical form, is the explanation of the behaviour of complex entities in terms of the properties of their parts, and some philosophers have taken this position to its logical conclusion and suggested that ultimately the world is, in principle at least, fully describable and intelligible in terms of the smallest microphysical particles it contains.

Reductionism has, however, been much criticized, including in the past by myself (Dupré 1993). I shall not explicitly pursue this critical project here. In opposition to reductionism I have, over a number of years, defended a quite different, pluralistic perspective (1993, 2002). According to this perspective there are many different kinds of things in the world, from physically simple things like electrons or quarks, to very complex things such as planets, elephants, or armies. Many or all these things, in my view, have equal claims to reality. As the basis of this position is the idea that many or all such entities have causal powers that are not simply consequences of the way their physical components are fitted together. This perspective gives biology, in particular, autonomy from the physical sciences. One objective of this chapter and the next will be to explain and defend this point of view.

Let me begin by pointing out what is perhaps the deepest difficulty with the reductionist hierarchy. Contrasting with the idea that life consists of a hierarchy of things, we may observe that it is more realistic to consider it as a hierarchy of processes. In a typical cell in a human body many thousands of chemical reactions are taking place every second. Molecules are constructed, reshaped, or dissolved. The cells in which they reside, divide, develop, and die. All of these countless events take place within a much longer process, the life cycle of the organism: conception, birth, death, and an exquisitely complex sequence of stages in between. And as these life cycles give rise to new life cycles through reproduction we begin to glimpse a much longer process still, evolution.[2] This reminds us that these life cycles are not a sequence of replicas but rather a sequence of similar but subtly different processes. Just as the process that is the life cycle of an organism changes constantly, partly in reaction to the demands put on it by its environment, so the sequence of life cycles changes in response to the longer term and greater changes to the environment—changes constituted most significantly by the changing patterns of life surrounding it.

Reductionism has, from its beginnings, been greatly inspired by our success in building machines, and even philosophers who have abandoned the epistemological dream of reductionism, the explanation of everything in terms of physics, still often adhere to versions of mechanism, the view that the functioning of complex systems, including biological systems, should be understood by analogy with

---

[2] I should emphasize that by 'reproduction' I include, for the case of organisms such as ourselves, much more than the biological process which is the primary referent of this term. Following so called 'Developmental Systems' theorists (see the Introduction, above, and passim), I take the concept of reproduction appropriate for evolutionary thinking to include everything that is required for the replication of the lie cycle. In the human case this might include, for instance, schools and hospitals. (See Oyama et al. 2001.)

machines.[3] So it is worth reflecting for a moment on how different the workings of a machine are from the hierarchy of processes that I have just sketched. The parts of a machine are not unchanging, of course, but their changes constitute a relentless and one-directional trend towards failure. A good machine starts with all its parts precisely constructed to interact together in the way that will generate its intended functions. The technical manual for my car specifies exactly the ideal state of every single component. As friction, corrosion, and so on gradually transform these components from their ideal forms, the functioning of the car deteriorates. For a while these failing components can be replaced with replicas, close to the ideal types specified in the manual, but eventually too many parts will have deviated too far from this ideal, and the car will be abandoned, crushed, and recycled.

Reductionism is almost precisely true of a car. We know exactly what its constituents are—they are listed in the manual—and we know how they interact: we designed them to interact that way. Reflection on the dynamic and interacting hierarchy of processes that constitute life should make us suspect that a very different picture is required.

An extreme reaction to this disanalogy might be that we should question the very idea of dissecting life processes into static things. I shall not take such an extreme position. One reason I shall not is that, most strikingly in the last few decades, mechanistic and even reductionistic explanations have provided extraordinary insights into living processes. Indeed, our understanding of the molecular mechanisms underlying living processes has been growing at a rate that perhaps exceeds any explosion of knowledge in the history of science. This growing understanding of the mechanical or quasi-mechanical interactions of molecules promises ever growing abilities to intervene in life processes, for example in combating disease. Certainly the processes of life are highly dependent on these mechanisms. It is even arguable that science is inescapably mechanistic; certainly its most impressive and uncontested achievements have been based on mechanical models. But even if this is all true, the great differences between living things and machines should tell us something very important about such scientific insights. Mechanical models, assuming fixed machine-like ontologies, are at best an abstraction from the constantly dynamic nature of biological processes. And it is this pervasive fact about biological science that is central to explaining the philosophical difficulties in characterizing the constituents of life that biologists hypothesize. If, indeed, science is essentially an examination of mechanisms, this points to ultimate

---

[3] Since writing this chapter I have had occasion to look more closely at an influential version of mechanism that has been promoted recently by a number of philosophers, especially in a series of recent papers by Carl Craver and collaborators. (See Machamer et al. 2000; Craver 2005.) This explicitly anti-reductionist mechanism is generally very congenial to the perspective developed in these chapters. The term 'mechanism' is used to stress the importance of distinguishing a set of interacting constituents that must be understood at several different structural levels. It seems to me that the disanalogies with machines that I stress in the text are sufficiently important to make the choice of term unfortunate, though this is, of course, a matter of no more than terminological taste.

limits in the ability of science to understand life. In the next chapter, however, I shall briefly consider some scientific ventures which promise a more realistic approach to biological processes.

Let me summarize the problem that I now want to address. The reductionist believes that in the end there is nothing in the world but the stuff of which things are made—let me call this basic physical stuff. Of course, the reductionist does not say, bluntly and absurdly, that houses, for example, don't exist. The claim is rather that a house is, ultimately, nothing but an aggregate of physical stuff, and all the properties of any house can, in principle, be fully explained by appeal to the properties and relations of basic physical stuff. So there is a possible, microphysically grounded, account of the world which would have no need to mention houses. I am insisting, on the contrary, that there is a whole hierarchy of increasingly complex things that really exist, and that have causal powers that are not reducible to the mechanical combination of the powers of their constituents. Yet I have also claimed that the things we distinguish in our descriptions of life, at least, are always to some extent abstractions from the dynamic processes that ultimately constitute life. This second claim may seem to undermine the reality of the members of the biological hierarchy to which the first claim attributes causal powers. I must now try to show how these theses can be reconciled.

Let me start with a very brief and abstract answer, and then illustrate what I mean with a homely example. The processes of life are of course massively heterogeneous. This heterogeneity is expressed, for example, when we inventory the thousands of chemical species to be found at any instant in a cell. Although such an inventory is a static snapshot of a dynamic entity—at best an idealized description of the cell, therefore—the molecules we distinguish are more or less transient foci of causal power, real nodes in the astonishingly complex causal nexus that drives the cellular processes. Crucially, they are not merely nodes in an upward flowing casual cascade from the microphysical, but equally in a downward flow of causal influence from complex things to simpler things. Now the homely example, from a very high level in the causal hierarchy.

Readers familiar with South Central Amsterdam will be familiar with Albert Cuypstraat. This street has an unusual capacity to attract people, a capacity which, I suggest, has significant similarities to the ability of a flower to attract bees, or the ability of a magnet to attract iron filings: all are causal powers of individual things. The particular causal power of Albert Cuypstraat will be obvious to anyone wandering around the streets in the immediate vicinity: while there will be a light scattering of people in these surrounding areas, immediately one reaches Albert Cuypstraat one will encounter a dense throng. The reason is no mystery, of course: this is a busy street market. The market could not exist without the people (and stalls, and products) that make it up, but equally there are properties of the market itself that attract the people to it.

The powers of this market are exactly matched to the powers of the people it attracts. They must know it is market, for instance, and how to get there. These are not difficult accomplishments: I myself managed to acquire them within a few days of arriving in Amsterdam. But of course I had acquired many of the necessary skills years ago: knowing what a market is, how to buy things, and so on. My return to the market to forage after my first accidental encounter with it is, however, a more complicated achievement than, say, returning to a place where I had previously discovered edible berries. I would not be similarly drawn to return to a place where I had seen delicious looking food through the window of a private house, for instance, and I would not return to the market at four o'clock on Sunday morning. The market is a social institution of a kind that I have learned to negotiate reliably. By learning this I have also become—willingly, I should add—susceptible to the attractive casual powers of this institution. The market depends for its existence on the people who go there to buy and sell; but it is simultaneously the power of the market that attracts the people that constitute its continued existence. And, insignificant though these may seem, the market effects changes in the people it attracts—it may determine, for example, what they eat for dinner. This is the sort of thing that I mean by a node in the causal nexus. I shall suggest that this model, incorporating the development of two-way causal interaction between a complex thing and its constituents, is the right model for interactions at many different levels of structural organization.

There is no better example of the consequences of the shift from a static to a dynamic view of life than the influence of Charles Darwin's revolutionary ideas on the subject of his most famous work, biological species. It may perhaps be thought that sorting organisms into species is more like constructing the automobile company's model catalogue than a parts list for one model. But, first, a majority of philosophers concerned with biology now hold that species should be seen as individual things, components of the evolutionary process.[4] And, second, sorting organisms into kinds raises many of the same issues as sorting, say, molecules or parts of molecules into kinds: classification is an essential part of scientific activity at any level of organization. The classification of organisms is both the most widely discussed and the most ancient such project—indeed a project that some believe was delegated to Adam when God invited him to name the animals. One crucial point that will emerge from consideration of this topic, and which should be less surprising viewed in the light of the general problem of abstracting objects from processes, is that there is no uniquely correct way of classifying organisms: different investigative interests dictate different and often cross-cutting modes of classification (see further Dupré 2002: chs. 1–4).

There is an ancient philosophical tradition that understands classification as involving the identification of the essence of things of a kind: the essence is a necessary and sufficient condition of being a thing of that kind and also the feature that most fundamentally explains the properties characteristic of that kind. So, for instance, a certain atomic structure might be both necessary and sufficient for a piece of stuff to be

---

[4] Classic statements of this thesis are Ghiselin (1974) and Hull (1976).

iron and, at the same time, provides an explanation of why that stuff has properties—being magnetic, being easily oxidizable, and so on—characteristic of iron. Whether or not such an idea works for chemistry, one thing that almost everyone now agrees on is that nothing similar works for biology.[5] A sufficient explanation of this failure is the agreement that one biological kind can evolve gradually into another. The identification of a kind of organism existing at this moment is an abstraction from a continuous process linking these current organisms through time to a long series of very different organisms and, indeed, if we trace evolutionary history back to a common ancestor and thence forward to the present, connecting any two currently existing kinds of organism. There is no way of understanding this link as consisting of a definite number of distinct types, each defined by its unique essence.

Just as evolutionary theory has put an end to certain traditional ideas about biological classification, so it underlies more contemporary views. What most contemporary theorists agree is that biological classification should reflect the evolutionary relationships between different kinds of organisms. Evolutionary history has traditionally been represented as a tree, with branches representing evolutionary divergences and the smallest twigs representing the most finely distinguished kinds, species. For a while the dominant view, the so-called Biological Species Concept especially associated with one of the twentieth century's most influential evolutionists, Ernst Mayr, reflected a theory about the mechanism of evolutionary divergence (e.g., Mayr 1963). The separation of branches of the tree, it was supposed, required that organisms on different branches be reproductively, and hence genetically, separated from those on other branches. Thus species were thought of as a reproductively connected group of organisms, reproductively isolated from all other groups. Unfortunately this idea often fits poorly with biologists' sense of what constitutes a species. Many groups of what seem to be well-defined species in fact show continuous reproductive links and, on the other hand, what seem like homogeneous species often divide into separate populations with little or no reproductive connection between them. In addition there is a major problem with asexual species, the members of which appear to be reproductively isolated from everything except their direct descendants and ancestors.

Since the 1960s an alternative programme has advocated a more direct relationship between the evolutionary tree and biological classification. So-called cladistic classification, or cladism, a version of this idea and increasingly the dominant school among taxonomists, aims directly to identify the branching points in the evolutionary tree.[6]

---

[5] A classic argument of this kind is Hull (1965). A number of recent commentators have suggested that the essentialism attributed to pre-Darwinian thinkers by recent anti-essentialists is something of a caricature, but this of course only strengthens the anti-essentialist position. See, e.g., Winsor (2003).

[6] The *locus classicus* for this idea is Hennig (1966). Cladism is generally understood as a form of phylogenetic classification that insists that all groups be monophyletic, which is to say that they must include all and only the descendants of an ancestral species. Less rigorous versions of phylogenetic classification, sometimes referred to as 'evolutionary taxonomy', relax this requirement so that it is possible to deny such apparently paradoxical claims as, for instance, that birds are a kind of dinosaur. The arguments below apply to both versions of phylogenetic classification.

Ideally, a distinct name would be given to any set of organisms lying between two branching points on the tree. The terminal branches will be the species. Because the patterns of branching in different parts of the tree can be very diverse, this often fails to reflect prior notions about how many species there are and how different they are from one another. But, cladists have tended to conclude, so much the worse for our existing notions about species.[7]

Before continuing with the discussion of classification, I must now introduce a topic that will be important throughout what follows. There is an English expression, 'the elephant in the room'. The elephant refers to a problem which, as is the way with elephants, is extremely obvious, but which, for whatever reason, all participants in a discussion decide to ignore. There is an elephant in the room of biological classification—indeed it is an elephant that can be found in many areas of biology and which I shall rudely point out at several points in my discussion. So let me now describe this elephant.

This elephant is not one large object, but a huge number of very small ones, the microbes. Microbes have been the only kinds of organisms on this planet for the majority, perhaps 80 per cent, of the history of life. And they continue to be the dominant life-form. It is calculated that even by sheer biomass microbes continue to constitute over half of contemporary terrestrial life. And the most extreme terrestrial environments remain too hot, cold, dark, or chemically hostile for other life-forms.

I should explain what I mean by a microbe. For now I shall think of microbes as including all single-celled organisms though I shall suggest later that this concept is not unproblematic. Two of the three branches of what is generally considered to be the most fundamental division among organisms consist of microbes. These are the super-kingdoms, or domains, Bacteria and Archaea. The third domain, the Eukarya, is also mostly composed of microbes, so-called protists, but also includes multicellular organisms, animals, plants, and some fungi. To emphasize their almost cameo role against the backdrop of microbial life, I and my collaborator on this topic Maureen O'Malley are attempting to popularize the word 'macrobe' to refer to those organisms, such as ourselves, that are not microbes. It seems absurd that we should have a word for the great majority of life-forms, but none for the small minority that this word excludes.[8]

I should now explain the relevance of this elephant to classification. Both the Biological Species Concept and cladistics have difficulties with asexual reproduction. The problem has already been noted for the Biological Species Concept. Cladistics is threatened in a somewhat different way. To see this we need to look more carefully at what is meant by asexuality. Sexuality is normally thought of, biologically, as a device through which two parents contribute genetic material in the production of a new individual. Asexuality, by contrast to this, is often thought of as parthenogenesis, the

---

[7] A variety of philosophical discussions of the main positions on the nature of species can be found in Ereshefsky (1991) and Wilson (1999).

[8] For more detailed elaboration of most of the points about microbes made here see Chapters 10 and 11.

production of offspring by a single parent. Sexual organisms sometimes abandon sexuality in favour of the latter method of reproduction, sometimes use it as an optional alternative. But even more than a device for facilitating genetic collaboration, sexual reproduction is part of a system for restricting the flow of genetic material. As the Biological Species Concept with its emphasis on reproductive isolation makes clear, sexual macrobes go to great trouble to make sure that their gene exchange takes place with very similar organisms. Indeed one influential descendant of the Biological Species Concept is called the mate-recognition concept, recognizing the diversity of mechanisms by which organisms, macrobes anyhow, ensure that they find the right partners for genetic collaboration (Paterson 1985). The asexuality typical of microbes should be seen by contrast to this aspect of sexuality.[9] As has become increasingly clear over the last several decades, from the perspective of genetic exchange, microbes are not so much asexual, as massively promiscuous. Microbes have a number of different mechanisms for exchanging genetic material, and they use them fully. They have mechanisms for so-called conjugation, exchanging genetic materials in a way analogous to macrobe sexuality; DNA is transferred from one organism to another by phages, viruses specific to microbes; and many microbes can pick up free, or 'naked', DNA from the environment. These mechanisms can facilitate DNA exchange between distantly related forms, even across the three domains at the base of biological classification. Because of the prevalence of these processes, typical microbes will include genetic material from numerous distinct lineages.

The problem with the phylogeny of microbes, then, and one reason that few if any microbial taxonomists endorse cladism, is that there is no unambiguous evolutionary tree on which to superimpose a taxonomic system: microbes have too many diverse ancestors.[10] Or, at any rate, they do if any past organism from which they derived genetic material is counted as an ancestor. Microbes for a long time seemed practically almost impossible to classify simply because of their dimensions. The development of tools capable of providing detailed inspection of genomes offered a solution to this problem. Comparison of microbial genomes would allow biologists to track the phlyogenetic histories of particular bits of microbial genome sequence, and infer the phylogeny, the evolutionary history, of microbes. In the early days of genomic classification of microbes a set of ribosomal genes was identified as particularly suitable for this purpose, and these continue to this day to provide an important resource for classificatory work. However, it is also becoming clear that the phylogenetic history

[9] In this and the following paragraph, my references to microbes apply mainly to the simpler organisms, the Bacteria and Archaea, lacking nuclear membranes, which are generally referred to as Prokaryotes. Matters are somewhat more complex and diverse for microbial Eukaryotes (protists). I use the term 'microbe' since it is much more familiar, and no serious confusion is likely to be engendered.

[10] This remains a controversial matter among microbiologists. A strong advocate of the impossibility of defining a microbial phylogeny is Ford Doolittle (see, e.g., Doolittle 1999). An influential resister is Carl Woese, the scientist responsible for distinguishing between the microbial super-kingdoms Archaea and Bacteria mentioned above. As will be clear, I find the former argument compelling.

produced using these genes is to an important extent an artefact of that choice. Using different genomic criteria the same organisms can appear in very different parts of the phylogenetic tree. This should be no surprise. What it indicates is merely that the genetic relations between microbes do not really form a unique tree at all, but rather a web. It may be useful for particular purposes to represent the evolutionary relations between microbes in the form of a tree, but we must remember that this is an abstraction from a much more complex reality.

I do not, in fact, believe that there is a uniquely correct way of classifying even macrobes (see my 2002: chs. 3 and 4), but the case is even clearer for microbes.[11] The failure of evolution to provide us with a unique and unequivocal method of biological classification enables us to see that there are many real discontinuities across the vast spectrum of different organic forms. And different discontinuities can ground different ways of classifying these, suited to different purposes, again both scientific and mundane. Certainly we can imagine that God, had he created the plants and animals, would have known how many distinct kinds he had come up with. Phylogenetic classification can be seen as a device that might have reconciled this ancient doctrine, to some degree, with post-Darwinian biology. But it cannot do that job. It may be an irreplaceable approach to biological classification, but it is not the only one possible, and it is an abstraction from the real complexity of biological relations. Once it is clear that only under quite special circumstances does evolution determine a unique way of classifying organisms, we should reject the cladist's indifference to the convergence of evolutionary theory on existing categories. Classifications serving different biological interests—ecology rather than evolution, for instance—and even more practical interests such as those of the forester, the herbalist, or the chef may equally be grounded in distinct natural discontinuities.

To mention one practical issue that is easily misunderstood by failing to understand this point, we might consider the problems of biological conservation. One might imagine that the aim of conservation is to save as many species as possible. Though I don't claim to know what the goal should be—I'd guess that it would be a mixture of aesthetic, utilitarian, ethical, and probably other aspects—the simple idea just mentioned surely won't do. Most fundamentally this is because it is incoherent: there is no unique way of counting the species. But even ignoring this, from any sensible conservation perspective not all species are equal. Apart from quite legitimate aesthetic arguments that the loss of tigers or gorillas would be more serious than the loss of one member of a large group of beetles, the former are plausibly far more biologically distinctive than the latter. There is, at any rate, no absolute conception of the species that contradicts this idea. Conservation of microbial diversity is an issue, and potentially a very important one, that has hardly been considered—perhaps because the actual objectives of conservationists typically are predominantly aesthetic.

---

[11] This argument is spelled out in greater detail in Chapter 10, below.

Does the denial that species represent a unique division of biological reality mean that they are unreal, or play no part in biological explanation? I have mentioned the widely held view among philosophers of biology that species are not kinds at all, but individuals. This view is linked to the idea that species are branches of the evolutionary tree and therefore inherits the limitations of that idea. However, to the extent that the evolutionary tree has branches at all it is sometimes useful to think of species as spatio-temporally extended individuals that can be identified with these branches. It is useful for theorizing much of macrobial evolution, and macrobial species should often be treated as individual things with significant causal powers. But macrobial species can be treated as individuals, because they do things: for example, they speciate, divide into two distinct species. Processes of macrobial speciation, the emergence of new biological forms, are often very real, and important for understanding biological diversity. Contrary to one popular idea, speciation is not always a slow or gradual process. About half of the species of flowering plants, for instance, appear to have arisen by a process of polyploidy, the doubling in size of the genome (Adams and Wendel 2005). Such a process creates instantaneous infertility with the ancestral species, and may produce immediate changes in the phenotype. Sometimes this is the doubling of the genome of a single parental organism, sometimes it happens through the hybridi-zation of two related plants. Because many plants are self-fertile, the appearance of a single polyploid individual may, if circumstances are propitious, found an entire new species.

The preceding case provides a nice reductive explanation of organismic diversity in terms of molecular processes. But species also play a part in explanations at their own level, and can be affected by their involvement in processes that could be thought of as at a higher level. Species interact with one another when, for instance, the members of one prey on or parasitize the members of another. This complex interaction will help to determine the dynamics of the size of the participant species. In the longer term, interactions between predator and prey species will direct the evolution of each—as is well documented in the phenomenon referred to, in an unfortunately common militaristic vein—as an evolutionary arms race. The lineages of cheetahs and gazelles, for instance, exhibit ever greater speeds as their lives depend on capturing or escaping one another. These may be interpreted as examples of large complex things—species—interacting with one another, though their significance does not depend on this interpretation. The more general point is that classifying a thing as a cheetah identifies a set of processes in which it can be involved. Classifying it in other ways might identify different processes. Such possibilities of multiple, perhaps cross-cutting, classification become more salient as classification becomes less determinate. This will be most clearly the case among the microbes.

Particular characteristics of human societies have also affected biodiversity in ways that are best described by identifying species, indicating a very different kind of interaction in which species (or merely organisms by virtue of being members of a species) may be involved. It may be that if tigers go extinct it is in part due to the belief, among significant proportions of the human species, that consuming tiger penises has great medical benefits. However decisive this factor may or may not be, it is certainly entirely possible for quite specific human beliefs to affect the trajectory of a non-human species, and there are surely many real instances of this happening. Beliefs about the relative desirability of rainforests and marketable timber are exterminating species as I write. The practice of selective breeding, now often involving targeted intervention at the molecular level, provides another obvious set of examples.

I want now to turn to a quite different biological concept, the concept of a gene, and again want to demonstrate the lack of any unique motivation underlying this concept, and the consequently distinct kinds of object that may serve these diverse concerns. Historically, the concept of a gene was introduced in the context of the experiments on breeding in the early twentieth century, deriving from the rediscovery in 1900 of Mendel's work. The gene was a hypothetical object that explained the distinctive patterns of inheritance of features of organisms discovered by Mendel. It was thus conceived as the transmittable cause of a specific phenotypic difference. For some time it remained a matter of debate whether genes should be thought of as material things at all, or rather conceived instrumentally as mere calculating devices. But by the time of the unravelling of the structure of DNA in 1953, it had become widely agreed that genes were material things and that they were located on chromosomes. This classical or Mendelian concept—the underlying cause of a difference—remains in use today, particularly in medical genetics, but as knowledge in molecular genetics has expanded exponentially in the last half century it has actually become more difficult to relate the classical gene to any particular molecular entity.[12]

Consider, for instance, the cystic fibrosis gene. This is a recessive gene, meaning that to suffer its effect, the severe congenital disease cystic fibrosis, one must receive the gene from both parents. The gene pretty accurately obeys Mendelian patterns of inheritance. But what is it? That is a harder question. Cystic fibrosis results from the failure of the body to make a particular protein, cystic fibrosis transmembrane conduc-tance regulator, involved in the production of channels that conduct salt through certain membranes. The cause of this failure is a defect (in both copies) of a bit of DNA sequence called the CFTR gene. However, there are more than a thousand known defects in this sequence that produce cystic fibrosis, though producing variably severe symptoms. So the cystic fibrosis gene is actually a large set of variations in a bit of DNA sequence. A set of variations is at least an unusual kind of object.

The gene CFTR, on the other hand, is a rather different kind of thing. It is generally defined as a sequence of 188, 698 base pairs on the long arm of human chromosome 7. This sounds a much more material kind of thing. However, it should be noted that there is no exact sequence of base pairs necessary to constitute a functioning CFTR gene. The genetic code, as is well known, is redundant, so that many changes will have

[12] Problems with the concept of a gene are discussed by Moss (2003) and in the essays in Beurton et al. (2000).

no effect at all on the functioning of the gene, and there are very likely to be changes that do make a difference to the transcription of the gene, but do not prevent its proper functioning. In short, then, the CFTR is a definite stretch of DNA sequence, though one that allows a good deal of variation; the cystic fibrosis gene is any of a large set of dysfunctional variations in the same part of the genome—or, perhaps, as we shall see, in quite different parts of the genome.

The CFTR gene, at any rate, looks a good deal more like the sort of thing people expect a gene to be in the age of genomic sequencing: a specific part of the genome with a specific molecular function. However, as so often in biology, things are not as simple as they may seem when we try to generalize this concept of the gene. A hint of the trouble can be seen in the fact that the number of genes in the fully sequenced human genome is currently estimated as being somewhere in the range of 20,000–25,000. It is often noted that this is a much smaller number than had been assumed necessary for the estimated number of gene-related human traits. But a more important puzzle for the moment is the vagueness of this estimate. Why, one may wonder, can they not just count them? The complexities that stand in the way of this task can only be sketched in the briefest way, but even such a sketch will be sufficient to make my overall philosophical point.

Properly molecular genes are often thought of as a part of the genome that codes for a particular protein. However, as a definition this raises numerous problems. Typically genes (in macrobes, at any rate) are composed of alternating sequences called exons and introns. After the gene is transcribed, into RNA, the introns are edited out, and the exons are then translated into protein molecules.[13] However, in many, perhaps most, cases there are alternative ways of splicing the exons into finished RNA sequence, and some bits may be left out. Further changes may be made either to the RNA sequence or to the subsequently produced proteins. In some cases elements of other genes may be incorporated. Thus the relation between molecular genes and proteins is not one to one, but many to many. Some genes are involved in making hundreds of distinct proteins.

It still might seem that the genes could be counted, even if they were then found to have much more diverse functions than might once have been supposed. But things get worse. First, genes can overlap. So a certain sequence can be part of two quite distinct primary RNA transcripts with quite different subsequent histories. Worse still, DNA is not always read in the same direction. So a sequence may be part of one gene read in the normal so-called 'sense' direction, but part of another when read in the opposite anti-sense direction. Philosophers Paul Griffiths and Karola Stotz have investigated empirically how many genes biologists claim to see in problematic bits of sequence and the answer, perhaps unsurprisingly, is that different biologists see different numbers of genes (Stotz et al. 2004). It would perhaps be possible to regiment the concept

---

[13] There is much to be said, and a good deal that has been said, about all these semantic metaphors—editing, transcribing, translating, coding—but that is not a topic I shall address here.

sufficiently so that the answer to such questions could be decided mechanically, but this would only conceal the real philosophical problem: nature has declined to divide the genome into a unique set of constituent entities. Different, overlapping, and non-contiguous elements of the genome are involved in different biological functions. A realistic conclusion is that a molecular gene is any part of a genome that a biologist has some reason to talk about. (Just as, indeed, it is sometimes said that a species is any group of organisms a competent taxonomist decides to put a name to.)

In fact this discussion has only scratched the surface of the diversity of entities that may legitimately be referred to as genes. The protein coding genes that I have been discussing make up only a few per cent of the DNA in many macrobial genomes, including our own. Until quite recently it used to be said that the remaining large majority of the genome was junk, a testament to the pernicious activity of genetic parasites.[14] It is becoming increasingly clear that much of this so-called junk serves important biological functions. At the very least it is essential for structural features of the genome. But it also appears that many parts of it are transcribed into RNA and that these RNA molecules play important roles in the functioning of the cell. It has also been known for a long time that non-coding sequences in the genome serve to regulate the expression of protein coding sequences, and a growing number of different kinds of such regulatory sequence are now distinguished. So, in short, there are many very different kinds of sequence that molecular biologists have reason to distinguish, and hence many different kinds of genes.

Nature, then, no more determines how to divide the genome into genes, than she does organisms into species. Particular parts of the genome can, however, as in the other examples I have considered, provide nodes on the causal nexus that are appropriate points of focus for particular investigative purposes. Reductionist explanation of the power of genes is familiar enough. Indeed, it was in large part the chemical explanation of the stability, complexity, and replicability of the DNA molecule that made the description of its structure such an extraordinary scientific achievement. A moment ago I pointed to the much more specific way in which various genetic anomalies help to explain a disease such as cystic fibrosis. Less familiar is the extent to which the DNA in a cell is in constant two-way interaction with other constituents of the cell. What was for a long time known (approvingly) as the Central Dogma of molecular biology was the view that information flowed in one direction only, from DNA to RNA to protein. In accordance with this dogma it was supposed that the function of RNA was primarily to carry information from DNA to proteins. But today the study of the vast number of different RNA molecules and their influence on gene

---

[14] These were, and still sometimes are, thought of as the truly selfish genes—merely competing with one another to occupy space in the genome. The transposable elements, the discovery of which eventually won Barbara McClintock the Nobel Prize, come closest to realizing this image, apparently concerned only with making more copies of themselves in the genome and often constituting a large proportion of the genome. Even these, however, are increasingly suspected of serving some more 'altruistic' purpose—of contributing something to the wider organism.

expression is one of the most rapidly developing fields in molecular biology. One class of these molecules, the so-called RNAi's, which can block the expression of a coding gene, are currently considered one of the most exciting prospects for molecular medicine. Many proteins, too, interact with nuclear DNA and affect the transcription of particular sequences.

It is still often supposed that the genome is the ultimate director of the process by which an organism develops, a supposition expressed in metaphors such as blueprints, recipes, or programmes. This is, in fact, an expression of the reductionist philosophy that I reject. For reductionism, a complex process such as organismal development can only be explained by causal influences from smaller constituents. DNA is seen as the largely unchanging structure that mediates this transfer of casual power from below. The vision of DNA as a node through which causal influence passes both upwards and downwards of course contradicts this picture, but does so on the basis of increasingly undeniable scientific evidence. Genes, in the end, are the diverse, nested, and over-lapping sites in the genome where these casual influences are focused, at different times, in different ways, and often in different ways at the same place.

A good way to get a sense of the implications of this picture is to contrast it with the reductionist picture of genetics that has grounded an extremely influential view of evolution but one that must now be seen as highly simplistic. If the genome were indeed an unchanging repository of information, then from the perspective of evolu-tionary theory one could see evolution as simply a temporal sequence of genomes. The organisms for which they were described as the blueprints, or the recipes, would develop as the genome dictated, and take their chances in the lottery of life, and the best ones would be selected. But all they would pass on to the next generation would be the most successful blueprints. Those familiar with the writings of Richard Dawkins should recognize this picture. Only the selfish genes in their immortal coils live on through evolutionary time.[15]

This picture should already look suspect when one sees that DNA is only one, admittedly very important, component of an interacting system of molecules, and that the whole system is passed on to the offspring in the cytoplasm of the maternal egg. But it now turns out that changes to the DNA itself that occur during the life of an organism can be transmitted to offspring. The best studied of such processes is methylation, a chemical modification of the DNA that prevents the expression of particular gene sequences. This is most familiar in stories about imprinting, the differential methylation of paternal and maternal DNA, claimed to reflect competing male and female evolutionary interests. But it is by no means restricted to this. One famous case is a study of the effects of maternal care on rats. Absence of such care,

---

[15] See Dawkins (1976) and many subsequent books. There is a good deal of more sophisticated theoretical work on evolution currently under way, though none unfortunately, that threatens to compete with Dawkins's sales volumes. Particularly important recent contributions include West-Eberhard (2003) and Jablonka and Lamb (2005). See also this volume, Chapter 9.

especially of licking by the mother, produced nervous, fearful offspring and, unexpectedly, these characteristics appear to be passed on to the offspring of the neglected animals (Weaver et al. 2004; Champagne and Meaney 2006). It has been verified that maternal care produces methylation of genes in the hippocampus, though the mechanism by which this change is passed on to subsequent generations remains obscure (Champagne et al. 2006). There is also a famous, though also still controversial case in recent human history, the Dutch famine of 1944–5. Unsurprisingly, mothers who experienced this famine tended to have small babies. Much more surprisingly, their generally well-fed children also tended to have small babies. Many have concluded that transmitted methylation patterns induced by the shock of malnutrition explain this phenomenon (see, e.g., Vines 1998). Students of such phenomena are even beginning to call themselves neo-Lamarckians, transgressing perhaps the most inviolable taboo of twentieth-century biology (Jablonka and Lamb 1995). The so-called epigenome, the set of inherited mechanisms that determine how genes are expressed, is another booming area of research. One of the successor projects to the Human Genome Project is the Human Epigenome Project that aims to map the methylation sites on the human genome. Epigenomics more generally, the study of the interactions between the cellular environment and the genome, is poised to become at least as significant a field of research as genomics itself.

The picture I have tried to sketch will not please those who are wedded to the crystalline clarity that the mechanistic vision of life offers. Shifting levels of organization with shifting, metamorphosing and even indeterminate constituents may seem like unlikely materials for understanding the exquisitely ordered and robust phenomena of life. And causal processes running upwards to exploit the diverse and specific capacities of countless chemicals and structures and downwards to provide externally enforced constraints on the actions of those structures and chemicals, may seem to be hopelessly intractable objects of real insight and understanding.[16]

Yet it is also worth considering how inadequate the mechanistic paradigm is for understanding these phenomena. As I explained with the example of a car, deterioration and failure are the inevitable history of a machine. Organisms, while perhaps all die in the end, show no such inevitable tendency. Some live for millennia with no obvious deterioration of vital functions, and it is now a matter of lively debate whether human ageing is an inevitable process of deterioration, or rather a biological function that we might, if we chose, find ways of subverting. There are powerful reasons for thinking that emancipation from the mechanistic paradigm is a precondition for true insight into the nature of biological processes.

---

[16] Carl Craver and William Bechtel, leading proponents of the contemporary mechanism mentioned in note 3, above, reject top-down causation but equally, and for similar reasons, deny bottom-up causation. They consider causation, strictu sensu, to be a concept only applicable within one structural level. Both top-down and bottom-up 'causation' they prefer to describe as 'mechanistically mediated effects'. Again, I suspect our substantive views are quite similar, though I am inclined to a much more catholic conception of causality, and am somewhat sceptical of any sharp divide between same and different levels (Craver and Bechtel 2006).

I am not, of course, the first person who has offered more complex and dynamic visions of life than are possible within the constraints of mechanism, and I shall end this chapter with one such vision that expresses, with a poetic elegance to which I can only aspire, a view remarkably congruent with much of what I have said today. The words are from Walter Pater, the British aesthete and philosopher, written in his conclusion to *Studies in the History of the Renaissance*, in 1873:

What is the whole physical life . . . but a combination of natural elements to which science gives their names? But those elements . . . are present not in the human body alone: we detect them in places most remote from it. Our physical life is a perpetual motion of them—the passage of the blood, the waste and repairing of the lenses of the eye, the modification of the tissues of the brain under every ray of light and sound—processes which science reduces to simpler and more elementary forces. Like the elements of which we are composed, the action of these forces extends beyond us: it rusts iron and ripens corn. Far out on every side of us those elements are broadcast, driven in many currents; and birth and gesture and death and the springing of violets from the grave are but a few out of ten thousand resultant combinations. That clear, perpetual outline of face and limb is but an image of ours, under which we group them—a design in a web, the actual threads of which pass out beyond it. This at least of flame-like our life has, that it is but the concurrence, renewed from moment to moment, of forces parting sooner or later on their ways . . . It is with this movement, with the passage and dissolution of impressions, images, sensation, that analysis leaves off—that continual vanishing away, that strange, perpetual weaving and unweaving of ourselves.

5

The Constituents of Life 2: Organisms and Systems

In the previous chapter I tried to explain some of the difficulties in defining central concepts in biology, and also offered a general hypothesis as to why these difficulties arise. The general hypothesis is that many of these difficulties stem from the conflict between, on the one hand, life itself as a hierarchy of dynamic and constantly changing processes and, on the other hand, our scientific understanding of living things as grounded on a picture of mechanistic interactions between fixed and statically defined components. While not wishing to deny the extraordinary insight that mechanistic models have provided into life processes, I tried to explain the deep differences between living systems and the machines that have been such a central source of inspiration for science generally. Mechanistic models have given us extensive knowledge of many of the elements of which living systems are composed, but they are inadequate to provide a full picture of life as a dynamic system.

Key concepts in biology, I suggested, are static abstractions from life processes, and different abstractions provide different perspectives on these processes. This is a fundamental reason why these concepts stubbornly resist unitary definitions. They specify, more or less, the level at which we are abstracting, but nature does not determine for us a unique mode of abstraction. This problem is central to explaining the philosophical difficulties that have been found in attempts to provide unique definitions of two central categories in biology, the species and the gene, difficulties which I summarized in the last chapter. While arguing that there is no unique and privileged way of dividing biological reality with these terms, I claimed nevertheless that there were many and diverse real biological entities falling under these concepts. In the most important cases, this reality consists in the more or less transitory focus that such entities, for example the particular parts of genomes sometimes identified as genes, provide for causal processes. But these entities must be understood not only as inheriting causal powers from their structural components, but also as recipients of causal influence from the larger entities of which they are part. This two-way flow of causal influence through a shifting and diverse array of entities presents a very different picture of life from the pristine mechanism which still influences so much scientific thinking. In the second part of this chapter I shall say something about how we might conceive the prospects for scientific progress when confronted with such a picture.

In the first part of the chapter, however, I shall enrich the general view being developed by looking at some crucial levels between the extremes of species and gene so far discussed. I shall begin with probably the most generally familiar kind of biological entity, the individual organism.

At first sight it will seem quite obvious that I, or my cat, or George W. Bush, are discrete biological entities whatever else is. To adapt US Supreme Court Justice Potter Stewart's famous remark about pornography, 'I know one when I see one.' But when we consider a little more closely what is to be included in these entities, matters become less clear. A natural way of describing the limits of the individual, John Dupré, would be to imagine the surface that includes all the parts that move together when John Dupré moves, and treat all the material included within that surface as part of John Dupré. This is a good legal definition: if someone violates that space, for example with a sharp instrument, they are considered grossly to have violated my rights.

In the previous chapter I introduced an 'elephant in the room'—the microbes, the overwhelming majority of living things. The elephant is still very much in the part of the room I am now describing. Within the surface I just mentioned, my own, 90 per cent of the cells are actually microbes. Most of these inhabit the gastro-intestinal tract, though within that, and elsewhere in the body, are a wide variety of niches colonized by microbial communities. Because of the diversity of these microbial fellow travellers, as many as 99 per cent of the genes within my external surface are actually bacterial (Xu and Gordon 2003).

We are sometimes told that the human body should actually be considered as a tube, so that the inside of my mouth or gut should rather be considered as part of the outside surface of my body. This will certainly reduce the microbial load in our own self-images, but it is somewhat counterintuitive. Considering the legal perspective just mentioned, it would be a very implausible defence against charges of various forms of serious sexual assault, for example. And it is biologically questionable. If we examine the inside surface of the gut we will discover complex and ordered communities of bacteria without which the interface between ourselves and the things we eat would be seriously dysfunctional. Our symbiotic microbes are essential to our well-being. Particularly interesting is the growing understanding that symbiotic bacteria are required for our proper development. It has been reported, for example, that environmentally acquired digestive tract bacteria in zebrafish regulate the expression of 212 genes (Rawls et al. 2004). In fact, for the majority of mammalian organism systems that interact with the external world—the integumentary (roughly speaking, the skin), respiratory, excretory, reproductive, immune, endocrine, and circulatory systems, there is strong evidence for the co-evolution of microbial consortia in varying levels of functional association (McFall-Ngai 2002). For these reasons, some biologists are now proposing a second human genome project—the human biome project—that will catalogue all the genetic material associated with the human, including that of their microbial partners.[1] At any

---

[1] [Note added to this edition] In fact in 2008 the United Sates National Institutes of Health launched the Human Microbiome Project, with just this goal.

rate, as a functional whole, there is much to be said for thinking of the whole community that travels around with me as a single composite entity.

I have mentioned that microbes remain by far the most versatile and effective chemists in the biosphere. The ability of multicellular organisms like ourselves to process food is entirely dependent on their cooperation. Being high on the food chain, we humans tend to consume highly processed foods that require less help from our microbial symbionts to metabolize than would more challenging inputs. It is worth recalling, though, that if we eat, say, cows, the amino acids we absorb were synthesized by microbes in one of the four stomachs of the animal, and if perhaps for moral reasons we prefer to get our amino acids from beans, this is only possible due to the nitrogen-fixing activities of bacteria associated with the roots of these plants. At any rate, as is familiar to users of powerful antibiotics, deficiencies in our gut bacteria are a serious problem even for human digestion.

Let me add a brief word about plants in this context. The best known metabolic capacity of plants is the ability of their long captive microbial symbionts, chloroplasts, to capture the energy of sunlight. But plants also feed through their roots of course. And the roots of a plant lie in the midst of some of the most complex multispecies communities on the planet. Bacteria and fungi not only form dense communities in the soil surrounding plant roots, but are also found within the roots themselves. These fungi put out nutrient-seeking tendrils, or hyphae, through the roots and into the outside soil. This is perhaps the most striking illustration of the idea that the interface between large multicellular organisms and their environments is typically mediated—and essentially so—by microbial communities. Given the permeation of the boundaries between ourselves and the external environment by comparably complex multispecies communities, and their essential role in managing the chemical and cellular traffic across these boundaries, we should at least question our intuitive sense that they are not part of us.

There is a theoretical ground for the assumption that an individual should be taken to exclude its obligatory symbionts. This might be stated as the thesis, one individual, one genome.[2] It is sometimes said, for example, that a group of trees, originating from the same root system, is *really* only one individual. The motivation for this stipulation is an evolutionary perspective, according to which evolution involves selection between genomes, but the stipulation is a problematic one. First, there are clearly important alternative perspectives that must sometimes be accommodated. From an ecological perspective, for example, we should surely prefer one trunk, one tree. We might also note that even complex animals are to some degree genomic mosaics. In extreme cases this may be the result of abnormal reproductive events, as is known to the great cost of a few human mothers who have failed genetic tests for parenthood of their children.[3] Transplant surgery, including blood transfusion, pro-

---

[2] For extensive criticism of this thesis, see Chapter 7.

[3] There are several cases of mothers having lost custody of children on the grounds that they were 'proved' not to be the biological mothers as a result of different parts of their mosaic genomes appearing in the child and the genetic test for parenthood. See Pearson (2002).

duces genomic mosaicism in some humans. More mundanely, mutation during development produces some genomic diversity, and in plants that reproduce vegetatively—for example by root suckers or rooting branches—this may provide material for natural selection.

It is also a familiar fact that the same genome may pertain to different individuals. Close to home, we do not consider monozygotic, or so-called identical, twins to be a single individual. This draws attention to the very important point, but not one I shall dwell on here, that there is much more to development than the unfolding of the genome.[4] But much more generally, we should again remind ourselves of the elephant. The vast majority of organisms do not produce an entirely novel genome in the process of reproduction. Although microbial genomes are extremely fluid over time, the basic process of reproduction is one of genome duplication. In short, the relationship between genomes and organisms is not one to one, but at least one to many. I want to suggest that it is a further but well-motivated step to admit this relationship as many to many: not only can one genome be common to many organisms, but one organism can accommodate many genomes.

I hope I have anyhow said enough to dispose of a simple criterion that might give a simple answer to the question, What is an organism? The correct answer, I suggest, requires seeing that there is a great variety of ways in which cells, sometimes genomically homogeneous, sometimes not, combine to form integrated biological wholes. The concept of multicellular organism is a complex and diverse one which, incidentally, provides no conceptual obstacle to the broader conception of the human individual sketched above.

The broader ramifications of this suggestion are once again best discerned by looking more closely at the elephant. I have described microbes as single-celled organisms, and this is how the organisms to which I wished to refer are generally conceived. However, there are compelling grounds for revising this view. Microbes are most commonly found as parts of communities, containing either one or many distinct types of microbe, communities which approximate many of the familiar features of multicellular organisms (Shapiro 1998; this volume, Chapter 10).

There are a number of converging types of evidence that support this perhaps surprising proposal. Probably the most important has been mentioned, the frequency of genetic exchange between microbes, particularly within associated communities of microbes. An increasing number of biologists are beginning to suggest that the genetic resources of a microbial community should not be thought of as partitioned into individual genomes in individual cells, but are rather a community resource, a genetic commons.

---

[4] This point, especially in the guise of 'Developmental Systems Theory', to which several chapters in this book advert, has been fundamental to recent critiques of classical models of evolution, from the mid-twentieth-century synthesis to Dawkins's gene selectionism. See Oyama (1985); Oyama et al. (2001). For application of this perspective to criticism of Evolutionary Psychological theories of human nature see Dupré (2001), and this volume, Chapter 14.

The project of metagenomics, the attempt to collect all the microbial genetic material in extended environments has become quite widely discussed, not least due to its association with Craig Venter, the leader of the free enterprise wing of the Human Genome Project, and a larger than life figure in contemporary biology. Venter embarked on a well-publicized expedition around the world's oceans collecting microbial genetic material from the water (Venter et al. 2004). This is sometimes seen as no more than a collecting or gene-prospecting exercise, and indeed Venter discovered an astounding quantity of unfamiliar genetic material and many previously unfamiliar types of microbes. But as suggested by my reference to a genetic commons, and by the great mobility of genetic material between microbial cells, it is but a small step from the metagenome, this totality of local genetic material, to the metaorganism, a multicellular organism composed of the community of microbes that shares this resource.

The clearest context in which to present the idea of microbial metaorganisms is with the phenomena of biofilms. Biofilms are closely integrated communities of microbes, usually involving a number of distinct species, which adhere to almost any wet surface. Biofilms are ubiquitous, from the slimy rocks and stones found under water and the chemically hostile acid drainage of mines, to the internal surfaces of drinking fountains and catheters; indeed biofilms are where most microbes generally like to be.[5] In addition to the genetic exchange already discussed, the constituents of biofilms exhibit cooperation and communication. These are most clearly exemplified by the phenomena of quorum sensing, in which microbes are able to determine the numbers of cells in their communities and adjust their behaviour—including reproduction—in appropriate ways. The general idea can perhaps be best appreciated by quoting a scientific paper on a very familiar kind of biofilm:

Communication is a key element in successful organizations. The bacteria on human teeth and oral mucosa have developed the means by which to communicate and thereby form successful organizations. These bacteria have coevolved with their host to establish a highly sophisticated relationship in which both pathogenic and mutualistic bacteria coexist in homeostasis. The fact that human oral bacteria are not found outside the mouth except as pathogens elsewhere in the body points to the importance of this relationship. Communication among microorganisms is essential for initial colonization and subsequent biofilm formation on the enamel surfaces of teeth and requires physical contact between colonizing bacteria and between the bacteria and their host. Without retention on the tooth surface, the bacteria are swallowed with the saliva. Through retention, these bacteria can form organized, intimate, multispecies communities referred to as dental plaque. (Kolenbrander et al. 2002)

For readers unconvinced that these should be counted as multicellular organisms, I invite reflection on the diversity of forms of multicellularity. Most familiar are the plants and animals, and it would certainly be possible to enumerate a series of major

---

[5] This is more clearly true of terrestrial than pelagic microbes, very large numbers of the latter being found as individual 'planktonic' cells. However, it also appears that a high proportion of these are in an inert state, and only become active in the context of microbial communities.

differences in the way these two prominent groups of organisms are organized. Or consider the third taxonomic group generally acknowledged to include multicellular organisms, the fungi. Fungi are generally divided into single-celled organisms, yeasts, and a variety of multicelled forms, such as mushrooms. But the multicellularity of fungi is a rather simple matter. Fungi form threadlike chains of cells, called hyphae, which generally exist in tangled mats, called mycelium. Some varieties occasionally organize their hyphae into much more ordered structures such as the familiar mushrooms that function to disperse fungal spores. This is a far less complex form of multicellularity than that exhibited by the many differentiated cell types of plants or mammals.[6] A mushroom is actually much more similar to the fruiting structures of such social bacteria as the myxobacteria, which also form colonial structures not unlike mushrooms for purposes of spore dispersal—though the cooperative hunting of other bacteria also reported in some species of myxobacteria perhaps suggests a more complex sociality than that of fungus cells. It is worth mentioning that some complex multispecies organisms have been familiar for a long time, most notably the lichens, symbiotic associations of photosynthetic algae or bacteria with a fungus. Anomalous from the perspective of a traditional dichotomy between unicellular organisms and monogenomic multicellular organisms, these seem quite unproblematic from the point of view of a more comprehensive understanding of multicellularity.

Multicellularity, even in the traditional sense just mentioned, is an enormously diverse phenomenon. Cells of different kinds organize themselves into a vast diversity of cooperative arrangements, with variously rigid structure, developmental trajectories, and so on. What is particularly surprising for traditional biological thought in the case of microbes, however, is that these cooperative ventures typically involve cells from quite diverse parts of the traditional phylogenetic tree—though as I have suggested this is not really unusual even for the more familiar multicellular organisms. The diversity of multicellular organization should be no great surprise. Leo Buss, perhaps the pre-eminent theorist of biological individuality, claimed over twenty years ago that multicellularity had evolved independently around seventeen times (Buss 1987). In fact it is fair to say that forming cooperative associations is very fundamentally what cells do. It is worth considering, as some biologists indeed have, that while no doubt evolution depends on competition among cells, it may be that what they primarily compete over is their ability to cooperate with other cells (Margulis 1998). Altruism, in its technical biological sense of assisting another organism at some cost to oneself, far from being a fundamental problem for evolutionary biology, may turn out to be ubiquitous in the living world.[7] From this perspective, the diverse communities that

---

[6] [Note added to this edition] Simpler in overall structure. On the other hand the cells of some fungi, which contain large numbers of nuclei with distinct genomes, are perhaps uniquely complex.

[7] A less radical but still controversial claim for the prevalence of altruism, grounded in broadly conventional evolutionary thinking, is Sober and Wilson (1998). The present discussion departs rather further from the widespread scepticism about cooperation.

make up microbial biofilms and even the diverse communities that constitute a properly functional plant or animal, including its mutualistic microbial communities, can quite properly be considered multicellular organisms.

Throughout this and the previous chapter, the one level of biological organization I have constantly referred to without qualification is the cell. And in fact it does seem that this is the most unproblematic such level. If there were any unique answer to the question, 'What are the constituents of life?', that answer would have to be cells. Cells are enormously diverse things, of course, but everything on the standard representation of the Tree of Life is a cell or is composed of cells. The problems I have indicated with a naïve conception of the organism derive from the complexity and diversity of the relations among very diverse sets of cells, but they do not problematize the idea of the cell as the basic constituent of these various associations.

I have to note, however, that hiding behind my now familiar elephant is yet another elephant. I have hardly mentioned the living forms that are not cellular and that are even more numerous than the cellular microbes, namely the viruses and related objects. Viruses have been found associated with every organism studied and they outnumber any other class of biological entities by at least an order of magnitude. Estimates of the numbers of viruses on Earth are in the range of 10 to the power of 31—a 1 followed by 31 zeros (Rohwer and Edwards 2002). This number is probably incomprehensible to non-mathematicians: it has been described by one virologist as amounting to 250 million light years of viral genes placed end to end (Hamilton 2006).

It is sometimes said that viruses are not living things at all. And it is true that they often exist in an entirely static state in an inert crystalline form that can hardly be said to be living. On the other hand, they are the most efficient replicators of their genetic material on Earth. It has been estimated that anything up to 50 per cent of marine bacteria are killed every day by pathogenic viruses, phages, and in this process the hostile virus produces thousands of replicas of itself for each bacterium it destroys (Breitbart and Rohwer 2005). It has been suggested that perhaps 10 to the power of 24 viruses are produced on Earth every second. And these massive replication rates are tied to high mutation rates and almost unlimited mutation mechanisms. Viruses can thus evolve at rates that are far beyond even what is possible for cellular microbes. It is a familiar observation that the HIV virus evolves significantly in the body of a single host, a fact that provides enormous obstacles to the development of effective therapies. When compared with our own 20–30-year generation spans, populations perhaps 22 orders of magnitude smaller, and handfuls of offspring, it is clear that viruses have abilities to explore the space of chemical possibility that organisms such as ourselves could hardly dream of. It is thus no surprise that viruses are the greatest producers and reservoir of genetic diversity on Earth.

Even more interesting than the viruses that kill their hosts are the ones that don't. Most viruses live in stable relations with their hosts but all viruses reproduce themselves by exploiting the chemical resources of their hosts, and many also insert their genetic material into the host genome. Since they may also incorporate DNA from their host's genome into their own, they can readily transfer DNA from one organism to another. Although viruses are generally quite specific in their hosts they can, as is well known, transfer to new hosts. When they do this they can transfer DNA from one species of organism to another. I mentioned earlier the so-called junk DNA that makes up most of the DNA of eukaryotes such as animals or plants. As well as being increasingly clearly not junk, this is in fact material mainly or entirely of viral origin (Villareal 2004b). It has been noted, for example, that the main differences between human and chimpanzee genomes are not, as might have been supposed, in coding sequence, but in non-coding regions derived originally from viruses (Villareal 2004b). The enormous powers of viruses to evolve and their ability to insert genetic material into the genomes of cellular organisms has led some biologists to speculate that it is viruses that are the prime movers of major evolutionary change or, at any rate, the main providers of novel biochemical resources. It is beginning to seem possible that, just as microbes are the expert metabolists in nature, so viruses are the leading evolvers. And as microbes provide us with indispensable chemical services, it may be that viruses provide us with comparably significant evolutionary services. At any rate, without disputing the fundamental importance of cells as foci of causal power and organization that make possible the complex biological structures and communities which we most naturally think of as biological objects, it is important not to forget the much larger number of non-cellular biological objects that spend their time moving genetic material into and between cells. The extraordinary biological capacities of microbes and viruses pose a very interesting question as to what it is that familiar multicellular organisms do well enough to exist at all unless, indeed, to adapt a well-known idea of Richard Dawkins, they are vehicles for carrying their microbial masters around, or niches constructed as residences for microbial communities.

My message so far may seem discouraging with regard to the prospects of real biological understanding. It is true that the very ability to discern the complexities I have sketched in these chapters displays the remarkable power of the instruments scientists have devised to explore the workings of life. Scientists have revealed in exquisite detail the structures of biological molecules and their modes of interaction with other molecules. And they are compiling comprehensive inventories of these molecules. Yet the sheer number of constituents thus discovered combined with two other problems that I have tried to emphasize throughout these discussions, presents a problem of almost inconceivable complexity. The first problem is that even as we discern these multitudinous constituents of living things, their biological significance cannot be fully discerned without a view both of the causal powers derived from their own structure and the causal powers of the larger systems in which they participate. We cannot properly appreciate the biological properties of a virus or a bacterium, say, without understanding both the chemical processes found within it and the much larger systems of which it is a vital constituent. The second problem is that even the inventory of causally significant objects at a particular level is not something fully determined by nature, but may vary according to the kind of question we want to ask.

Nature is not divided by God into genes, organisms, or species: how we choose to perform these divisions is theory relative and question relative.

It is, then, possible to achieve remarkable insights into life processes, but is there any way we can ever hope to fit them together into an integrated understanding of how a living thing, even a single living cell, functions? There is an exciting project that is currently receiving a lot of attention and investment and which does have this aspiration, integrative systems biology. I shall now say a few words on this topic.

First, what is systems biology?[8] It is often said that it is nothing new. General systems theory is generally traced to Karl Ludwig von Bertalanffy (1950) in the mid–twentieth century, and a number of biologists, perhaps most notably the American theoretical biologist Robert Rosen (1970), have developed ideas that were at least important precursors of contemporary systems biology. At another extreme I have heard biologists say that systems biology is no more than a new name for physiology. The difference, I think, and perhaps this is a case where a difference of degree becomes a difference of kind, is the vast quantity of data, especially molecular data, that is available to the current theorist. Indeed it is not too cynical to say that a major motivation for this entire project is the question, now that we have all these terabytes of molecular data, what do we do with them?[9]

The proposed answer, very broadly put, is that we employ teams of biologists, mathematicians, and computer scientists to work out how we can create adequate mathematical representations of the multitude of diverse objects discerned in a biological system, and then use these representations to explore and better understand the real processes. The envisaged subjects of such models range from chemical subsystems within cells, to whole cells, complex organisms, and even organismal communities. Microbial communities of the kind described earlier are one attractive target, and are indeed currently one of the major areas to which the very large contemporary investments in systems biology are being directed. Optimists about this project envisage that we may eventually have good enough models of human cells and systems that we can use them for example to test drugs in silico. Apart from likely economic advantages, this prospect looks an attractive way, especially in the UK, to avert the unwanted attentions of animal rights activists. In light of the problems I have been describing, one very clear advantage of this project is that in principle, at least, it seems possible for such models not only to represent systems as dynamic, but even, in principle again, to represent the interactions of different scales of dynamic systems.

There are two extreme perspectives on how systems biology should proceed—though no doubt most actual attempts at its implementation will lie somewhere in between the extremes. At one extreme is a broadly reductionist approach that simply aims to find ways of representing as much molecular data as possible, including of course interactions between molecules, and calculate what happens when all of this is

---

[8] For a more detailed discussion of systems biology, see O'Malley and Dupré (2005).
[9] Since writing this we have moved into the era of petabytes (a petabyte is 1,000 terabytes).

put together. At the other extreme, many biologists hold that the only chance of making such a project work is to appeal to some much more general principles as to how systems work in order to have some way of deciding, among the vast mass of biological data, what is important and what is not.[10]

My own sympathies lie some way towards the latter end of the spectrum. Scientific modelling is not like building a scale model of a ship, where the ideal outcome is to produce an exact miniaturization of the original object. Rather, scientific models are successful to the extent that they identify the factors, or the variables, that really matter. I have emphasized throughout this work that the objects we distinguish in biological investigations are generally abstractions from the complexities of dynamic biological process. The models we are now considering, then, are abstractions of abstractions—selections among the first level of abstractions that we hope may provide us with approximations of the full functioning of biological objects. While the complexity of biological phenomena is forcing us to develop new kinds of models capable of including very large numbers of factors, it is unimaginable that this project can be extended to the extreme of in silico miniaturization. Some theoretical principles or assumptions will surely be needed to guide this second level process of abstraction. Moreover, in opposition to the extreme reductionist understanding of systems biology, and as I have emphasized throughout these chapters, one cannot understand the biological significance of a molecule without appeal to larger structures with which it interacts.

On the other hand, I am extremely sceptical of the idea that there are general laws of systems that can be applied equally to economic systems, weather systems, biological systems, or even, for that matter, all biological systems. I don't claim to have an argument that there couldn't be such laws, I just see no reason why there should be. Apart from their various levels of complexity, different systems work in very different ways—I'll say a bit about this in a moment. I suggest that what we need as a theoretical infrastructure for modelling biological systems—and here is an argument that the topics I have discussed in these lectures really matter—is ontology. We cannot expect to understand the behaviour of molecules in a cell unless we have a clear idea of what sorts of molecular objects there are and, even more importantly, what kinds of larger structures they are particularly suited to interact with. Let me try to make this clearer by returning to an example that, though it may well be as complex as, or even more so than, the molecular economy of the cell, is in certain respects much more familiar, the behaviour of people.

I shall limit myself to a very modest aspect of human behaviour, one I touched on in the last chapter, the movement of people round a city. I mentioned then the Albert Cuyp market and its curious capacity to attract large throngs of people. I noted that this is a capacity that it only has at specific times—a matter I myself confirmed with a

---

[10] See again O'Malley and Dupré (2005).

pointless trip there on Queen's Day, a major public holiday. It also differentially attracts certain people. Some households still exhibit a division of labour between wage earners, domestic workers, and free-riders (children). A market has a much stronger capacity to attract people belonging to the second of these categories. For reasons that could in principle be explored, even among those who regularly buy food, some are much more drawn to street markets, others to supermarkets. And so on. The point I want to make is only that we would have no chance of modelling the human movements around a city merely by a detailed inventory of the dispositions even of every individual person. These movements are constrained and promoted by a huge number of physical and institutional structures: roads, shops, schools, playgrounds, city councils, and so on. And importantly, my visit to the Albert Cuyp will make a difference to me and, however minimally, the market. Depending on my experience there I will be more or less likely to return and may eat different foods. And my activities will make marginal differences to the experience of the traders with whom I interact. Aggregated with the activities of many other visitors, this will ultimately affect the continued participation of traders in the market, and so on.

It is not entirely fanciful to compare this scenario with the inside of a cell. Millions of molecules move about this microscopic space, and their movements are also constrained by a complex cellular infrastructure of organelles, membranes, and so on. Ribosomes, for example, structures that host the production of protein molecules, are sites where amino acids and messenger RNA molecules congregate—not, of course, because they have intentions or plans but either because they are transported there by molecules with that particular function, or simply because they tend to stick when they bump into these structures. Just as my movement around a city has to be understood as a relation between my internal dispositions and the infrastructure that surrounds me, so the traffic of molecules around a cell is jointly determined by the capacities that derive from their molecular structure and features of the cellular environment, including most notably the density of other molecular species and the membrane topography of cellular infrastructure.

I do also want to keep in view that disanalogies are important too. We should not assume that the same principles will emerge from a system connected by social conventions and language as one based on chemical interactions. And we might usefully remember that there are simpler systems, for example the weather, based on purely physical interactions. Though we are becoming quite good at modelling such simpler systems, they are hardly simple: the best models used to predict global warning take several months to run on our fastest computers.[11] And useful information from such models, information, that is, that is understandable to the many and diverse non-expert consumers of meteorological information, must also be conveyed in terms of abstractions from the flux of process, for example hurricanes, cold fronts, droughts, or showers.

---

[11] Richard Betts, talk to the Migrations and Identity Symposium, Exeter University, November 2005.

In earlier critiques of reductionism I have suggested that what reductive explanations do is explain the causal capacities of things (Dupré 1993). Where they go wrong is when it is supposed that this is sufficient to explain what they actually do—which capacities are exercised when—something that will typically require detailed knowledge of the context in which they are placed. But I now want to say that this is too simple. My capacity to deliver the Spinoza Lectures required both internal capacities of mine (I hope!) and institutional facts about the University of Amsterdam without which no one could have such a capacity. Similarly, the important capacity of messenger RNA molecules to adhere to ribosomes requires both the chemistry of RNA and the presence, and salient features, of ribosomes. Even the capacities of things are produced jointly by internal structures and features of the context in which those capacities are to be exercised.

So an adequate model of, say, a cell, must at least be rich enough to include the mutual determination of properties of objects at different structural levels. If this is true, it may seem to imply that there can be no stopping place short of the entire biosphere. If cells have properties partially determined by, at least, the organisms of which they are part, and the relevant properties of organisms are in part determined by the larger associations of which they are part, then everything mutually determines everything else. But though this has to be admitted in theory, in practice things are not necessarily this bad. What these points do indicate is the importance of deciding what is a sufficiently isolated system to be a plausible target for modelling—this, of course, is part of the process of abstraction that I just mentioned was inevitably central to such modelling projects. And as many theorists have pointed out, partial isolation of systems and subsytems, or modularity, is very probably a necessary feature of any partially stable system above a fairly low level of complexity, so we can expect some help from nature in providing appropriate boundaries for our abstractions. At the lower end there is little doubt that intrinsic chemical properties can be taken as brute facts from the point of view of biology—little benefit is likely to accrue from trying to explain variations in the behaviour of biological molecules by appeal to quantum mechanics, though no doubt such explanations are also possible. If we are trying to model a complex organism there are probably many features of the behaviour of cells that can be treated as given, though which these are will be a difficult part of the work of creating such models.

The upper end is harder. The clearest example of a plausible target for a system model is the individual cell. As mentioned earlier, this seems the most unproblematic layer of organization into individuals in the biological hierarchy, though even here the difficulties consequent on the sensitivity of cells to biological context should not be underestimated. A strictly reductionist approach to cellular differentiation, for instance, would not easily have appreciated the importance of geometrical distortion by neighbouring cells as clusters of cells divide. Defining the outer limits of an organism, for reasons already discussed, is much harder. For many aspects of organismic function—

digestion, immune response, or development, for example—it may prove that an adequate model requires treatment of the whole biome, and in the case of ourselves, and all or almost all other plants and animals, this would include not only traditionally human cells but those of our vast array of microbial, including viral, commensals. This is a good point at which to introduce the final question I want to discuss.

The question is, granting that ontology matters, Why is this a matter for philosophers? Isn't biological ontology a matter for biologists? With some important qualifications, I would answer Yes to this last question. The substantive claims I have made in this chapter and the last have mostly depended for their plausibility on my, no doubt limited, understanding of biological issues. Why not leave it to those with less limited understanding?

Part of the answer that I want to stress is that ontology is much less simple a matter than might appear, even to those with a deep understanding of the facts. Indeed my earliest work in the philosophy of biology, concerned with showing the deep differences between the taxonomies of organisms required for biological investigations, and those required for everyday life, was part of a project of showing that biological ontology is seldom simple or obvious. One way in which it is less simple than generally assumed is that it is equivocal. My example of the human biome can illustrate the point. To understand human development, human susceptibility and resistance to disease, or human digestion, this may be the narrowest sufficient system to consider. But for examining human behaviour it is likely that a more traditional conception of the human, ignoring microbial symbionts, will be appropriate (this will perhaps be a relief to psychologists, philosophers, and others who study human behaviour). This is not only because—though this is a fascinating fact—the nervous system is the only part of the human biological system that is not currently believed to have co-evolved with a commensal microbial community. It is more simply because from the point of view of behaviour, most of biology can be taken as given. The question why my arm goes up when I decide to raise it is an enormously difficult one for physiology, and a perpetually intriguing one for philosophy, but it can be taken for granted by most scientific students of behaviour. Of course my microbial associates will sometimes directly affect my behaviour: when my gut flora are unhappy, my behaviour will be much restricted. But it will usually be sufficient to note that I have an upset stomach without going into cellular details.

A central concept for addressing the ontological issues I have been considering is that of a boundary. The boundary to which I have just alluded is one that screens behaviour from the details of cellular chemistry and thereby intercommunity cooperation. But this is a boundary that screens behaviour but not, for instance, disease. So ontological boundaries are relative to the issues with which we are concerned, which is a central part of the reason why there is no unique ontology. To return to another of my favourite examples, how we divide organisms into kinds or species (which, in some instances, coincides with dividing them into individual things), depends on why we are doing it.[12] Note also that division of organisms into species amounts to discrimination of things—evolutionary lineages—just to the extent that there are real biological boundaries in place, that is to say, the boundaries that block the flow of genetic information. Microbial lineages are less plausibly treated as things than are some macrobial lineages, exactly because the boundaries of the former are so permeable, especially to genetic material. When I suggested earlier that the cell was the most unequivocal constituent of life, I might also have said that the cell membrane is, for a very wide range of theoretical questions, an effective boundary.

The clearest example of the importance of ontology to biology is in the theory of evolution. The issue that has been most extensively discussed by philosophers of biology for the last thirty years or so is the so-called units of selection problem: given that evolution is driven by natural selection, what are the things that selection selects? An idea that has been enormously popular in this regard has been that promoted so effectively by Richard Dawkins (1976), that the units of selection are bits of DNA. For a variety of reasons philosophers have almost uniformly rejected this idea, most importantly on the ground that it assumes a simplistic view of the relation between genes and organisms. Prior to the popularity of gene selectionism, it was assumed that the targets of selection were organisms. Nowadays it is widely held that the answer must be that there are a variety of levels—genes, organisms, and probably groups of organisms.[13] I think it may be possible to reinstate something like the idea that organisms are the primary target of selection, but with three very important qualifications. First, as I have argued here, we should not take it as obvious what the organisms are. It may be that the typical 'organism' is really a community of co-evolved cell types. Second, as a number of biologists have argued in recent years, organisms do not evolve in passive response to their environment. The evolutionary 'niche' to which an organism is adapted is as much a product of the organism as a cause of the organism's adaptation (Odling-Smee et al. 2003). This should seem a natural idea in the context of these chapters. Central to the niche of a bird, for instance, is its nest; but it didn't evolve to occupy the nests that happily turned out to be lying around, but rather modified its environment to provide the resources to which it is adapted. This is most obvious of all for humans for whom the environmental niche includes schools, hospitals, and such-like, all of which play an essential role in the life cycles of individual humans. And, third, as is indicated by my reference to life cycles, we must avoid seeing the organism as a static thing with a fixed set of properties. Organisms are generated, develop, reproduce, age, and die. All of these stages are adapted in different ways to the niche with which the organism is co-evolved. Thus if I say that the organism is the normal

---

[12] And, in this case, what kind of biological entities we are doing it for: there is no reason to assume that the best principles for classifying birds will also be well suited to bees or bacteria, and many reasons for doubting it.

[13] For a good survey of recent thinking on the so-called units of selection problem, see Sterelny and Griffiths (1999: esp. chs. 3–5).

unit of selection, it should be understood that the concept of organism involved is far removed from a naïve and static conception of a living individual. This organism is a process—a life cycle—rather than a thing; it may be a community of distinct kinds of organisms rather than a monogenomic individual; and it must be understood as conceptually and of course causally linked to its particular environment, or niche, which both contributes to the construction of the organism in development, and is constructed by the organism through its behaviour.

Simplistic understandings of evolution, often based on naïve views of deterministically understood genes as units of selection, can underlie bad, and even dangerous, science. I have argued in some detail to this effect about the distressingly influential project of so-called Evolutionary Psychology (Dupré 2001; see also Chapter 14, this volume). This is the view that our basic behavioural dispositions are best understood by reflecting on evolutionary forces that acted on ancestral humans in the Stone Age. Among the fundamental failures of this programme is its grounding in an antiquated view of evolution based on a crude ontology of genes with deterministic developmental capacities, and isolated, self-interested individuals.

This suggests another reason why biological ontology may be fit work for philosophy. Most biologists, and for good reasons, are strongly focused on very specific problems. The best biologists often do concern themselves with ontological issues, and it is their work in these moods that is often most valuable for the kibitzing philosopher. But even these biologists may be constrained by their disciplinary expertise. It is rare to find someone professionally expert in microbiology, vertebrate evolution, and immunology, say. One might recall Ernst Mayr, one of the most distinguished evolutionists of the twentieth century, dismissing Carl Woese's revolutionary reformulation of taxonomy into the three domains I described in the last chapter, with the remark that 'Woese was not trained as a biologist and quite naturally does not have an extensive familiarity with the principles of classification' (Mayr 1998: 9721).

And, finally, the discussion of ontology, what there is, can benefit from the availability of a set of conceptual resources that have been (and are still being) hammered out for centuries by philosophers—natural kinds, individuals, causation, and so on. I would argue that one of the best tests of the value of such tools and the value of the very abstract modes of argument philosophers have used to discuss them, is their ability to throw useful light on the much more concrete and specific issues that concern different societies at different times. And no set of issues, I suggest, should concern our own society, at this time, more than the remarkable insights into nature being offered by contemporary biology.

In the last two chapters I have tried to sketch a synoptic view of biology—a survey of the kinds of things that constitute the biological world and the kinds of relations they have with one another. My only consolation for my undoubted inadequacy for this task is the thought that perhaps no one is properly adequate. Our scientific understanding of life processes is growing at a breathtaking rate, and our ability to synthesize and assimilate this understanding will have fundamental effects not only on everyone's understanding of life, but even on the future trajectory of human life. I, at least, am convinced that the task is of sufficient importance to outweigh the undoubted risks of getting things wrong.

6

Understanding Contemporary Genomics[1]

Recent molecular biology has seen the development of genomics as a successor to traditional genetics. This chapter offers an overview of the structure, epistemology, and (very briefly) history of contemporary genomics. A particular focus is on the question to what extent the genome contains, or is composed of, anything that corresponds to traditional conceptions of genes. I conclude that the only interpretation of genes that has much contemporary scientific relevance is what I describe as the 'developmental defect' gene concept. However, developmental defect genes typically only correspond to general areas of the genome and not to precise chemical structures (nucleotide sequences). The parts of the genome to be identified for an account of the processes of normal development are highly diverse, little correlated with traditional genes, and act in ways that are highly dependent on the cellular and higher level environment. Despite its historical development out of genetics, genomics represents a radically different kind of scientific project.

An ancestor of this chapter was written for a symposium on 'Proof and Demonstration in Science and Mathematics'. This presented the immediate difficulty that I was unsure whether there was anything to be said on these topics relating to my current areas of study, genetics and genomics. Proof, at any rate, is not a concept I often encounter anywhere in biology. Certainly there is plenty of evidence for some biological claims, but I'm not sure these generally amount to anything that would count as proof. Proof in mathematics is a much more familiar idea, but in so far as it points to a logical relation between axioms and theorems, its application to contemporary science suggests an antiquated philosophy of science which, at least in the context of biology, has been almost entirely discredited. There is, of course, a tradition in the philosophy of science of thinking of explanations as derivation from laws and initial conditions, which raises obvious parallels with deduction from axioms. But nowadays philosophers of biology are much more inclined to talk about models than

[1] A version of this chapter was presented at the Athens-Pittsburgh Symposium in the History and Philosophy of Science and Technology in Delphi in June 2003, and benefited from the comments of several members of the audience there. I am especially grateful to Richard Burian, who saved me from some significant historical errors.

about laws, and are generally quite sceptical even of the existence of biological laws (see, e.g., Lloyd 1994).

A quite different connection might be through the use of biological evidence in juridical contexts. It is often claimed for instance that the analysis of DNA found at crime scenes provides proof of guilt or innocence. Interesting though such contexts undoubtedly are, and interesting though the issues they raise in understanding juridical proof may be, I doubt whether they raise profound philosophical issues in the study of science.

Demonstration is, I think, a broader concept, and perhaps holds more promise. It can, of course, be more or less synonymous with proof, as again in the context of mathematics and as illustrated by the initials QED which schoolchildren used to be required to add to the conclusion of what they fondly hoped were mathematical proofs. In simpler times, demonstrations were an important part of science pedagogy. A physics teacher might hang weights of various sizes on a spring, measure the length of the spring, and plot the length of the spring against the numbers stamped on the shiny brass weights. Subject only to our confidence that these numbers corresponded to a real property of the weights, something rendered intuitively plausible by their visible sizes—the one marked 5kg looks quite a few times as big as the one marked 1kg, for instance—this might reasonably be taken as a demonstration of Hooke's law. If one attends carefully to the performance, the demonstration, one can more or less see that Hooke's law is true (or for the very sceptical, true here, today, anyhow).

In striking contrast to this simple demonstration, I recently had the good fortune to receive a tour of the Sanger Centre, outside Cambridge, where a large part of the Human Genome Project, the sequencing (more or less) of the entirety of the genetic material in a human cell, was carried out. Over the reception desk an electronic display flashes a stream of Cs, Gs, As, and Ts which, we are informed, constitute a real-time readout of some DNA that is being sequenced somewhere on the premises. Touring the building where the sequencing actually takes place, the first stop is a room in which large robots stick tiny probes into Petri dishes and then into rectangular arrays of test tubes. Spots on the nutrient gel in the Petri dishes, we are told, contain bacteria infected by viruses with pieces of human DNA. We next peer through a window in the door of a room containing small but expensive machines that perform the polymerase chain reaction, the process that multiplies the quantities of DNA generated in the first process to the quantities required for the sequencing machines. These latter, finally, occupy a warehouse-sized space in which conversation is rendered difficult by the hum of the powerful cooling systems. There are perhaps a hundred of these machines, each connected to a familiar looking desktop computer, all busily sequencing genomes. The room is largely devoid of human activity, except for the occasional lab assistant carrying trays of material to be fed into the machines. A separate building, which I did not see inside, houses the bioinformatics operation, in which the output of all these machines, and others like them around the world, are chewed over by powerful computers.

My hour or two touring the sequencing centre might perhaps be referred to as a demonstration of a state-of-the-art genomics laboratory. But it is very clearly far removed from the simple demonstration of Hooke's law. By contrast with my modest faith that the weights have been accurately labelled, in the Sanger Centre everything is taken on trust. If the entire operation was a mock-up by Lucasfilms, I'm sure I would be none the wiser.

Perhaps the most interesting moral of this comparison is the way in which it points to the division of labour in much of modern science. Though there are of course plenty of biologists who understand the basic biological principles underlying the various bits of machinery in the sequencing lab, it's a fair bet that few or none of them know in any detail how all of these machines work. Moreover, even those who do know how they work surely don't normally have the expertise in operating them possessed by experienced technicians familiar with their quirks and occasional malfunctions. And even those technicians surely don't have the expertise of the engineers who design and construct the machines or who repair them when they malfunction in serious ways.

To cut a long story short, a project such as the sequencing of the human genome involves the collaboration of thousands of people with hundreds or thousands of different forms of expertise, not to mention requiring many years of work by this large and diverse group of people. Clearly no one could offer a demonstration that the human genome was . . . One is reminded of Descartes's concern that a proof should be compact enough for all the steps to be held in the mind at the same time, though presumably Descartes never dreamed of anything quite this far from meeting this optimistic ideal. If one has confidence that the published drafts of the human genome bear some close relation to something in reality this is based not on proof or demonstration, but on trust. And this is as true for Sir John Sulston or Craig Venter as it is for the casual reader of *Nature* or tourist in the Sanger Centre.

No doubt this is all too ambitious. Surely within the practice of genetics and genomics, as within any human practice with even a minimal intellectual content, there are arguments. For instance:

This gene codes for the Bacillus thuringiensis toxin
If we insert it into the genome of this plant, the plant will produce BT toxin
BT toxin poisons insect pests
Therefore, if we insert this gene into this plant, the plant will poison insect pests.

This argument is plausible, if a bit enthymematic. One premise that might start to flesh it out is:

If we insert a gene for x into a (living) genome then that genome will produce x.

This premise shows us that the argument, whether or not plausible, is not sound. For the missing premise is certainly false. There are lots of reasons for this falsity. One of the most interesting involves the familiar redundancy of the genetic code. Amino acids, the constituents of proteins, are coded for by as many as six different base-pair triplets.

However, different organisms tend to use different triplets preferentially and will be disproportionately equipped with extra-nuclear equipment for reading the preferred codons (Ikemura 1981). Consequently they may be very bad at transcribing a gene from a distantly related organism. More simply, whether a sequence is transcribed will depend very much on where it ends up in the genome, on its spatial relations to other genes, especially promoter and suppressor sequences, and even to other structures in the cell. Current techniques for inserting genes into alien genomes are thoroughly hit or miss as far as where the genes end up.

Another reason that the gene may fail to produce the toxin is that the plant may die before it has a chance to do so. If the inserted gene should land in the middle of a sequence of the genome vital for the plant's functioning then the plant will not function. Inserted genetic material may also have a range of effects on the host organism distinct from those intended (pleiotropy), and these may be harmful or fatal.

The relevant moral of these genomic factoids is that genomic events are diverse and specific. One familiar model of scientific argument, that most closely connected to mathematical ideas of proof and demonstration, essentially involves generalizations—traditionally thought of as scientific laws—and generalization is a risky business in biology generally and genetics in particular. The simple example just discussed illustrates the difficulty. The attempt to convert such simple generalizations into exceptionless laws would be extremely difficult if not impossible. Such considerations lead naturally to the conclusion that there are few if any laws that apply to genes.

And there is an even more basic reason for the lawlessness of genes: as I shall explain shortly, it is doubtful whether there are any genes at all. My point so far is not, of course, that no one engaged in genetics or genomics ever deploys any kind of argument. My thesis is rather that argument has no special role in genetics beyond that which it plays in any other intelligent human activity. There are no general patterns of argument to be found, certainly no premises that recur across indefinitely many different genetic arguments.

One of the reasons for the lack of such recurring premises is that genes, the apparent subject matter of genetics, if they should be said to exist at all, are highly diverse entities and not the kinds of things that might be the subjects of broad generalizations. We can, perhaps, refer relatively unproblematically to whole genomes. A biologist colleague likes to define the genome as 'a space in which genetic things happen'.[2] Genes are then, perhaps, the things that things happen to in genomes. But all kinds of different things happen in genomes, and they happen to different kinds of things. Generality in genetics and genomics applies to some interesting extent to the tools, techniques, and instruments that can be used to provide insight into genetic events. But genetic events themselves are hardly more homogeneous than, say, things seen through a telescope, or a microscope.

---

[2] Thanks to Steve Hughes for this illuminating idea.

A more positive way of stating the point is the following. Traditional philosophy of science sees central concepts as opening up the possibility of discovering laws of nature or, at any rate, general knowledge of nature. The example of genes suggests something quite different: the function of this concept is rather to allow us to talk about lots of different things (see Rheinberger 2000 for a related account of the term 'gene'). Such a concept facilitates communication between people with different but related concerns, and facilitates continuity between successive historical enquiries. It may also provide a risk of serious misunderstanding. This risk is probably minimal in the case of working scientists communicating their results to one another, as far more specific, local interpretations of a word such as 'gene' will be expected and provided. Misunderstanding arises rather as scientific results disseminate to different areas either of science or to other domains of human life, and such dangers may be exacerbated by obsolete philosophy of science. At any rate, there is a likely role for philosophers of science in attempting to delineate the diversity of meanings of such complex concepts and even in insisting on the fact that such meanings are diverse.

Are There Genes?

Understanding the problems with the concept of a gene requires a brief excursion into scientific history, though I'm afraid the present excursion will be somewhat Lakatosian in character.[3] We might begin the Lakatosian enterprise by imagining that Mendel invented the term gene (the footnote attributes this to Wilhelm Johannsen, in 1909). At any rate, the tradition of transmission genetics generally supposed to have been inspired by Mendel's work, and epitomized by the famous *Drosophila* experiments of Morgan and Müller, was concerned with genes as hypothetical factors responsible for differences in phenotypes. The gene for red eyes was whatever caused some flies but not others to have red eyes. Of course, nobody supposed that this was the complete cause, as if the gene was something that you could dump in the laboratory disposal bin, and the bin would grow red eyes. But it was the factor that caused the difference in the developmental process that led to the animal having its distinctive eye colour.

Inevitably this programme inspired an interest in the question, What (if anything) is the physical instantiation of these hypothetical factors? Attention quite quickly focused on chromosomes as the likely location for genes, and in 1927 Müller provided evidence for this hypothesis by establishing that X-ray damage to chromosomes could produce genetic changes in flies. In 1944 Oswald Avery argued that the physical basis of heredity was DNA on the basis of experiments in which DNA was transferred from pathogenic bacteria into a harmless related species, and thereby transmitted the

---

[3] Lakatos (1980) famously suggested that history of science should mainly be concerned with 'rational reconstruction' of what should have happened to best explain the current state of our knowledge, with the actual history confined to the footnotes. Perhaps unlike Lakatos, I do not mean to imply any disrespect for the very important business of real history. Unfortunately I am not equipped to provide it.

pathogenicity. This, however, remained controversial. In 1953, as we all know, Crick, Watson, and others disclosed the chemical structure of DNA, and the basis of its capacity for replication. This quite quickly established consensus on the identification of DNA as the genetic material. In the 1960s, following the 1961 breakthrough by Marshall Warren Nirenberg, finally, the genetic 'code' was 'cracked', and the basis of the ability of strands of DNA to determine the production of specific polypeptide chains was understood.

In an obvious sense the elucidation of the structure of DNA was the culmination of the project of transmission genetics. But it was also, in a less obvious sense, the beginning of the end of that project—for it initiated the process of seeing that there really weren't any Mendelian genes, or anyhow not many. This was in fact the conclusion of the discussion that might perhaps be said to have inaugurated contemporary philosophy of biology, the question whether the Mendelian gene could be reduced to the molecular gene. The argument that it could not was stated in David Hull's classic introduction to the philosophy of biology in 1974. The already uncontroversial central premise of Hull's argument was that the relations between molecular genes and phenotypic traits were many/many. A typical molecular gene would have a variety of effects on the phenotype, and any phenotypic trait would require numerous molecular genes for its realization. So the characterization of genes in terms of their phenotypic effects seems drastically underdetermined.

This is, to put it mildly, a simplified story. As I already noted, classical geneticists did not typically have crassly naïve views of the role of genes and intended only to refer to differentiating causes (see, for example, the sophisticated discussion of the developmental relations between gene and trait in Morgan et al. 1915). The point about many/many relations does not show that a molecular gene may not, under normal conditions perhaps, make the difference between one phenotypic condition and another.

But there is a more important point. The initiation of the programme of molecular genetics inevitably directed attention towards development: What are the processes by which genes affect the development of the organism? From the point of view of development, the question of genetic difference is only of tangential interest. It's of more fundamental interest to see how the eye develops at all than to know why it should be red, purple, orange, or crimson. And many or most of the genes involved in normal development aren't even candidates for Mendelian genes, because they are required for development, and the only distinguishable phenotypic state connected to variations in such genes is non-viability.

The importance of the relation of Mendelian and molecular genes is a concern that the Mendelian tradition might seem to license a general program of identifying genes in terms of their phenotypic effects. (This is a version of what Lenny Moss [2003] has described as the preformationist conception of the gene—the gene as carrying the information necessary and sufficient for the production of a particular trait.) Of course, this is the only way that the tradition could possibly identify genes and it would be quite unfair to accuse its exponents of making any such ungrounded universal claim.

Still, what we do see is why there is a strong discontinuity between the Mendelian and molecular traditions. This is just that while the tradition gave access to some of the molecular phenomena, and motivated the search for the molecular phenomena, the phenomena that that search ultimately revealed were not even generally the kinds of things that Mendelian genetics had investigated.

So what is a molecular gene? The natural move in the light of the first decade or so of information about the actual function of DNA was to suggest that a gene was a bit of DNA that contained the code for producing a functional polypeptide, or protein. Thus the connection would be maintained with some product for which the gene was responsible, but the product would be identified much nearer to DNA itself in the causal chain. As a matter of fact, it is quite common to hear this conception of the gene defended to this day (for a sophisticated version of this sort of view, see Waters [1994]).

This conception of the gene is, however, of little general use for analysing the genome. To begin with, possibly as much as 95 per cent of the human genome (proportions vary for different species) doesn't appear to code for anything or even to have any function at all, and this is often referred to as 'junk' DNA. Of course having no known function is not the same as having no function, and it remains possible that all kinds of further functions may be discovered. It seems increasingly likely that the three-dimensional structure of the whole genome may be functionally important, in which case some or most 'junk' DNA will be functionally relevant to maintaining this structure.[4] If there is genuine junk in the genome, it is an interesting speculation that this is the DNA to which Richard Dawkins's notorious conception of 'the selfish gene' may really apply: this is DNA that exists because it has successfully competed for space in the genome. From the point of view of the organism it is a mere parasite. All this is, however, perhaps a rather minor issue. If, to use a standard abusive expression, the genome were composed of genes 'like beads on a string', then all the presence of junk DNA would show would be that there turns out be a lot of string and not so many beads.

Even within the 5 per cent of the genome that seems definitely to be functional, only about 60 per cent is both transcribed into RNA and translated into polypeptide chains. There are, in addition, sequences involved in a variety of ways in promoting, suppressing, terminating, and activating other sequences. So only about 3 per cent of the genome even holds out the hope of fitting the definition under consideration. But even this modest target can quickly be seen to be unreachable. Sequences identified as 'genes', it now appears, are typically composed of alternating coding sequences and sequences which, while often functional, are not part of the sequence for which the

[4] [Note added to this edition] Writing today, I would put these points much more strongly. It is quite clear that far less of the genome lacks function, and the notion of junk DNA is in rapid decline. At least 70% of the genome is transcribed to RNA, and functions for these transcript are being discovered at an impressive rate. (A brief but more up-to-date discussion of these topics is provided in Chapter 4.) Recent work in epigenetics has made the importance of genome structure, specifically as modulated by modifications to the histone core, uncontentious.

gene as a whole codes. These are known, respectively, as exons and introns. All or some of the exons, finally may be transcribed and then assembled into a variety of distinct and often functionally different proteins, sometimes employing in addition coding sequences from other parts of the genome.

A final point, the importance of which is increasingly being realized is that there are variably transient, but heritable, changes to the genome, that can have major functional consequences. Most important of these is the process of 'methylation', a modification of the cytosine molecule, one of the bases in the DNA sequence, that affects the activity of a particular coding sequence. The importance of this process is currently being explored in the 'Epigenome Project', one of the main successors to the Human Genome Project. This phenomenon emphasizes the extent to which DNA is increasingly perceived as interacting with other elements in the cell, and indeed indicates the accelerating demise of the one-time 'central dogma' of molecular biology that postulated a strictly unidirectional flow of information from DNA to RNA to protein.[5]

Without going too deeply into these complexities, what emerges can also be seen as a recurrence of the many/many problem that derailed reductionist aspirations for the relation between phenotypes and genotypes. Even between DNA sequences and polypeptides there are many/many relations: a DNA sequence may be involved in the production of a variety of polypeptides, and the production of a polypeptide will normally involve a variety of often spatially distinct DNA sequences. One thought—more likely to occur to a philosopher than a biologist, I suspect—is that one might still maintain the principle one polypeptide, one gene (though not vice versa) and simply recognize that genes had proved to be overlapping and spatially discontinuous entities.[6] But even apart from the rather serious objection that this will overturn most or all existing genetic nomenclature, it fails for more technical reasons. The processes of polypeptide assembly do not necessarily end with translation from RNA to amino acid chain. One salient case is that of chains that split into smaller units after translation. Sometimes these units are identical, sometimes different. In the latter case it appears that we must find some way of avoiding the conclusion that all the polypeptide fragments are products of the same gene. Any way we find of doing this is likely to force us to say that the identical products in the former case all are products of different genes. The point is just that the diversity of the processes intervening between DNA sequence and functional protein is such as to make it an unpromising venture to look for some uniform relation between the latter and some privileged part of the former held to have a canonical causal responsibility for it. So, it appears that we cannot use the protein products to base a taxonomy of bits of the genome, and the problem of dividing the genome into genes remains unanswered.

---

[5] [Note added to this edition] The modifications to histones mentioned in note 3 are also considered an important class of epigenetic change.
[6] This idea is discussed, and the difficulties explained, in detail by Fogle (2000).

To recapitulate: Mendelian genes, postulated causes of differences between conspecific organisms are, at the molecular level, scarce and equivocal. They are scarce not only because a large proportion of the genome does not even contain candidates for Mendelian genes, but also because much of that which does cannot vary in functionally significant ways without fatally derailing the development of the organism. And they are equivocal because genes are pleiotropic, having a range of different effects on the organism. If we think of genes as 'made molecular', as the components of the genome, then Mendelian principles are of little use in delimiting genes.

If, on the other hand, we start with the concrete physical genome, we might perhaps think of genes as the functional constituents of the genome. Unfortunately from this perspective the delineation of genes appears to be massively underdetermined. There are many different kinds of such functional constituents and, moreover, functional constituents themselves have smaller functional constituents. Is an exon a gene? An intron? For that matter, why not a base pair?

If the Mendelian gene concept is largely inapplicable and the molecular gene concept hopelessly indeterminate, it begins to look as if we would do well not to talk about genes at all.

One reason this may sound surprising is that we not only talk about and hear about genes on a daily basis, but we even learn with considerable regularity that scientists have discovered them. We are naturally inclined to attempt to make sense of this talk. I suggest that most of this talk assumes a concept that has not been sufficiently recognized, what might be called the developmental defect concept.[7]

It is not uncommon for discussions of behavioural genetics to establish the credentials of their subject by referring to the genetic disorder phenylketoneuria (PKU). This condition involves the inability to metabolize phenylalanine. The accumulation of this amino acid leads, in turn, to various physical problems and a degree of mental retardation. The disorder is caused by any of a range of mutations in both alleles of the sequence that codes for the enzyme phenylalanine hydroxylase. The pathological condition, PKU, is commonly thought of as a monogenic disease. This, in turn, is naturally interpreted as meaning 'a disease caused by a single gene'. But we can immediately see that this isn't quite right. The disease is caused by the disfunctionality of a particular gene. And the various mutations that lead to such disfunctionality are not in any natural sense genes for PKU but dysfunctional variants in a genetic region that codes for phenylalanine hydroxylase. The referent of the phrase 'gene for PKU' therefore is not a physical object at all, but a set of defects in another object, a coding region involved in the production of a particular protein. There is, I suppose, a

---

[7] An anonymous referee correctly pointed out that the following discussion considers only deleterious germline mutations, whereas the concept of genetic disease also extends to disease caused by somatic cell mutations. It would be possible to insist that somatic cell mutations can be included under a sufficiently broad conception of development. However, given the clear distinction between these cases it is no doubt better to distinguish them, and recognize a wider range of applications of the term 'genetic disease'. I do not think this correction significantly affects the philosophical argument here (and nor, I am pleased to say, did the referee).

technical interpretation of the phrase 'gene for x' according to which any of these defective regions is a gene for PKU: the defective allele makes a difference to the developing phenotype. No one of them, however, is *the* gene for PKU. But there is not even a technical sense in which the functionally unimpaired version of the gene is a gene for the prevention of PKU. (It does of course function in the production of phenylalanine hydroxylase, but we have already considered the reasons why it would be misleading to call it the gene for phenylalanine hydroxylase.) Compare with this the idea that the heart is an organ for preventing oxygen deprivation-induced brain damage.

It is striking that perhaps the most familiar roughly Mendelian human physiological trait is a nice example of the developmental defect gene concept, namely eye colour.[8] Blue eyes are, roughly speaking, the result of a recessive genetic defect in the production of the pigment that gives eyes their proper brown colour. Since in Northern latitudes this defect has no serious consequences, such defects have accumulated in some populations to the point where blue eyes have become the norm. But there is no gene for blue eyes in the rather strong sense that the cause of this trait is a pure absence. There is a little more to be said for talking about a gene for brown eyes, though certainly all the standard problems of pleiotropy, polygeny, and so on will make the terminology liable to mislead.

When we refer to a 'gene for x' it is natural to suppose that we are referring to a gene the physiological function of which is to produce x. If behavioural genetics is the study of forms of behaviour caused by identifiable genes, then PKU is completely irrelevant to the subject. The same is equally true for familiar physiological disorders. We can, of course, insist on using the phrase 'gene for x' in a different technical sense derived from Mendelian genetics and also adopted by some evolutionists. The problem then is that behavioural genetics will have only the slimmest connection with the causes of behaviour. My own reading of the evidence is that there is, in fact, little reason to expect that genetic differences will be useful in explaining behavioural or mental differences beyond the cases of serious incapacity caused by malfunctioning genes. This would, indeed, make the developmental defect gene concept the appropriate one in this context, but would also undermine most of the publicly expressed pretensions for this field of study. To take one example, it was recently widely reported in the press that the gene for human arts and culture had been found. The consilient evidence for this claim was, first, that a gene had been isolated with a mutation that occurred subsequent to the split of the human lineage from that of the great apes; and, second, that damage to this gene caused people to be deficient in artistic and cultural skills. I hope it is clear that this does not provide the slightest shred of an argument for the discovery of a gene for arts and culture in any normal interpretation of that expression.

---

[8] As usual, this example is really much more complicated as there is also a gene for green eyes, and not all colour variation has been genetically explained.

Let me offer one more simplistic summary of the simplistic historical narrative. For much of its history genetics was driven by a hypothetical kind that it saw itself as investigating, the gene. As we gradually identified the material referent of this hypothetical kind and were able to learn something about how its instances worked and what they did, it became increasingly clear that they were not a kind at all but a diverse set of molecular objects and processes. There is perhaps a legitimate kind, DNA sequence, and some instances of this kind do indeed do something interesting: they are transcribed into RNA sequences, some of which are translated into proteins.

I won't go into much detail here about positive accounts of the gene. A number of accounts have been offered by people who have come to terms with the sort of complexities just discussed. Most of these, in my view, have the fatal defect of legislating a concept much narrower than historical conceptions of the gene and a concept too closely tied to a particular theoretical idea.[9] At the opposite extreme, and rather more promisingly, Hans-Jorg Rheinberger (2000) has suggested that central scientific concepts, like the gene, function precisely by remaining sufficiently vague to allow communication between all the various groups that have an interest in talking about such things, but very diverse accounts of what it is they are talking about.

I do think there must be something right about this last view. However, it does at least need supplementation to account for the great precision with which particular genes are referred to in narrow scientific contexts. I shall offer a rather different tentative suggestion. There are interesting parallels with a topic I have been interested in for many years, the so-called species problem. It seems almost indisputably impossible to find a definition of the species that is applicable across the whole range of biological diversity. There are partisans for a variety of species concepts, and these supporters take various attitudes to the bits of the biological world they don't adequately cover (they're not important; more research is needed; they don't form species at all; etc.). An inevitable, and in my view correct, reaction to all this is pluralism: there is no definition of species and groups of various kinds should qualify. Most pluralists nevertheless try to hang on to some theoretical core to the concept, generally that a species have some kind of phylogenetic coherence.[10]

My own preference is for total abandonment of such theoretical commitment. One reason for this is the insistence that 'species' is not primarily a theoretical concept at all, but a classificatory concept. (Certainly this accords with the principle of priority, which is an important one within scientific taxonomy.) It is naturally assumed that these will coincide, as seems to be the case, for example, with the classification of chemical elements. But the path to pluralism reveals that this is not in fact the case for biological kinds. One is driven to pluralism by the realization that theoretical principles that seem to work nicely for some domains of biological classification turn out to be wholly inapplicable to others, so that attempts to provide a monistic account of what a species

[9] For example, the interesting account of Beurton (2000) seems to me to suffer from this defect.
[10] My views on this topic are explained in detail in Dupré (2002: chs. 3 and 4).

is leave us unable to classify large areas of biological diversity. When theoretical conceptions of the species are applied to practical taxonomy, theory can even become an enemy of classification. Changes in theory will lead to changes in classification and stability is an obvious desideratum of a classificatory scheme. Taxonomic conservatism must be recognized as an important criterion in assigning species names to groups of organisms, and even more so when it is recognized that there is no universally adequate theoretical conception to which classification should be answerable.

Turning now to the naming of genes, I was struck recently by the following sentences on a major bioinformatics website:

Keeping stable names for "things," such as genes, in databases is very important. This allows scientists in different labs around the world to be confident they are all referring to the same thing.

Ensembl goes to great lengths to try to maintain stable names for genes and other features in the genome.[11] (Ensembl Naming Conventions. From Ensembl website database.)

The diversity of kinds of entity and the desirability of taxonomic stability seem exactly to mirror the issues that arise for the case of species. So my proposal is for an atheoretical pluralism similar to that which I advocate for species: a gene is any bit of DNA that anyone has reason to name and keep track of. Genes may be proper parts of other genes; they may overlap; they may have non-contiguous parts, perhaps on two or more chromosomes. And, as illustrated for the case of developmental defect genes, 'gene' may even refer to a functionally connected class of DNA segments. My conclusion is that there are genes—an important point given how much people talk about them— but that the price of this is conceding that it doesn't take much to be a gene. Not much, but not nothing either. I am assuming that genes are real material entities. Many of the genes discussed by behavioural geneticists for instance, may well not even meet this minimal condition.

Some consequences

One conclusion I would like to draw from this recognition of the diversity of the referent of the term 'gene' is a familiar one in contemporary philosophy of biology. It is that the traditional philosophy of science that sees science as ultimately concerned with the articulation of wide-ranging and fundamental truths—laws of nature, for in- stance—has little relevance to biology. Genetics and genomics offer little project of such general truths because of the diversity of their subject matter. Recalling Rhein- berger's suggestion mentioned above, it may be that the function of the most general terms in such sciences is precisely to compensate for the lack of such general truths by allowing some degree of communication between people with varying interests in the workings of, for example, the genome. Similarly, I suggest, the term 'species' is useful

---

[11] http://www.ensembl.org/. Thanks to Dick Holdsworth for drawing my attention to this statement.

in allowing people with different interests in the classification of organisms, different principles for accomplishing this, and consequently different groups of organisms to which they need to refer, nevertheless to understand when reference is being made to a group of organisms at a certain important level in the taxonomic hierarchy. The great diversity of the subject matter of biology calls for the most central terms not to be those in terms of which laws can be formulated, but rather those which are tolerant enough in their reference to bridge the divides between the various phenomena in which local communities of researchers may be interested. There are, I suppose, some general truths about DNA that make it possible for DNA to constitute genes, but there are lots of ways for bits of DNA to be genes of various kinds, and all of these depend on the relations between bits of DNA and other things to which they are related.

It is, as I have noted, hardly a novel suggestion that the view of science as the search for universal laws is of little or no relevance to biology, but the extent to which this suggestion has been reinforced by recent developments in genetics has not yet been fully appreciated. Indeed, it is still sometimes imagined that the annoying failure of biology to generate law-like generalizations is a consequence merely of its continuing concern with complex and variable structures, and its concomitant failure to get down to the real action at the molecular, and ultimately even more fundamental, levels.

One moral of my preceding remarks is just that no such consequences result as we investigate the inner structures of biological things. On the contrary, what we find as we become more familiar with molecular processes is a diversity of structure and action quite comparable with that which we find at more complex levels. We are far from approaching the few simple laws that earlier theorists imagined might reduce complexity and diversity to order and uniformity.

My argument here is not in any simple way anti-reductionist. It is clear in genetics that enormous illumination and insight has come from our ability to investigate and describe molecular processes. It is, however, anti-reductionist in the sense of rejecting the hierarchical view of nature often associated with reductionism. Knowledge of different levels of organization is complementary, not competing. The molecular view is not a superior view to, say, the cellular view, and one that in principle should render the latter obsolete. And the reason for this is simply that the molecular view is not even separable from the cellular view. There is no possibility of specifying the behaviour or function of bits of DNA independently of a detailed description of the biological context in which they exist. Minimally this context will include further genomic and cytological information. Sometimes the relevant context will be much broader, including physiology, ecology, and even sociology. And of course this dependence on context is a large part of why what may look very similar—strings of DNA—may nevertheless prove to be so diverse.

There is a vision of the cell as a nugget of information suspended in a soup of dumb and formless goo, a notion that still seems common in popular presentations of biology, and this vision perhaps best represents the remaining aspirations of hierarchical reductionism.

The extra-nuclear goo, in this vision, is no more than the minimal context necessary for the expression of the structure inherent in the DNA.

But in reality the extra-nuclear goo is as structured, as rich in information, as is the nuclear DNA. The sorts of things bits of DNA can do involve diverse reactions with particular chemicals and structures in the cell. Biochemistry only becomes molecular biology when it is embedded in cytology. Lower-level knowledge cannot possibly displace higher-level knowledge.

And this, as one final important philosophical moral of our growing understanding of the cell and the genome, should also make clear the futility of seeing causality as something always elusively located in lower levels of order, ultimately filtering up from the most basic constituents of matter. What was once the controversial thesis of 'downward' causation is a commonplace in biology. One striking example is the differentiation of cells in development. All the diverse varieties of cells in multicellular organisms, the liver cells, blood cells, hair cells, and so on, trace their origin to the same ancestral cell. The explanation of the different developmental paths leading to these diverse outcomes does not reside in differences in the DNA, but in the ability of the spatial relations between cells and the spatial distribution of relevant biologically active substances in the egg and, later, in different locations in the body to affect differentially the behaviour of the DNA within different cells. This seems as clear as possible a case of the behaviour of a low-level entity being caused by higher-level entities of which it is part. The prejudice in favour of the causal priority of the small, visible in a range of weak reductionisms and supervenience theses is, I think, just that, a prejudice.

I don't know how useful it is to read scientific models as political allegories. But it is remarkable how naturally a common picture of the cell fits with a hierarchical model of social organization. Command and control inheres in a central administration, the genome, and orders are carried out by messengers, clerks (transcribing, translating, and so on). The construction work takes place at various sites decently removed from the seat of power. Contemporary molecular genetics takes us away from this Stalinist model towards something more Smithian. The efficiency of the cell is unimaginable, from this perspective, without the distribution of specialized capacities across a very large range of different agents. Command and control do not descend from the central administration building, but emerge spontaneously, as if guided by an invisible hand. On the Smithian model order at the lower level is an order of teleological mechanism: events fit together in efficient ways to produce valued outcomes. Broad generalizations—like this last one—emerge if at all at higher levels. Empirical evaluation of the attempt to provide a science of economics as a set of axioms and their consequent theorems shouldn't encourage us to hold out a great deal of hope for these higher-level generalities.

Genetic things, genes in the catholic sense I have advocated, are unquestionably real. They cause things to happen at the phenotypic level and intervening levels of biological structure, just as those things cause the activation and specific action of particular genes. Hence only at many levels simultaneously can we begin to get a full account of the nature of an organism.

One final question, the answer to which I hope has been illuminated to some degree by the foregoing, is why so much contemporary discourse is replacing the term 'genetics' with 'genomics'. Genetics, a science of hypothetical entities held to be responsible for inheritance, can be caricatured, but not altogether unfairly, as a science developed in accordance with a reductionist epistemology and a law-seeking methodology. Over the course of a century genetics led us to a remarkably detailed view of the genome. Among many remarkable properties of the genome is its total unsuitability for both this epistemology and this methodology. Genomics, I am tempted to suggest, is the successor science to genetics that rejects this obsolete epistemology and methodology.

A more positive way of stating the point recapitulates and largely endorses Rheinberger's idea. Traditional philosophy of science sees central concepts as opening up the possibility of discovering laws of nature or, at any rate, general knowledge of nature. But as I have tried to show, genomic events are diverse and specific and there are few if any laws that apply to genes. The example of genes suggests something quite different: the function of this concept is rather to allow us to talk about lots of different things. Such a concept facilitates communication between people with different but related concerns, and facilitates continuity between successive historical enquiries. It also provides a risk of serious misunderstanding. This risk is probably minimal in the case of working scientists communicating their results to one another, as far more specific, local interpretations of a word such as 'gene' will be expected and provided. Misunderstanding arises rather as scientific results disseminate either to different areas of science or to other domains of human life, and such dangers may be exacerbated by obsolete philosophy of science. At any rate, there is a likely role for philosophers of science in attempting to delineate the diversity of meanings of such complex concepts.

7

The Polygenomic Organism

Introduction: genomes and organisms

Criticisms of the excessive attention on the powers of genes, 'genocentrism', have been common for many years.[1] While genes, genomes, or more generally DNA are certainly seen as playing a fundamental and even unique role in the functioning of living things, it is increasingly understood that this role can only be properly appreciated when adequate attention is also paid to substances or structures in interaction with which, and only in interaction with which, DNA can exhibit its remarkable powers. Criticisms of genocentrism are sometimes understood as addressing the idea that the genome should be seen as the essence of an organism, the thing or feature that makes that organism what it is. But despite the general decline not only of this idea, but of essentialism in general,[2] the assumptions that there is a special relation between an organism and its distinctive genome, and that this is a one-to-one relation, remain largely intact.

The general idea just described might be understood as relating either to types of organisms or to individual organisms. The genome is related to types of organism by attempts to find within it the essence of a species or other biological kind. This is a natural, if perhaps naïve, interpretation of the idea of the species 'barcode', the use of particular bits of DNA sequence to define or identify species membership. But in this chapter I am interested rather in the relation sometimes thought to hold between genomes of a certain type and an individual organism. This need not be an explicitly essentialist thesis, merely the simple factual belief that the cells that make up an organism all, as a matter of fact, have in common the inclusion of a genome, and the genomes in these cells are, barring the odd collision with a cosmic ray or other unusual accident, identical. It might as well be said right away that the organisms motivating this thesis are large multicellular organisms, and perhaps even primarily animals. I shall not be concerned, for instance, with the fungi that form networks of hyphae connecting the roots of plants, and are hosts to multiple distinct genomes apparently capable of moving around this network (Sanders 2002). I should perhaps apologize for this narrow focus. Elsewhere in this book (Chapter 10) I criticize philosophers of biology

---

[1] For a recent example, see Barnes and Dupré (2008).

[2] At any rate among philosophers concerned with the details of scientific belief. Essentialism has had something of a resurgence among more abstractly inclined metaphysicians (Ellis 2001; Devitt 2008).

and others for a myopic focus on a quite unusual type of organism, the multicellular animal. Nonetheless it is unsurprising that we should have a particular interest in the class of organisms to which we ourselves belong, and this is undoubtedly an interesting kind of organism. And in the end, if my argument is successful for multicellular animals it will apply all the more easily to other, less familiar, forms of life.

At any rate, it is an increasingly familiar idea that we, say, have such a characteristic genome in each cell of our body, and that this genome is something unique and distinctive to each of us. It is even more familiar that there is something, 'the human genome', which is common to all of us, although, in light of the first point, it will be clear that this is not exactly the same from one person to another. The first point is perhaps most familiar in the context of forensic genomics, in the realization that the tiniest piece of corporeal material that any of us leaves lying around can be unequivocally traced back to us as its certain source. At any rate, what I aim to demonstrate in this chapter is that this assumption of individual genetic homogeneity is highly misleading, and indeed is symptomatic of a cluster of misunderstandings about the nature of the biological systems we denominate as organisms.

Organisms and clones

A clone, outside Star Wars style science fiction, is a group of cells originating from a particular ancestral cell through a series of cell divisions. The reason we suppose the cells in a human body to share the same genome is that we think of the human body as, in this sense, a clone: it consists of a very large group of cells derived by cell divisions from an originating zygote. A familiar complication is that if I have a monozygotic ('identical') twin, then my twin will be part of the same clone as myself. Although this is only an occasional problem for the human case, in other parts of biology it can be much more significant. Lots of organisms reproduce asexually and the very expression 'asexual reproduction' is close to an oxymoron if we associate biological individuals with clones. For asexual reproduction is basically no more than cell division, and cell division is the growth of a clone. If reproduction is the production of a new individual it cannot also be the growth of a pre-existing individual. Indeed what justifies taking the formation of a zygote as the initiation of a new organism, reproduction rather than growth, is that it is the beginning of a clone of distinctive cells with a novel genome formed through the well-known mixture between parts of the paternal and the maternal genomes.

As I have noted, it is common to think of genomes as standing in one-to-one relations with organisms. My genome, for instance, is almost surely unique and it, or something very close to it, can be found in every cell in my body. Or so, anyhow, the standard story goes. The existence of clones that do not conform to the simple standard story provides an immediate and familiar complication for the uniqueness part of this relation. If I had a monozygotic ('identical') twin, then there would be two organisms whose cells contained (almost) exactly the same genomes; we would both, since originating from the same lineage-founding zygote, be parts of the same clone. And lots of organisms reproduce asexually all or some of the time, so this difficulty is far from esoteric.

Some biologists, especially botanists, have bitten the bullet here. They distinguish ramets and genets, where the genet is the sum total of all the organisms in a clone, whereas the ramet is the more familiar individual (Harper 1977). Thus a grove of trees propagated by root suckers, such as are commonly formed, for instance, by the quaking aspen (*Populus tremuloides*), in the deserts of the south west United States, is one genet but a large number of ramets. Along similar lines it has famously been suggested that among the largest organisms are fungi of the genus Armillaria, the familiar honey fungi (Smith et al.1992). A famous example is an individual of the species *Armillaria ostoyae* in the Malheur National Forest in Oregon that was found to cover 8.9 km^2 (2,200 acres).[3] To the average mushroom collector a single mushroom is an organism, and it would be strange indeed to claim that two mushrooms collected miles apart were parts of the same organism. There is nothing wrong with the idea that for important theoretical purposes this counterintuitive conception may be the right one; there is also nothing wrong with the more familiar concept of a mushroom. The simple but important moral is just that we should be pluralistic about how we divide the biological world into individuals: different purposes may dictate different ways of carving things up.

It's pretty clear, however, that we cannot generally admit that parts of a clone are parts of the same individual. Whether or not there are technical contexts for which it is appropriate, I doubt whether there are many interesting purposes for which two monozygotic human twins should be counted as two halves of one organism. Or anyhow, there are certainly interesting purposes for which they must be counted as distinct organisms, including almost all the regular interests of human life. An obvious reason for this is that most of the career of my monozygotic twin (if I had one) would be quite distinct from my own. And for reasons some of which should become clearer in light of the discussion below of epigenetics, the characteristics of monozygotic twins tend to diverge increasingly as time passes. The careers of monozygotic twins may carry on independently from birth in complete ignorance of one another; but it is hardly plausible that if I were now to discover that I had a monozygotic twin, this would drastically change my sense of who I was (i.e., a spatially discontinuous rather than spatially connected entity). Some kind of continuing connection seems needed even to make sense of the idea that these could be parts of the same thing. Being parts of the same clone is at any rate not a sufficient condition for being parts of the same biological individual.

However, we should not immediately assume that the concept of a genet encompassing a large number of ramets is generally indefensible. A better conclusion to draw is that theoretical considerations are insufficient to determine unequivocally the

[3] See http://www.scientificamerican.com/article.cfm?id=strange-but-true-largest-organism-is-fungus (accessed 2 November 2009).

boundaries of biological objects. Sometimes, perhaps always, this must be done relative to a purpose. There are many purposes for which we distinguish human individuals and for the great majority of which it would make no sense to consider my twin and myself part of the same entity. My twin will not be liable to pay my debts or care for my children, for instance, if I should default on these responsibilities, though it is interesting in the latter case that standard techniques for determining that they are my children would not distinguish my paternity from my twin's. This may even point to an evolutionary perspective from which we are best treated as a single individual. And when it comes to the trees, this is surely the right way to go. For the purposes of some kinds of evolutionary theory the single genet may be the right individual to distinguish, but if one is interested in woodland ecology, what matters will be the number of ramets. If this seems an implausible move, this is presumably because of the seemingly self-evident individuality of many biological entities. I hope that some of the considerations that follow will help to make this individuality a lot less self-evident than it might appear at first sight. But whether or not the pluralism I have suggested for individual boundaries is defensible, the assumption of a one-to-one relation between genomes and organisms is not. I will explain the objections to this assumption in what I take to be an order of increasing fundamentality. At any rate, as the next section will demonstrate, the various phenomena of genetic mosaicism suffice to demonstrate that genotypes will not serve to demarcate the boundaries of biological individuals. Or in other words, genomic identity is not a necessary condition for being part of the same biological individual.

Genomic chimeras and mosaics

The general rubric of genomic mosaicism encompasses a cluster of phenomena. An extreme example, sometimes distinguished more technically as chimerism, is of organisms that have resulted form the fusion of two zygotes, or fertilized eggs, in utero. The consequence of this is that different parts of the organism will have different genomes—the organism is a mosaic of cells of the two different genomic types from which it originated. A tragic consequence of this has been the occasional cases of women who have been denied custody of their children on the basis of genetic tests that appeared to show that they and the child were not related. It has turned out that the explanation of this apparent contradiction of a connection of which the mother was in no doubt was that she was a genomic mosaic of this kind, and the cells tested to establish the parental relation were from a different origin than were the gametes that gave rise to the child (Yu et al. 2002). With the exception of a modest degree of chimerism found in some fraternal twins who have exchanged blood and blood cell precursors in utero and continue to have distinct genotypes in adult blood cells, such cases are generally assumed to be very rare in humans. However, chimeras do not necessarily experience any unusual symptoms, so the prevalence of full chimerism, chimerism derived from multiple zygotes, is not really known, and may be much higher than suspected.

Probably more common than chimerisms resulting from the fusion of two zygotes are those resulting from mutations at some early stage of cell division. One well-known example of this is XY Turner syndrome, in which the individual is a mixture of cells with the normal XY karyotype, the complement of sex chromosomes found in most males, and XO cells, in which there is no Y chromosome and only one X chromosome (Edwards 1971). Turner syndrome is a condition of girls in which all the cells are XO (i.e., with one X chromosome missing, as opposed to the standard XX); people with XY Turner syndrome generally have normal male phenotypes, though a small percentage are female and a similar small percentage are intersexed. The large majority of foetuses with either condition are spontaneously aborted. The phenotype displayed by XY Turner cases is presumably dependent on exactly when in development the loss of the X chromosome occurs.

Chimerism is quite common in some other organisms. When cows have twins there is usually some degree of shared foetal circulation, and both twins become partially chimeric. This has been familiar from antiquity in the phenomenon of freemartinism, freemartins being the sterile female cattle that have been known since the eighteenth century invariably to have a male twin. This is the normal outcome for mixed-sex bovine twins, with the female twin being masculinized by hormones deriving from the male twin.[4] This has occasionally been observed in other domesticated animals. Even more than for the human case, the prevalence of this, and other forms of chimerism, in nature is not known.

The chimeras mentioned so far are all naturally occurring phenomena. Much more attention has lately been attracted by the possibility of artificially producing chimeras in the laboratory. And unsurprisingly, the most attention has been focused on the possibility of producing chimeras, or hybrids, that are in part human. Recent controversy has focused on the ethical acceptability of generating hybrid embryos for research purposes by transplanting a human nucleus into the egg cell of an animal of another species, usually a cow.[5] Since all the nuclear DNA in such a hybrid is human, it can be argued that this is not a chimera at all, at least in the genetic sense under consideration. On the other hand, such cells will contain non-human DNA in the mitochondria, the extra-nuclear structures in the cell that provide the energy for cellular processes.[6] No doubt the mixture of living material from humans and non-humans is disturbing to many whether or not the material in question is genetic, as is clear from controversy over the possibility of xenotransplantation, use of other animals to provide replacement

---

[4] Exactly to what extent this is the normal outcome remains as with so many phenomena in this area somewhat unclear, however (Zhang et al. 1994).

[5] Although research involving hybrid embryos is generally thought unacceptable unless there are clear potential medical benefits, opinion in the UK is quite finely balanced on this topic (Jones 2009).

[6] As a matter of fact the mitochondria are now known to be descendants of bacteria that long ago became symbiotically linked to the cells of all eukaryotes, or 'higher' organisms. This may suggest a further sense in which we are all chimeric, a suggestion I shall elaborate shortly.

organisms. But this will not be my concern in the present chapter (but see Parry 2010 and Twine 2010).

Modern laboratories, at any rate, are well able to produce chimeric organisms. At the more exotic end of such products, and certainly chimeras, are such things as 'geeps', produced by fusing a sheep embryo with a goat embryo. The adults that develop from these fused embryos are visibly and bizarrely chimeric, having sheep wool on parts of their bodies and goat hair on others. Much more significant, however, are the transgenic organisms that have caused widespread public discomfort in the context of genetically modified (GM) foods (Milne 2010). These are often seen as some kind of violation of the natural order, the mixing together of things that nature or God intended to keep apart (Barnes and Dupré 2008). Whatever other objections there may be to the production of GM organisms, it will become increasingly clear that this is not one with which I am sympathetic: organisms do not naturally display the genetic purity that this concern seems to assume.

The chimeric organisms discussed so far in this section have been organisms originating to some degree from two distinct zygotes. (The exception is the XY Turner syndrome, which should strictly have been considered in the context of the following discussion.) Other cases relevant to the general topic of intraorganismic genomic diversity, but generally referred to by the term mosaicism rather than chimerism, exhibit genomic diversity but deriving from a single zygotic origin. Such mosaicism is undoubtedly very common. One extremely widespread instance is the mosaicism common to most or all female mammals that results from the expression of different X chromosomes in different somatic cells. In the human female, one of the two X chromosomes in each cell is condensed into a cellular object referred to as the Barr body and is largely inert. Different parts of the body may have different X chromosomes inactive, implying that they have different active genotypes. This phenomenon will apply to most sexually reproducing organisms, though in some groups of organisms, for example birds, it is the male rather than the female that is liable to exhibit this kind of mosaicism.[7] The most familiar phenotypic consequence of this phenomenon is that exhibited by tortoiseshell or calico cats, in which the different coat colours reflect the inactivation of different X chromosomes. Although there are very rare cases of male calico cats, these appear to be the result of chromosomal anomaly (XXY karyotype), chimerism, or mosaicism in which the XXY karyotype appears as a mutation during development (Centerwall and Benirschke 1973).

Returning to chimerism, mosaicism deriving from distinct zygotes, a quite different but very widespread variety is exhibited by females, including women, after they have borne children, and is the result of a small degree of genomic intermixing of the maternal and offspring genomes. Though scientists have been aware of this

---

[7] Curiously, however, it appears that birds find less need to compensate for the overexpression of genes on the chromosome of which one sex has two (in birds the male has two Z chromosomes). So this kind of mosaicism will be less common, or may not occur at all (Marshall Graves and Disteche 2007).

phenomenon for several decades it has recently been the focus of increased attention for several reasons. For example, recent work suggests that the transfer of maternal cells to the foetus may be important in training the latter's immune system (Mold et al. 2008). Another reason for increasing interest in this topic is the fact that it opens up the possibility of genetic testing of the foetus using only maternal blood, and thus avoiding the risks inherent in invasive techniques for foetal testing such as amniocentesis (Lo 2000; Benn and Chapman 2009). It should also be noted that maternal cells appear to persist in the offspring and vice versa long after birth, suggesting that we are all to some degree genomic mosaics incorporating elements from our mothers and, for women, our offspring.

One final cause of chimerism that must be mentioned is the artificial kind created by transplant medicine, including blood transfusions. Very likely this will continue to become more common as techniques of transplantation become more refined and successful. A possibility increasingly under discussion is that this will eventually be extended, through the development of xenotransplantation, to include interspecific mosaicism. At any rate, any kind of transplantation, except that involving cells produced by the recipient himself or herself, will produce some genomic chimerism. So, in summary, both natural and artificial processes, but most commonly the former, generate significant degrees of chimerism in many, perhaps almost all, multicellular organisms including ourselves. The assumption that all the cells in a multicellular organism share the same genome is therefore seriously simplistic and, as mentioned above, conclusions drawn from this simplistic assumption, for example about the violation of nature involved in producing artificial chimeras are, to the extent that they rely on this assumption, ungrounded.

Epigenetics

The topics of chimerism and mosaicism so far discussed address the extent to which the cells that make up a body are genomically uniform in the sense of containing the same DNA sequences. This discussion runs a risk of seeming to take for granted the widely held view that, given a certain common genome, understood as a genome with a particular sequence of nucleotides (the As, Cs, Gs, and Ts familiar to everyone in representations of DNA sequence), the behaviour of other levels of biological organization will be determined. Perhaps a more fundamental objection to the one genome, one organism doctrine is that this common assumption is entirely misguided. The reason that the previous discussion may reinforce such an erroneous notion is that the comparisons and contrasts between genomes were implicitly assumed to be based entirely on sequence comparisons. But to know what influence a genome will actually have in a particular cellular context one requires a much more detailed and nuanced description of the genome than can be given merely by sequence. And once we move to that more sophisticated level of description it becomes clear that, even within the sequence-homogeneous cell lineages often thought to constitute a multicellular organism, there is a great deal of genomic diversity. These more sophisticated descriptions are sought within the burgeoning scientific field of epigenetics, or epigenomics.

A good way of approaching the subject matter of epigenetics is to reflect on the question why, if indeed all our cells do have the same genome, they nevertheless do a variety of very different things. It is of course very familiar that not all the cells in a complex organism do the same things—they are differentiated into skin cells, liver cells, nerve cells, and so on. Part of the explanation for this is that the genome itself is modified during development, a process studied under the rubric of epigenetics or epigenomics.[8] The best known such modification is methylation, in which a cytosine molecule in the DNA sequence is converted to 5-methylcytosine, a small chemical addition to one of the nucleotides, or bases, that make up the DNA molecule. This has the effect of blocking transcription of the DNA sequence at particular sites in the genome. Other epigenetic modifications affect the protein core, or histones, which form part of the structure of the chromosome, and also influence whether particular lengths of DNA are transcribed into RNA. It is sometimes supposed that these are not 'real', or anyhow significant, alterations of the genome, perhaps because we still describe the genome sequence in the same way, referring to either cytosine or 5-methylcytosine by the letter C. But all this really shows is that the standard four-letter representation of genomic sequence is an abstraction. As a matter of fact there are about twenty nucleotides that can occur in DNA sequences, and it is only our choice of representation that maintains the illusion that some chemically fixed entity, the genome, can be found in all our cells. If we were to change the representation to a more fine-grained description of chemical composition, we would find a much greater genomic diversity than is disclosed by the more abstract and familiar four-letter code.

It is true that part of the value of the abstraction that treats the genome as consisting of only four nucleotides is that this does represent a very stable underlying structure. This has provided extremely useful applications that use stable genome sequence to compare or identify organisms, applications ranging from phylogenetic analysis to forensic DNA fingerprinting. Phylogenetic analysis, the investigation of evolutionary relations between kinds of organisms, here depends on the stability of genomes as they are transmitted down the generations, and DNA fingerprinting depends on the admittedly much shorter term stability of genome sequence within the life of the individual. Methylation, on the other hand, is reversible and often reversed. However, overemphasis on this stable core can be one of the most fundamental sources of misunderstanding in theoretical biology.

---

[8] It appears that the phenomenon in question may not be fully explicable at all, however, as gene expression is also importantly affected by random processes, or noise (Raser and O'Shea 2005). But there is also growing evidence that noise of this kind may be adaptive, and hence this effect may have been subject to natural selection (Maamar et al. 2007).

Such misunderstanding is sometimes expressed in the so-called Central Dogma of Molecular Biology.[9] This is generally interpreted as stating that information flows from DNA to RNA to proteins, but never in the reverse direction. I don't wish to get involved in exegesis of what important truth may be alluded to with this slogan, and still less into the vexed interpretation of the biological meaning of 'information' (Maynard Smith 2000; Griffiths 2001). What is no longer disputable is that causal interaction goes both in the preferred direction of the Central Dogma, and in the reverse direction. Epigenetic changes to the genome are induced by chemical interactions with the surrounding cell (typically with RNA and protein molecules). A reason why this is so important is that it points to a mechanism whereby even very distant events can eventually have an impact on the genome and its functioning. The classic demonstration of this is the work of Michael Meaney and colleagues, on ways in which maternal care can modify the development of cognitive abilities in baby rats, something which has been shown to be mediated by methylation of genomes in brain cells (Champagne and Meaney 2006). The most recent work by this group has provided compelling reason to extrapolate these results to humans (McGowan et al. 2009). Whether epigenetic research shows that genomes are diverse throughout the animal body of course depends on one's definition of 'genome' and one's criterion for counting two as the same. It needs just to be noted that if we choose a definition that, *pace* the points made in earlier sections, counts every cell as having the same genome, we will be overlooking differences that make a great difference to what the cell actually does.

Symbiosis and metaorganisms

In this section I want to make a more radical suggestion. So far I have considered the diversity of human (or other animal) cells that may be found in an individual organism; and the phenomena I have described are generally familiar ones to molecular biologists. In this section I shall propose that there are good reasons to deny the almost universal assumption that all the cells in an individual must belong to the same species. This may seem no more than tautological: if a species is a kind of organism then how can an organism incorporate parts or members of different species? The resolution of this paradox is to realize that very general terms in biology such as species or organism do not have univocal definitions: in different contexts these terms can be used in different ways. For the case of species, this is quite widely agreed among philosophers of biology today (for discussion see various essays in Wilson 1999). I am also inclined to argue something similar for organisms. Very roughly, I want to suggest that the organisms that are parts of evolutionary lineages are not the same things as the organisms that interact functionally with their biological and non-biological surroundings. The latter,

---

[9] This phrase was introduced originally by Francis Crick, and I have no wish to accuse Crick himself of misunderstanding. Indeed the use of the word 'dogma' suggests a degree of irony.

which I take to be more fundamental, are composed of a variety of the former, which are the more traditionally conceived organisms. But before explaining this idea in more detail I need to say a bit more about the facts on which it is based. I shall introduce these with specific reference first to the human.

A functioning human organism is a symbiotic system containing a multitude of microbial cells—bacteria, archaea, and fungi—without which the whole would be seriously dysfunctional and ultimately non-viable. Most of these reside in the gut, but they are also found on the skin, and in all body cavities. In fact about 90 per cent of the cells that make up the human body belong to such microbial symbionts and, owing to their great diversity, they contribute something like 99 per cent of the genes in the human body. It was once common to think of these as little more than parasites, or at best opportunistic residents of the various vacant niches provided by the surfaces and cavities of the body. However, it has become clear that, on the contrary, these symbionts are essential for the proper functioning of the human body. This has been recognized in a major project being led by the US National Institutes of Helath, that aims to map the whole set of genes in a human, the Human Microbiome Project.[10]

The role of microbes in digestion is most familiar and is now even exploited by advertisers of yoghurt. But even more interesting are their roles in development and in the immune system. In organisms in which it is possible to do the relevant experiments it has turned out that genes are activated in cells by symbiotic microbes, and vice versa (Rawls et al. 2004). Hence the genomes of these cells and the symbiotic microbes are mutually dependent. There is no obvious reason not to extrapolate these findings to the human case. And it seems plausible that the complex microbial communities that line the surfaces of the human organism are the first lines of defence in keeping out unwanted microbes.[11] Since the immune system is often defined as the system that distinguishes self from non-self, this role makes it particularly difficult to characterize our symbiotic partners as entirely distinct from ourselves. Finally, it is worth recalling that we are not much tempted to think of the mitochondria that provide the basic power supply for all our cellular processes as distinct from ourselves. Yet these are generally agreed to be long captive bacteria that have lost the ability to survive outside the host cell.

These phenomena are far from being unique to the human case, and arguably similar symbiotic arrangements apply to all multicellular animals. In the case of plants, the mediation of the metabolic relations between the plant roots and the surrounding soil is accomplished by extremely complex microbial systems involving consortia of bacteria as well as fungi whose webs pass in and out of the roots, and which are suspected of

---

[10] See http://nihroadmap.nih.gov/hmp/ (accessed 28 October 2009).

[11] More traditional views of the limits of the human organism might make it seem strange that a strong correlate of infection with the hospital superbug, *Clostridium difficile* is exposure to powerful courses of antibiotics, though this correlation is not quite as pervasive as was earlier thought (Dial et al. 2008).

transferring nutrients between diverse members of the plant community, suggesting a much larger symbiotic system (Hart et al. 2003).

My colleague Maureen O'Malley and I (this volume, Chapter 12) have suggested that the most fundamental way to think of living things is as the intersection of lineages and metabolism. The point we are making is that, contrary to the assumption that is fundamental to the one genome, one organism idea, the biological entities that form reproducing and evolving lineages are not the same as the entities that function as wholes in wider biological contexts. Functional biological wholes, the entities that we primarily think of as organisms, are in fact cooperating assemblies of a wide variety of lineage-forming entities. In the human case, as well as what we more traditionally think of as human cell lineages, these wider wholes include a great variety of external and internal symbionts. An interesting corollary of this perspective is that although we do not wish to downplay the importance of competition in the evolution of complex systems, the role of cooperation in forming the competing wholes has been greatly underestimated. And there is a clear tendency in evolutionary history for entities that once competed to form larger aggregates that now cooperate.

Conclusion

It should be clear that there is a continuity between the phenomena I described under the heading of chimerism and mosaicism and those discussed in the preceding section. Living systems, I am arguing, are extremely diverse and opportunistic compilations of elements from many distinct sources. These include components drawn from what are normally considered members of the same species, as illustrated by many of the cases of chimerism, but also, and more fundamentally, by the collaborations between organisms of quite different species, or lineages, which have been the topic of the preceding section. All of these cases contradict the common if seldom articulated assumption of one genome, one organism.

One plausible hypothesis about the attraction of the one genome, one organism assumption is that it represents an answer to the question, What is the *right* way of dividing biological reality into organisms? But, as I have argued throughout this essay, there is no unequivocal answer to this question. From the complex collaborations between the diverse elements in a cell, themselves forming in some cases (such as mitochondria) distinct lineages, through the intricate collaborations in multispecies microbial communities, to the even more complex cooperations that comprise multi-cellular organisms, biological entities consist of disparate elements working together. Different questions about, or interests in, this hierarchy of cooperative and competitive processes will require different distinctions and different boundaries defining individual entities. As with the more familiar question about species, in which it is quite widely agreed that different criteria of division will be needed to address different questions, so it is, I have argued, with individuals. This is one of the more surprising conclusions that have emerged from the revolution in biological understanding that is gestured at by the rubric, genomics.

Returning finally to the distinctively human, the capacities that most clearly demarcate humans from other organisms—language, culture—are the capacities that derive from our increasing participation in ever more complex social wholes. A further extension of the argument sketched in the preceding paragraph would see this as the next stage in the hierarchy of collaboration and perhaps, as has often been speculated, genuinely marking the human as a novel evolutionary innovation. Rather less speculatively, it is arguably a striking irony that the often remarked centrality of individualism in the last 200 years of social theory has perhaps been the greatest obstacle to seeing the profoundly social, or anyhow cooperative, nature of life more generally.

8

It is not Possible to Reduce Biological Explanations to Explanations in Chemistry and/or Physics[1]

This chapter was originally written as a debate with the well-known historian and philosopher of biology Evelyn Fox Keller (Keller 2010) though the differences between our views turn out not to be very great. I argue that it is a mistake to imagine that complex systems, such as those found in biology, can be fully understood from a sufficiently detailed knowledge of their constituents. My central claim is that the properties of constituents cannot themselves be fully understood without a characterization of the larger system of which they are part. This claim is elaborated through a defence of the concepts of emergence and of downward causation, causation acting from a system on its constituent parts. Although much of this argument can be read as having only epistemological or methodological force, the final section of the chapter defends a more robust metaphysical reading: even purely metaphysical understandings of reductionism such as are commonly represented by supervenience theses are misguided.[2]

Introduction: no need for special biological laws?

Kenneth Waters (1990) has referred to the 'antireductionist consensus' in the philosophy of biology, so it is perhaps not too surprising that I find myself in agreement with most of what Evelyn Fox Keller says. At the beginning of her paper, she says that she 'could as easily have gone the other way', and I would say that, perhaps, she might better have chosen the other side. However, there are some passing statements in her paper with which I am inclined to differ, and I shall explore a few of these to see whether there may be some significant disagreement after all.

---

[1] I am grateful to Maureen O'Malley, Alex Powell, Robert Arp, and Evelyn Fox Keller for comments that led to substantial improvements in the final version of this chapter.
[2] This paragraph was added in this edition.

Like Keller, I am a materialist. That is to say, I do not believe there is any kind of stuff in the world other than the stuff described by physics and chemistry. There are no immaterial minds, vital forces, or extra-temporal deities. Keller writes, however, that as a materialist she is 'committed to the position that all biological phenomena, including evolution, require nothing more than the workings of physics and chemistry'. Even as a materialist, I'm not sure I feel committed to this; but, of course, that depends upon exactly what the title question means. A little unpacking of this question may help to reveal where (if anywhere) there is a serious difference between Keller's position and my own.

One could start with a trivial interpretation of this sentence. If 'the workings of physics and chemistry' meant no more than the workings of things that were made of physical or chemical stuff, then a materialist, such as myself, could hardly deny it. But an ambiguity immediately appears in the phrase 'physics and chemistry.' It could be read—as I just have read it—ontologically, as referring to the things of which physics and chemistry speaks. And for a materialist, therefore, it refers to the entire material world.[3]

But on a quite different reading, one might more naturally think of physics and chemistry as scientific domains, traditions of enquiry, or suchlike. Then, it is far from trivial to claim that biology requires no more than the workings of physics and chemistry. Why should there *not* be biological workings that are quite different from those of physics and chemistry? And here, I appear to be in agreement with Keller when she denies that biology could be derived from the *theories or laws* of physics and chemistry.

As a matter of fact, much of the recent history of reductionism, as a philosophical doctrine, has addressed the relation between theories or, perhaps, laws of nature. Classical versions of the doctrine held that the relation in question was logical: laws of biology should follow deductively from the laws of physics or chemistry. Within the philosophy of biology, something that has surely received the status of a consensus is that no such derivations are plausible. One holdout against this consensus might be Alexander Rosenberg (2006), who seems to suggest that the laws of physics, supplemented by the principle of natural selection, would suffice to derive the whole of biology. I discuss this position critically elsewhere (Dupré 2007). Even ignoring the rather serious problem that, as Keller notes, there do not seem to be many, or any, laws in biology, there is a further problem that the concepts employed in different sciences seem to be incommensurable. As David Hull observed over thirty years ago (Hull 1974), the relations between Mendelian genes and molecular genes are many/many. So Mendelian genetics, a scientific project still very much alive in medicine and

---

[3] Incidentally, the disjunction here and the inclusive disjunction in the title are somewhat problematic. If strong doctrines of reductionism are true, then chemistry is reducible to physics. Hereafter, I shall generally refer just to physics, though, of course, any practical scientific project will surely be happy with a reduction to chemistry.

agronomy, employs concepts that are incommensurable with those that pertain to the molecular entities that, in some sense, underlie the Mendelian phenomena. And subsequent developments in molecular genetics have suggested that the problems are more severe even than Hull could have known (Barnes and Dupré 2008).

If the question of reductionism were merely a question of whether all of biology could be derived from the laws of physics, then we could confidently assert that the issue had been resolved. As I noted in the preceding paragraph, it cannot. However, this is clearly not what Keller has in mind and, despite her ambivalence on the topic, she has some willingness to be counted as a reductionist. So, again, what are we to make of 'the workings of physics and chemistry'?

The reductionist principle

Suppose we are interested in the ecological system that includes lynxes and hares. Everyone can agree, I assume, that, among the constituents of this system are lynxes; and everyone can agree (or everyone I am concerned to argue with) that the lynxes are made entirely of physical stuff. There is nothing else to be made of. The first of these propositions is the one that the anti-reductionist will tend to emphasize, while the reductionist will be more inclined to stress the second. But where should we look for a definite disagreement between these opposing camps?

We have agreed that the lynx is made only of physical stuff. Sometimes this is expressed as the claim that the lynx (or anything else) is 'nothing but' an array of physical parts. While this claim might be endorsed by most reductionists, even the moderate reductionists who have abandoned deductive relations between successive theories, it is liable to be treated with more suspicion by anti-reductionists. So we might try to separate the claim about constitution from the 'nothing but' claim. What more is there to what a thing *is* than what it is made of?

Of course, one answer to the last question that, again, everyone can agree on is that it matters how the physical (or chemical) constituents are put together. In fact, to make a lynx they have to be put together in a stunningly intricate way; and a pile of chemicals that happened to be the very same molecules that could, properly assembled, constitute a lynx would be no more than an inert heap of stuff. So the reductionist's claim should be that the lynx is nothing but a collection of physical parts *assembled in a certain way*. So here, finally, is a proposition that we might expect the reductionist and anti-reductionist to disagree about: if we knew everything about the chemicals that make up a lynx, and the way they are assembled into cells, organs, and so on, we would, in principle, know everything about the lynx. Reductionists will generally endorse something like this, whereas anti-reductionists will deny it. Let me call this, with the specific biological system lynx replaceable by any system we care to investigate, the *reductionist principle* (RP).

An important feature of RP is, of course, the phrase 'in principle'. Certainly, no one knows how to explain all the properties of a complex organism in terms of the properties and arrangements of its parts; the question is whether this is simply a reflection of the underdeveloped state of our current biology, or whether there are deeper obstacles, obstacles in principle, that will continue to prevent us from doing this. The kinds of principles involved will distinguish a variety of different versions of anti-reductionism or, as it is often called, *emergentism*, the belief that there are emergent properties, properties that could not have been predicted (even 'in principle') from a complete knowledge of the constituents of a thing and their internal relations. (I should note, however, that this probably still does not capture any disagreement between Keller and me, as she acknowledges explicitly the existence of emergent properties. It may be that we differ as to what they are.)

Strong emergence

One conception of emergence, championed recently by Mark Bedau (2003), proposes the obstacle to explanation as the lack of a general principle connecting features of the constituents to features of the whole, but holds on to the reductionist intuitions with the idea that a fully detailed *simulation* of the interactions among the constituents would generate the behaviour of the whole. The behaviour of the whole is fully determined by the behaviour of, and interactions between, the parts; but the only way to get from the latter to the former is by a complete simulation. Bedau calls this kind of emer-gence—which shares with reductionism the insistence on the dependence of the whole on the properties of the parts—*weak emergence*. Strong emergence, in contrast, denies such dependency.

I propose, here, to defend strong emergence. That is to say, I want to deny that the behaviour of the whole is fully determined by the behaviour of, and interactions between, the parts. And hence, the elements of behaviour that are not so determined are what we don't know when we know everything about the parts and the way they are assembled; and thus, finally, what violates RP.

At this point, we need to be rather more careful with the relations between *dispositions to behaviour* and *behaviour*. No one believes that the behaviour of a whole is, in general, determined solely by the properties of its parts, even for the most paradigmatically mechanical systems. A properly functioning grandfather clock, say, the action of which is powered by a slowly falling weight, will not function if the weight is supported so as to disconnect its gravitational force from the action of the clock. With few, if any, exceptions the properties of parts translate into the behaviour of wholes under specific circumstances. So, the most any reductionist should claim is that dispositions of the whole are determined by properties of constituents, together with appropriate surrounding conditions. With the important qualification that these dispositions may be probabilistic rather than deterministic, let me, for the sake of argument, concede this much. Have I, then, conceded what is important about reductionism? In earlier work, I have distinguished further between the case in which probabilistic dispositions involve determinate probabilities of specific behaviour, and the case in which the behaviour is possible, but in which there is no reason to think that it occurs with any determinate probability (Dupré 1993: ch. 9). As a matter of fact, I see no reason why the second case should not be characteristic of much that happens, an idea to which I shall return briefly and tangentially in the conclusion to this chapter.

One way of glossing the previous remarks about context is to note that many, at least, of the dispositional properties that appear to fall within the range of the preceding discussion are relational. An elephant gun has the capacity to kill elephants. This is a property that depends on many features of the gun and many features of elephants. One could deprive the gun of this capacity by fitting all elephants with suitable armour plating or, indeed, by killing them all so that nothing any more has the capacity in question. But it is natural to think that there is an intrinsic capacity of the gun, the capacity perhaps to project a lead pellet of a particular mass at a particular velocity in a determined direction, that underlies the relational capacity. Relational capacities of a thing are quite obviously not reducible to any information about the parts of that thing alone, since they depend also on features of the other party to the relation. But the reductionist might reasonably propose that all the intrinsic properties of the thing are reducible to properties of the parts, and that the relational properties were deducible from a knowledge of the intrinsic properties of the things related. The hardness of an elephant's skin, the distance from the skin to organs necessary for the maintenance of life, and so on, combined with the intrinsic capacity of the gun to project a leaden missile, together entail the capacity of the gun to inflict fatal damage on elephants.

So, here is a possible ontological picture for the reductionist. Imagine arranging entities in a traditional ontological hierarchy: elementary particles, atoms, molecules, cells, organisms, and so on.[4] At each level of the hierarchy, we can determine a set of intrinsic properties of the relevant entities. From the intrinsic properties of entities at one level, say atoms, and the relations between the atoms, we can infer the intrinsic properties of molecules. Thus, ultimately, the intrinsic properties of everything are consequences of the intrinsic properties of their constituents and, ultimately of their smallest (physical) constituents. Or anyhow, this will follow as long as we can take care of the relational properties smuggled into the story. (Keller probably would not endorse this reductionist picture. For she writes that biological explanations assume 'the dependence of the identity of parts, and the interactions among them, on higher-order effects'. As will emerge very shortly, I take this to capture a fundamental deficiency in the reductionist picture.)

---

[4] I won't worry here exactly what are the appropriate steps in the hierarchy. As a matter of fact, I am increasingly sceptical whether there really is such a hierarchy rather than a number of points at which we have found it useful to abstract objects of particular degrees of complexity. For example, we typically think of multicellular organisms as distinct from the multitudes of microbes with which they are in an obligate symbiotic relationship, but this is not appropriate for all purposes. The argument below concerning the contextual dependence of the identity of biological entities points more generally in this direction, but I won't develop the point any further here.

Complex relations in biology

Unfortunately, of course, taking care of the relational properties is not an easy matter. Sometimes these are simply a matter of location, as with the relative positions of the elephant and the gun, and spatial relations presumably belong comfortably in the realm of the physical. But many relational properties in biology are not that simple. Consider, for example, the characteristic properties of enzymes. An enzyme is a catalyst, generally a protein but in some cases an RNA molecule, which facilitates a generally highly specific biochemical reaction in a cell. The mechanisms by which many enzymes work are well understood and involve a variety of spatial and electrochemical interactions between the enzyme and its substrate. Enzymes typically have an active site, a small part of an often very large and complex protein, which binds to the substrate and changes its spatial configuration or electrical charge pattern in a way that reduces the activation energy of the reaction the enzyme catalyzes.

We may, no doubt, assume that intrinsic properties of the substrate and the enzyme are sufficient to explain the capacity of the latter to act as a catalyst on the former. However, the intrinsic properties of a large and complex molecule such as a protein will very likely allow it to catalyze many different reactions. And as a matter of fact it has become clear that many proteins do, in fact, have multiple functions. Ramasarma (1999) lists over fifty proteins (or groups of closely related proteins) with more than one known function (also see Jeffery 1999). These alternate functions include a range of activities in addition to serving as enzymes, such as binding or transporting various molecules, inhibiting various cellular processes, or forming subunits of larger proteins. Glyceraldehyde-3-phosphate dehydrogenase (GAPDH), for example, a common 'housekeeping' enzyme, is believed to act as an acyl phosphatase, an esterase, a protein kinase, and a Uracil-DNA glycosylase, in ADP-ribosylation, microtubule-binding, t-RNA binding, amyloid protein binding, and membrane binding (Ramasarma 1999). The number of *possible* functions of a protein molecule seems, in principle, quite indefinite.

Hence, finally, a complete knowledge of the physical and chemical properties of a protein will certainly not tell us what a protein does. When we know what the protein does, chemistry may certainly be able to explain how it does it; but that is a different matter. The distinction between explaining *how something does what it does* and explaining *what it does* was central to my earlier critique of reductionism (Dupré 1993). The idea is developed further by Pigliucci and Kaplan (2006).

Note, also, that it is common practice to say that a protein *is* an esterase, a protein kinase, etc. In fact, the primary name of the protein just mentioned, GAPDH, indicates its role in catalysing the transformation of glyceraldehyde-3-phosphate; most proteins are named by reference to (one of) their functions. But being a GAPDH, a molecule with that particular catalytic function, requires not merely a particular chemical structure, but an environment in which there is glyceraldehyde-3-phosphate to transform.

I take it that the point I have just been trying to make is part of what Keller means when she writes of 'the dependence of the identity of parts, and the interactions among them, on higher-order effects'. But this dependence points, I believe, to a fundamental *objection* to reductionism. Chemistry, alone, cannot tell us that a particular protein is a GAPDH rather than one of the countless other functionally defined things it might have been. To be a GAPDH requires, in addition, an environment that includes the other elements that make the performance of its specific function possible.

The point can perhaps be more intuitively illustrated by thinking of the quite different case of human capacities. As is most definitively argued in Wittgenstein's private language argument (1953/2001), the ability most humans have to speak a language is one that would be impossible, in principle, without the existence of a linguistic community of which they were part. Countless human capacities—to write a cheque, make a promise, play chess, and so on—depend for their possibility on the existence of a social context in which conventions or rules create the conditions for such activities. No amount of knowledge of my physics, chemistry, neurophysiology, or the like, could determine whether I was able to write a cheque. But the point also applies, in important ways, to less socially embedded physical activities. My physical properties do not determine, for example, whether I am able to move vertically through buildings. If my legs function in the standard way, my ability to do this will depend on the availability of staircases. If I use a wheelchair, my vertical mobility will depend on the provision of ramps or elevators. Again, the capacities of a thing, as opposed to the countless merely possible capacities, can be seen to depend on the relationship between the thing and the environment in which it exists.

It may seem that by conceding that the actual capacities—the capacities that become actual rather than merely possible in an appropriate environment—can be explained in physical or chemical terms, I have conceded everything the reductionist really cares about. Certainly, I hope I have conceded what is necessary to account for the extraordinarily successful practices of scientists engaged with molecular aspects of biology. Here, however, it is useful to recall a banal point, though one that occasionally gets overlooked in such discussions, that we are talking about science, not nature. Biological explanations are part of biology, not part of the world, and biology, like any other science, is an articulated conceptual structure and not a repository of things-in-themselves. I shall move into metaphysics and attempt to say a little about nature at the end of this chapter, but, for now, I shall continue to address science. And the fact that biology—a science—works with *concepts* that depend on the larger systems of which they are part, as well as on their constituents, is a fatal objection to the claim that 'it is possible to reduce biological explanations to explanations in chemistry and/or physics'.

A misinformed slogan and its contributions

The preceding point can be developed, as well as some of the main dangers of reductionist thinking illustrated, by considering the view that the genome contains *all* the information required to build an organism. In fact, this view is still too often promulgated by scientists who should know better, and widely asserted in popular science writing. Probably not a lot of people, biologists or otherwise, who have thought seriously about such things still believe this. However, a close look at why it is so deeply mistaken will be a useful way of elaborating the difficulties with reductionism. I won't dwell too long on the tricky concept of information. As developmental systems theorists, in particular, have pointed out for some time, the concept faces an impossible dilemma (an excellent introduction to development systems theory is Griffiths and Gray 1994; Oyama 1985 is the *locus classicus*). If it is interpreted in the everyday semantic sense, then it is obviously false. Genomes contain meaning only in a highly metaphorical sense. But if it is interpreted in the technical sense of information theory, according to which, very roughly speaking, a source conveys information about a target when it reduces the uncertainty about the state of the latter, then there is no sense in which the genome carries information that does not apply equally well to everything else that is necessary for features of the genome to have their normal effects on the cell. Without a complete transcription mechanism, for instance, the genome carries no information; and taking the genome as part of the channel through which the information flows, the transcription machinery carries information about the same targets. In what follows, I shall occasionally use the concept loosely to refer simply to casual determination.

But this rather technical issue points directly to a more fundamental difficulty even with less pedantic interpretations of the idea under discussion. Strings of DNA, or even real genomes replete with histones and other structural elements that make up real physical chromosomes, do nothing on their own. Their involvement in the production of proteins is as part of a system that includes a very large number of additional molecules and cellular structures, and although there are very special and biologically important features peculiar to nucleic acids, singling them out in the way suggested by the reductionist slogan makes no sense. There are several simplistic ideas that contribute to the continued popularity of the slogan, however, and I shall briefly discuss three of these.

The first is what was named by Francis Crick (1958), presumably with a trace of irony, the Central Dogma of Molecular Biology. According to the dogma, information flows from DNA, to RNA, to protein sequence, but never in the opposite direction. Although one should question whether merely specifying a list of amino acid sequences that constitute proteins would be sufficient to specify the entire organism, the doctrine that DNA is only a source of information, entirely immune to the influence of signals from its cellular environment, provides a powerfully reductive perspective on the economy of the cell. But, at any rate, interpreted this way the dogma is entirely misguided. Apart from its dependence, already referred to, on the transcription mechanism that produces mRNA sequence from DNA sequence—not to mention the mechanisms including complex cellular structures, ribosomes, that facilitate the production of proteins in accord with mRNA sequence—the behaviour of the genome is affected by countless other molecules in the cell. It does certainly remain the case that the important activity of the genome is providing a template for transcription of RNA molecules, but which RNA molecules are transcribed and in what quantities is dependent on interactions with many other constituents of the cell, and the actual structure of the DNA in the genome is constantly being modified by these other constituents. Information, then, flows constantly to the genome from RNA and protein molecules.

The importance of RNA molecules in modifying genome behaviour is a field still in its infancy, but one that is thought by many to be likely to revolutionize cell biology. A second simplistic view that is relevant here is the idea, once widely cited, that most of the DNA in the genome, perhaps up to 98 per cent of it, is 'junk'. It has been realized for some time that only a very small proportion of the DNA in the genome provided sequence that ended up translated into protein structure. In accordance with the central dogma, it was concluded that most of the rest lacked any function at all and that, perhaps, its presence reflected no more than competition for space in the genome among genetic elements that played no part in the functioning of the organism (Dawkins 1982). However, it now appears that the large majority of the genome, at least 70 per cent, is transcribed, and as more is understood about the variety of RNA molecules in cells and their diverse functions, it becomes increasingly imprudent to assume that these RNAs—and, therefore, the genomic DNA from which they derive—may not play essential roles in the economy of the cell.

At this point it might be said that, even acknowledging that the function of the genome is affected by numerous RNA and protein molecules, the latter are derived from genomic DNA sequence, so that all that has been added to the idea of the DNA as controlling molecule is a few feedback loops. Everything still begins with the genome. This leads to the third simplistic view I want to mention. Sometimes, it is imagined that all that passes from one generation to the next, through reproduction, is the genome. Clearly, this presupposes the idea that the genome contains all the information necessary to build an organism, since the organism is built, and nothing is there to build it from but the genome. But this view is quite wrong, as well. The smallest 'bottleneck' in the developmental cycle is a single cell, the zygote. And this cell contains all the machinery necessary for the functioning of the DNA, and the rest of a normal cellular complement of molecular constituents and subcellular structures. The cell is an evolved structure which, far from being assembled through instructions contained in the DNA, is a product of several billion years of evolution. Naturally occurring DNA is always and everywhere found as part of such exquisitely complex evolved systems.

This brings me back to the central point that occupied the earlier parts of this chapter. The capacities of DNA are not merely consequences of its molecular constitution, but are simultaneously determined by the systems of which DNA molecules are part. The best way to illustrate this point is by considering the ways that we divide the genome into functional parts, genes.

Genes

The concept of a gene has two rather different traditions of use. Its origin is in the breeding experiments, especially on fruit flies, of the first half of the twentieth century, though the classic experiments of Mendel in the 1860s are generally considered canonical precursors of this work, and this research tradition is often referred to as *Mendelian genetics* (Sturtevant 2001). In this context, the gene was a theoretical term used to track inherited features of the organism under study. The 'gene for red eyes' was the hypothetical cause of a pattern of inheritance of red eyes. Flies with red eyes were assumed to possess this gene, and this, through Mendelian patterns of inheritance and theories of dominance and recessiveness, explained the quantitative characteristics of the pattern. Such a gene made the difference between having and not having a particular trait, in this case a particular eye colour, and it is an important point about this research tradition that genes were always, and only, difference makers. Where a kind of organism showed no variation in a trait, there were no genes for that trait to investigate.

It is uncontroversial that, in this tradition, genes were what they were only in a very complex context. Even a quite deterministic view of the action of genes would need to allow that only deep within the body of a fly would any molecular entity actually make a difference in eye colour. The same entity (if indeed it is an entity) might appear in another organism with a quite different effect, or indeed might appear as a production stage in a chemical factory making parts for organic computers. Being a gene for red eyes is very far from being an intrinsic feature of a bit of chemical.

Of course, the reason for this is clear enough: we have identified the gene in terms of its effect in a much larger system. Nowadays, we are inclined to think of genes rather as sequences of nucleotides, and conceived that way surely they are simply chemical objects. And indeed, we might suppose that the genes for red eyes and suchlike could now be identified with sequences of nucleotides describable quite independently of the biological context in which they appeared. As a matter of fact, however, this turns out not generally to be possible. One of the major areas in which Mendelian genetics remains a thriving tradition is in the medical genetics that addresses single gene disorders. A standard example is cystic fibrosis, which is caused by a recessive gene with very serious health consequences. However, there is no particular sequence that corresponds to the gene for cystic fibrosis. Rather, cystic fibrosis results from a large range of mutations in a certain transcribed DNA sequence (known as the *cystic fibrosis transmembrane conductance regulator gene*) any of which render the gene incapable of producing a functional product. If both copies of the gene are dysfunctional, the disease state ensues. Currently, over 1,000 such mutations have been catalogued. But not every possible mutation of this stretch of the genome will render the gene dysfunctional, and in fact the severity of the disease will vary according to the precise mutation. So, whether a piece of DNA sequence is a cystic fibrosis gene is determined only by how it functions in the entire organism.

It might be supposed that things would be much more straightforward from the perspective of the second tradition, molecular genetics. In this research tradition, which developed out of the identification of DNA as the material out of which genes were made and the resolution of the chemical structure of that molecule, genes are generally identified as specific sequences of nucleotides. Surely being such a sequence is something that occurs quite independently of any context? I have already explained part of the problem with this thought. In so far as molecular genes are identified as templates for particular proteins, and proteins are distinguished by (one of) the specific function(s) they carry out in the cell, then the identity of the gene, as it is conceived as a gene with the function of providing a template for a particular protein, is, again, tied to a molecular context. And it is under some such description that we are motivated to distinguish a particular sequence as a discrete entity, a gene. But the difficulty goes deeper than this, arising from fundamental limitations to the gene concept itself.

In the early days of molecular genetics, and nicely encapsulated in the Central Dogma, it was supposed that genes were discrete DNA sequences that specified precisely the amino acid sequence for a particular gene product. Subsequently, however, things have proved far more complicated. First of all, it was discovered early on that genes (in eukaryotes, anyhow, the division of life that includes, among other things, animals such as ourselves and plants) were not typically unbroken coding sequences, but parts of the coding sequence (known as *exons*) were interspersed with non-coding parts (known as *introns*). The whole sequence, exons and introns, is transcribed into RNA; but, subsequently the introns are excised by further processing machinery. Subsequently it was discovered that in many (probably most) genes this excision process could be carried out in different ways, resulting in different RNA products, so-called *alternative splicing*. These RNA products were also liable to undergo further 'editing', alteration of details of their sequence, prior to being translated into amino acid sequences. And these amino acid sequences might subsequently be assembled into a variety of different functional proteins. So, different and discontinuous parts of the sequence, initially thought to be a gene, might end up in a range of different functional proteins.

And this is not the end of the relevant complications. The early picture had discrete genes, separated by non-coding (either regulatory or 'junk') sequences in a reasonably orderly sequence. But it now seems that coding genes may overlap one another; they may be embedded within the intron of another gene; and they may sometimes be read in both directions, as so-called sense and anti-sense genes, and a particular sequence might be part of both a sense gene and an anti-sense gene. A particular part of the genome might be part of several different genes. In fact, it now seems that something like 23,000 'genes'—in the sense originally assumed in molecular genetics—are involved in the production of perhaps as many as a million proteins. It is hardly surprising that the process from the former to the latter is not straightforwardly linear.

I have presented these complications in terms of genes, but the fact is that they raise serious worries as to whether there is really a coherent concept of the molecular gene at all. The philosophers Paul Griffiths and Karola Stotz carried out a research project in which they presented scientists with DNA sequences involved in various complexities of the kinds just described, and asked them how many genes there were in these sequences (Stotz and Griffiths 2004; Stotz et al. 2004). There was little consensus as to the right answer. The best way to understand this finding is certainly open to debate, but my own view would be not that it showed that there were no such things as genes, but that distinguishing part of the genome as a gene only makes sense in relation to some function that particular bit of sequence serves in the general functioning of the cell. There is, therefore, no objectively unique division of the genome into genes. Again I conclude, the conceptualization of the genome, as an object of study and as divisible into discrete functional constituents, requires that it be placed in the wider *context* with which it interacts.

Causation

We can also approach the question of the reducibility of the biological by looking at intuitions about causation. One of the intuitions underlying reductionist thinking is that, whereas it is natural to think of parts of an entity as causally explaining the behaviour of the whole, the reverse, causal explanation of the behaviour of the parts in terms of features of the whole, so-called *downward causation*, is somehow considered mysterious. (Downward causation has been the subject of philosophical debate for some time, generally dated from a proposal of Donald Campbell [1974]. Interest has been greatly increased recently in the context of systems biology, of which more below.) So it seems natural to explain, for example, the movement of my arm in terms of a series of biochemical processes leading to the contraction of bundles of fibres attached to parts of the skeleton. This constitutes a classic causal/mechanical explanation in terms of pushes, pulls, hinges, and suchlike. It seems strange to many, on the other hand, to suppose that the whole organism of which the muscle tissue is part could somehow cause the necessary molecular activities. Of course, philosophically untutored intuition may find the second possibility quite natural. The naïve explanation of my arm's going up is that I intended to reach for a book, say, which explains the bodily movement in terms of a feature of the whole, its intention. If the whole person is capable of raising the arm, and raising the arm is caused by (among much else) calcium being pumped into the sarcoplasmic reticulum, then it appears that the person is capable of causing calcium to be pumped into the sacroplasmic reticulum.

Downward causation seems a very natural way to think of much of what I have been saying about molecular biology. What causes the human genome to behave in the particular ways it does—for example, various sequences being transcribed or not at varying rate, changes in conformation and spatial relation of chromosomes, and so on—is a variety of features dispersed over the surrounding parts of the cell. The behaviour of the part is to be explained by appeal to features of the whole.

Another example that fits naturally into this picture is the phenomenon of protein folding. A major problem in molecular biology is to explain the transition from an amino acid sequence to the baroquely complex structure that results as this sequence folds into a three-dimensional shape. The topology of this structure is essential to the proper functioning of the protein, yet in many cases it appears to be strongly under-determined by the chemical properties of the links between successive amino acids. It is known that many proteins require specific collaboration from other proteins, known as chaperones, to accomplish this complex feat. One might argue that this was simply another interaction, between the folding protein and the chaperone, fully compatible with a traditional reductionist perspective.

However, even if, as is probably a great oversimplification, interaction with the correct chaperones were all that was required for correct folding, the kind of argument considered with regard to protein function applies equally here. There is a very specific environment, in this case one replete with appropriate chaperones, which endows the amino acid sequence with the capacity, or disposition, to fold in a particular, function-ally desirable, way. And hence it is a specific environment that disposes the various relevant parts of the genome to produce, in the end, an appropriately folded protein. And again, this environment is not something that could possibly be generated *de novo* by the genome but, on the contrary, it is one that took a few billion years to evolve. The cell, I think we must say, with all its intricate structure and diverse contents, is what causes these contents to behave in these life-sustaining ways (Powell and Dupré 2009).

Systems biology

A scientific development that has brought these issues of downward causation, emer-gent properties, and reductionism to the fore is the rapid growth of *systems biology*. Systems biology can be seen as a response to the growing realization that the accel-erating avalanche of molecular data from ever-faster gene sequencing and comparable technologies for assaying RNAs and proteins had not been matched by similar growth in the ability to assemble these data into adequate models or explanations of larger-scale phenomena. Systems biology was conceived as a collaborative effort among molecular biologists, mathematicians, and computer scientists to attempt to provide such integra-tive understanding. In earlier work (O'Malley and Dupré 2005), my collaborator and I distinguished between top-down and bottom-up tendencies in current systems biology. The former, generally reductive tendency hopes to build up more global understanding by gradual integration of information from molecular censuses and knowledge of molecular interactions. Top-down systems biologists doubt whether this can be done, and insist on the need for more general principles that emerge at higher levels of organization, and constrain the behaviour of constituents.

This is a relatively crude dichotomy, of course, and the consensus among biologists involved in these projects is that some combination of the two will be needed for systems biology to succeed (Krohs and Callebaut [2007] offer criticism of the dichoto-my just mentioned). This is exactly what should be expected in the light of the preceding discussion: the capacities of parts will require explanation through reductive, bottom-up approaches; but a top-down approach is required to understand their actual behaviour, and to identify the capacities that need to be explained. However, whereas some top-down systems biologists hold that this is a matter of identifying laws that govern complex systems, my own prejudice is that the top-down part is more a matter of higher-level description of particular systems. One cannot infer the behaviour of a cell by treating it as a bag of chemicals; one might begin to make progress by describing the intermediate structures—viz., ribosomes, Golgi apparatuses, and so on—and the heterogeneous distribution of various molecules in relation to such things.

Metaphysical coda

Reductionism is inspired both by observations of the methods of science and by more purely metaphysical reflections or intuitions. Though, as I have explained, I don't doubt that reductionist methods have a vital role in science, I think this role can be overstated, especially in biology. A science such as molecular biology tells us not only how particular entities come to have the complex capacities they do, but also how complex systems enrol some of these capacities to create stability, order, and function. In doing so, I have suggested, those systems constrain and causally influence the behaviour of their molecular constituents.

An influential movement in recent philosophy of science has attempted to describe biological systems in terms of mechanism (e.g., Bechtel 2006; Machamer, et al. 2000). I am generally sympathetic to this movement, and these accounts have strong parallels, for example, with the view of top-down systems biology mentioned at the end of the preceding section. Although these recent accounts of mechanism do not rest a great deal on the implicit parallel with machines, this parallel does have serious disadvantages, as well as some virtues.

On the positive side, machines, like organisms, exploit capacities of their constituents to create order and predictable behaviour. But there are important differences. The machines we construct typically have a fixed set of parts, and those parts are invariably subject to decay and failure over time. Organisms, in contrast, constantly renew and replace their parts, often with different ones. Organisms have life cycles; machines have only a linear progression towards decay.

However, both machines and organisms illustrate one very important point. Order is difficult to achieve. Machines achieve it with all kinds of ingenuity, and auxiliary devices that anticipate, and sometimes prevent, the common causes of failure. Organisms maintain order with stunningly complex arrays of interacting parts, the 'Bernard machines' eloquently described in Keller's paper, and much else besides. Also, I maintain, these order-preserving systems work by creating synergies of mutual deter-mination between different levels of organization. Although this last point may indicate a fundamental difference between machines and organisms, there is a crucial point in emphasizing their similarity, namely, to indicate the dubiousness of an intuition that underlies much reductionist thinking. This is the idea that order is everywhere, i.e., that everything is determined by the unvarying capacities of microscopic constituents; and it is not at all borne out by a close study of the systems that do manage to maintain order and predictability. Biological order, I argue, is the extraordinary achievement of systems honed by billions of years of evolution. It is not something that comes for free with the determinism of the physical and chemical worlds.

Throughout this chapter, I have been concerned to engage with reductionism as a serious aspect of scientific methodology. I have tried to produce an account that does justice to the undoubted importance that working scientists attach to reductionist methods, while avoiding philosophical conclusions that go beyond what such a concern requires. However, a great deal of philosophical discussion involves much weaker notions that have no such connections with scientific methodology, actual or even imaginable. Perhaps the most widely discussed, and certainly of no threatened relevance to science, are various theses of 'supervenience', a form of reductionism sometimes considered so weak that any sane person must accept it. Indeed, super-venience is often thought to be a paradigmatic form of *anti-reductionism*. A thesis of the supervenience of the biological on the physical asserts that, however inaccessible are principles connecting lower levels to higher levels, nevertheless, the biological depends on the physical in the sense that for any biological system there is a physical state that constitutes it, and wherever we were to find an identical physical state we would find an identical biological state.

It should be obvious what my worry with such a position will be. Perhaps this would be true for any closed biological system, but then there are no closed biological systems. This is one way of understanding the dependence of the identity of biological entities on context that I have emphasized in this chapter. Bounded biological systems do not supervene on their physical parts because aspects of what they are depend on the context with which they interact, a context always extending beyond any predeter-mined boundaries. Perhaps I should concede that everything in the universe supervenes on the total physical state of the universe? Perhaps. But, here, we are so deeply into the domain of purely speculative metaphysics that I more than happy to remain agnostic.

9

Postgenomic Darwinism

Introduction

I might perhaps have better called this chapter post-Darwinian genomics. One point I want to make is that it is time we disconnected our discussions of evolution from an unhealthily close connection with the name of Charles Darwin. Darwin, after all, wrote his most famous work 150 years ago, and rapidly advancing sciences do not generally rest directly on work a century and a half old. Darwin knew nothing of genetics or genomics and, as I shall especially emphasize, there have also been remarkable advances in microbiology that he could not have known about and that fundamentally affect our understanding of evolution.

I do not, of course, have any wish to deny Darwin's greatness as a scientist. It is impossible to read his extensive scientific writings without being struck by the powers of his observation, the encyclopaedic breadth of his knowledge, and a remarkable ability to move between detailed observation and the grand sweep of theory. Moreover, the fact that it was Darwin who convinced the learned world of the fact of evolution, of the common descent of humans and other forms of life, gives him an uncontestable place in the history of ideas. This has provided a cornerstone of the naturalistic world-view which, if hardly the universal perspective of the human race, has increasingly become the dominant perspective among its most educated and reflective minorities.

But this is not just a quibble about an anomalous degree of deference to a distinguished and influential dead scientist. I think this deference can act as an obstacle to the advance of the science. At its most extreme—and here one cannot help seeing an ironic defeat for biology in its debate with religious creationists—Darwin takes on the role of scriptural authority, and his words are subject to detailed exegetical analysis as if this was a way to better understand the biological world. It sometimes seems that Darwin, like God in war, appears on both sides of most major biological debates. One of the great epistemic virtues of science is that it constantly attempts to revise itself and advance its understanding as new information or insight accumulates. Excessive deference or even reverence for past authorities is the antithesis of this epistemic commitment.

But more subtle and specific misunderstandings are also associated with the excessive reverence for Darwin. It is sometimes forgotten that whereas Darwin quite rapidly convinced the learned world of the truth of evolution, the transformation between distinct species, after the publication of the *Origin of Species*, conviction for the process commemorated in his subtitle, natural selection, was not achieved widely until well into the twentieth century, with the synthesis of Darwinian natural selection with Mendelian genetics. The real target of this chapter is not so much Darwin's own views, but the cluster of ideas that emerged at that time as the 'New Synthesis', and has evolved today into what is often called neo-Darwinism. There is a popular view that Darwin got just about everything right that was possible for someone deprived of an adequate understanding of genetics, and the New Synthesis filled in this final gap. And it is this vision, lent weight by the towering authority of Charles Darwin, which I suggest is becoming an obstacle to the advancement of our understanding of evolution and its ability to take account of the very remarkable advances in our biological understanding over the last few decades.

Neo-Darwinism

By 'neo-Darwinism' I mean the New Synthesis as modified by the emergence of molecular genetics in the 1950s and beyond. From the New Synthesis it maintains (in addition to the core commitment to natural selection) the Mendelian idea of inheritance as particulate, the concept of genes that are transmitted to offspring in their entirety or not at all, and the concept, following August Weismann, of a sharp division between germ cells, which carry the transmitted genes, and somatic cells. Neo-Darwinism can be defined, for my present purposes, in terms of two core theses and one important corollary. The first thesis is that overwhelmingly the most important cause of the adaptation of organisms to their environment, or conditions of life, is natural selection. This is the heart of the Darwinism in neo-Darwinism. The second thesis is that inheritance, at least as far as it is relevant to evolution, is exclusively mediated by nuclear DNA. This thesis could be seen, if a little simplistically, as a blend of Mendel and Weismann seen through the lens of Crick and Watson.

The corollary, especially stemming from the Weismannian ingredient of the second thesis, is the rejection of Lamarckism. Lamarckism here has perhaps less to do with the actual opinions of Jean-Baptiste Pierre Antoine de Monet, Chevalier de la Marck even than do contemporary understandings of Darwinism with the ideas of Charles Darwin. Lamarckism now has come to mean the inheritance, or bequeathal to descendants, of somatic characteristics acquired in the lifetime of an organism, and this has become the ultimate taboo in Darwinian theory. The significance of the taboo is that it presents a powerful restriction on the variations that can be the targets of natural selection, the differences between which nature selects. These differences are now assumed to be, or to be direct causal consequences of, randomly generated changes in the genes or genome of the organism.

In the following pages I shall describe some developments in recent biology that show that neo-Darwinism, if not entirely obsolete, is at least severely limited in its ability to encompass the full range of evolutionary processes. My suggestion is that the association with a long-dead hero can convey the message that in general outline the problems around evolution have been solved long ago, and only the details, perhaps of evolutionary history, need to be sorted out. This message is sometimes explicitly promoted in opposition, particularly in the United States, to the powerful voices of creationists opposed to the very idea of evolution. As I have already suggested, the response is surely a counterproductive one. We should celebrate the fact that the exploration of evolution is an exciting scientific project and, far from being essentially complete, it is one of which we are still only at the very early stages. Those who insist on having the whole 'truth' whether or not we have any serious grounds for believing it are perhaps closer to the religious fundamentalists they so vehemently oppose than they would like to believe. At any rate, what I shall do in the main body of this essay is look at some areas of biological research that are radically altering our views of evolution and challenging neo-Darwinian orthodoxy. As should by now be clear, I take this as illustration of the excitement and dynamism of evolutionary science, certainly not any indication of its vulnerability.

Revisionist Darwinism 1: the Tree of Life

The first topic I want to address that will indicate the shakiness of the neo-Darwinian orthodoxy is the concept of the Tree of Life. The Tree of Life is the standard neo-Darwinian representation of the relatedness of organisms. As a tree, crucially, it constantly branches, and branches always diverge, never merge. Species are represented as small twigs; larger branches represent larger groups of organisms. By following down from the branches towards the trunk of the tree it is possible in principle to work backwards through all the ancestors of a group of organisms to the earliest beginnings of life at the tree's base. Darwin's imprimatur for this divergent evolutionary structure is often secured by a picture in the notebooks that seems to represent a divergently branching structure, accompanied, to the delight of philosophical commentators, by the legend 'I think' (Figure 9.1). More significant still, though, is the sole illustration in the *Origin of Species* representing with a branching diagram the formation of new species through the divergence of varieties within a species, an illustration that follows a chapter adumbrating the benefits of divergence by analogy with the division of labour (Figure 9.2).

But this image of the Tree of Life has been rendered at least partially obsolete by recent developments, especially in microbiology, where so-called lateral gene transfer, the passage of genetic material not from ancestors, but from sometimes distantly related organisms on widely separated branches of the Tree of Life, is common. One reason for the importance of this phenomenon is that it threatens to undermine the pattern of explanation of features of biological organisms that is universally mandated by the divergently branching structure of the tree. Neo-Darwinism, it will be recalled, attributes the adaptation of organisms to natural selection, working on variations in the genetic material. These variations are generated endogenously and transmitted within the narrow confines of the species, understood as groups of organisms sharing access to the same gene pool. Embedding this idea within the wider frame of the Tree of Life, we can see that the explanations for all the characteristics of an organism are to be sought in the sequence of ancestors traceable down the branches of the tree, and in the evolutionary process, namely natural selection, to which these ancestors had been subject. Explanation of the characteristics of an organism by lateral gene transfer, on the other hand, puts no limit in principle on where in the history of life a particular aspect of a lineage may have originated. This is immediately obvious when we note that if lateral gene transfer is common, the overall structure of relations between organisms will take the form not of a tree, but of a web, or net. And in a web, unlike a tree, there are many paths from one point to another.

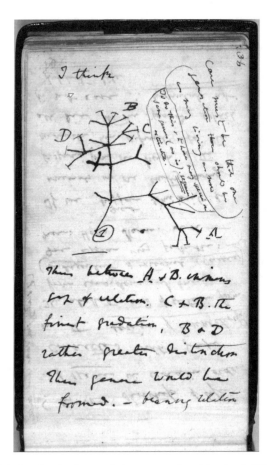

Figure 9.1 Darwin's first sketch of an evolutionary tree from Notebook [B], the first notebook on *Transmutation of Species* (1837). Reproduced by kind permission of the Syndics of Cambridge University Library.

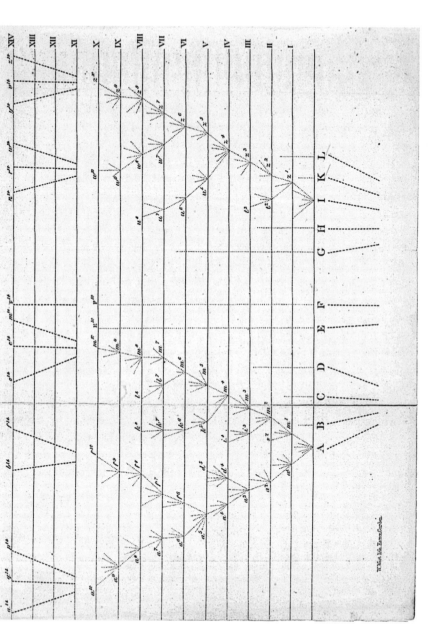

Figure 9.2 The sole illustration from *The Origin of Species*, showing the divergence of ancestral species, first into varieties and eventually species. Species G and H, for instance, have gone extinct, whereas species I eventually gives rise to six descendant species. Reproduced by kind permission of the Syndics of Cambridge University Library.

Lateral gene transfer is widely recognized to be endemic among microbial life-forms (see, for example, Doolittle 1999). Microbes transfer bits of DNA from one to another by a process sometimes likened to sex called *conjugation*, in which a tube down which the genetic unit passes is inserted by one cell into another; by *transformation*, the uptake of free DNA from the environment; and by *transduction*, in which the transfer is mediated by viruses. These processes can result in genetic transfers between the most distantly related organisms, even organisms from different domains, the threefold classification now taken to be the most fundamental division of living organisms.[1] This, in short, removes the presupposition that the evolutionary exigencies of linear ancestors explain the features of their living descendants. Lateral gene transfer allows features to have come from, more or less, anywhere in the biosphere.

Questioning the Tree of Life remains, none the less, a controversial business.[2] Although many microbiologists have accepted that there is no unique tree for microbes, some still resist this conclusion, and insist that there is a core genome, resistant to lateral transfer, and in terms of which a microbial phylogenetic tree can be reconstructed (Lawrence and Hendrickson 2005; but see Charlebois and Doolittle 2004). There are serious problems with this, however. First we might wonder, even if the claim can be sustained in some sense, whether the tree based on the core genome is very useful. Or in other words, what is the Tree of Life for? If, as I have been suggesting, its function is to underpin evolutionary explanations of organismic features, then the more prevalent is lateral transfer, the less will any tree be able to serve this end. This is even more so as the genes that are likely to form the constant core will inevitably be ones with fundamental, and therefore substantially invariant, functions across a very wide range of organisms. They will, for that reason, be the least useful in tracking differences between organisms. This leads naturally to the question, why track phylogeny using these genes rather than some others? Because of lateral transfer not all trees of genes will coincide. And it may be that different gene trees will be useful for answering different questions. Perhaps the defenders of the core genome have in mind that what they should attempt to construct is the cell tree, the tree that traces the sequence of (vertical) cell divisions back to the beginning of cellular life. The trouble then is that

---

[1]   These three domains, probably still not widely familiar to lay readers, are the Bacteria, the Archaea, and the Eukarya. Most people are still probably more familiar with the outdated fivefold classification, Animalia, Plantae, Fungi, Protista, and Monera. But in fact the first four of these are now understood to belong within just one of the currently distinguished fundamental domains, the Eukarya. The remaining kingdom, Monera, is now divided between the domains Archaea and Bacteria (and some miscellaneous organisms now seen as Eukarya). Evidently this change reflects a vastly increased awareness of the diversity of microbes, which include the latter two domains as well as the Protista and many of the Fungi (yeasts). As will be emphasized below, microbial life is the dominant form of life on Earth in terms of its antiquity, its diversity, and even its sheer mass.

[2]   One reason for this briefly alluded to above is that, especially in the USA, any dissent from Darwinian orthodoxy can be seen as providing ammunition for creationists. I have occasionally had the dubious pleasure of finding my own work cited with approval on creationist websites. As also noted above, however, for science to insist on orthodoxy in defence against theological fundamentalism could provide only the most pyrrhic victory.

this seems just to assume what is at issue, that vertical inheritance is what really matters. If this position is to be maintained regardless how much the contents of the cells may be changed by other interacting, non-vertical processes, one might wonder whether ultimately it would end up as little more than a fetishism of the cell membrane.

Eukaryote[3] biologists are generally much more confident of the Tree of Life, and with good reason.[4] Lateral gene transfer seems less common among eukaryotes, and there is little question that the tracing of vertical ancestral relations is a powerful and useful way of classifying these organisms.[5] Even here, though, there is reason to be cautious. For a start, hybridization seems to be much more common than was once thought (Mallet 2008). But perhaps more important, the transfer of genetic elements by viruses certainly does continue in eukaryotes, and may well prove to be an important factor in evolution. About half of the human genome, for instance, consists of material that is thought to have originated in transfers from viruses. Much of this, it is true, consists of highly repetitive sequences that have seemed unlikely to be function-ally significant. When the idea of 'junk DNA' was fashionable, these were prime candidates for junk. However, it now appears that at least 70 per cent of the genome is transcribed into RNA, and investigation of the roles of various kinds of RNA fragment in regulating the genome is one of the fastest growing fields in molecular biology. It would be premature to assume that sequences of viral origin may not play crucial roles in such regulatory systems. And finally, there are examples of significant functional features of cells that do appear to involve protein-coding sequences of viral origin. The best example here is of the evolution of placental mammals. The tissue that provides the barrier between foetal and maternal circulations, a multinucleated cell-like struc-ture called a syncytium, is believed to be coded for by genes of viral origin (Mallet et al. 2004). There may surely be other equally significant cases. It is at any rate clear that, even among eukaryotes, lateral origins play some role in explaining the current features of organisms. The always branching, never merging, tree of traditional phylogeny is not enough.

Revisionist Darwinism 2: evolution by merger

Lateral gene transfer can be seen in a rather different light as an example of something much broader, evolution by merger. This gets to one of the most general points I want to make about the limitations of neo-Darwinism. The first thesis mentioned above, the overwhelming emphasis on natural selection, has encouraged neo-Darwinian evolutionists

[3] Eukaryotes, it will be recalled, are the animals, plants, and fungi, as well as a diverse and miscellaneous collection of microbes. They are distinguished by a more complex cell structure which includes a distinct compartment, the nucleus, which houses the genetic material.

[4] As an important qualification, there is no consensus about the evolutionary origins of eukaryotes.

[5] Notice that the tendency to lateral gene transfer, or mechanisms to prevent it, are contingently evolved features of particular classes of organisms, so there is no reason why its prevalence among microbes should imply anything about multicellular eukaryotes.

to think a great deal about competition, but very little about cooperation. Indeed, the latter appears mainly in the guise of a problem—the 'problem of altruism'. The problem of altruism is, crudely put, the problem of understanding why it is that, in a 'Darwinian' world in which the only survivors are the most ruthless and self-interested competitors, some organisms are actually nice to one another. But looked at from a quite different perspective, life is a massively cooperative enterprise and 'altruism' should hardly be surprising. The elements in a cell or the cells in a multicellular organism must obviously work in a highly coordinated way and subordinate their own 'interests' to those of the whole of which they are part. It will be objected at once that this is cooperation within an organism, not between organisms, and so of course not a problem. But this reply assumes that we know exactly what constitutes an organism and what is merely a part of an organism, an assumption I shall suggest is highly problematic.

It is perhaps hardly controversial to note that natural selection will frequently select the organisms that are best at cooperating with the organisms with which they interact. This is just one way of adapting to the environment, the most salient part of which is typically the other organisms that inhabit it. I want to go a step beyond this, however, and suggest that merger with other organisms (or suborganismic biological entities) is a central process by which biological organisms evolve. Such a process is referred to as endosymbiosis, and is most widely familiar from the ideas of Lynn Margulis (1970) about the origins of the eukaryotic cell.[6] It is now universally acknowledged that the mitochondria that provide the energy source for all eukaryotic cells, and the chloroplasts that effect photosynthesis in plants, were both originally free-living organisms but are now more or less independently reproducing but wholly dependent constituents of larger cells. Although the details are much more controversial, it is also believed by many that the eukaryotic cell itself derived from a merger between two prokaryotes, perhaps a bacterium and an archaeon.

The examples just mentioned are instances of fully obligate endosymbiosis: mitochondria are parts of eukaryotic cells, and there is no more question of why they are acting altruistically towards the containing cell than of why my liver acts altruistically towards me. However, it is important to note that endosymbiosis is something that may evolve over a long period of time, and in the meantime may consist of a range of degrees of interdependence from conditional and reciprocal cooperation to full endosymbiosis. There are, for example, well-studied cases of varying degrees of endosymbiosis between insects and bacteria. *Buchnera aphidicola*, endosymbionts of aphids, have been associated with their partners for up to 200 million years, and have lost the ability to carry out various essential metabolic functions on their own. *Wolbachia*, on the other hand, a genus of bacteria associated with a very wide range of arthropod species including perhaps half of all insect species, is generally referred to as a parasite. *Wolbachia* are particularly interesting for their ability to control the reproductive behaviour of

---

[6] Margulis and Sagan (2002) present a broader argument for the central importance to evolution of what I am calling evolution by merger.

their hosts. Some can kill or feminize males, or induce parthenogenesis. They can also induce reproductive incompatibility between insects infected with different *Wolbachia* strains, possibly playing a determinant role in speciation.

It is generally supposed that the manipulation by *Wolbachia* of their hosts' reproduction contributes to their own rather than their hosts' reproductive interest. However, as some host species appear unable to reproduce without the assistance of *Wolbachia*, and as *Wolbachia* are obligatorily symbiotic, it is not always clear how these interests are to be separated. *Wolbachia* are involved in transfers of DNA between insect species, raising questions about genetic differentiation of insect species (Whitworth et al. 2007), and a whole *Wolbachia* genome has been found embedded within a *Drosophila* genome (Dunning Hotopp et al. 2007). It has also been found that *Wolbachia* may reduce the vulnerability of their hosts to viral infections (Teixeira et al. 2008). It would be difficult to assess the ratio of costs and benefits to the parties in these intimate associations, but it seems likely that this balance will vary from case to case, and that in some cases the relationship has moved to full mutualism or even symbiosis.

One reason I have spent a little time on this example is that it begins to introduce a fundamental question, namely how we determine the limits of an organism. No one doubts that mitochondria are parts of the organisms in which they are found whereas, on the whole, everyone takes *Wolbachia* and their insect hosts to be distinct organisms. But what is the basis of this different treatment? It will be recalled that discussions of altruism tend to assume that this question is unproblematic. If, as I shall suggest, it is a thoroughly indeterminate matter, settled as much by our interests as investigators as by anything in nature, it will clearly be necessary to rethink the question of altruism or, more broadly, competition and cooperation.

What is an organism?[7]

Although philosophers have for many years questioned some of the key concepts of biology, such as the species or, more recently, the gene, on the whole they have not seen much to worry about with the concept of an organism. According to the orthodox view, there are two kinds of organisms: single-celled, or microbes, and multicellular, or (as I have elsewhere suggested we call them (this volume, Chapter 10)) macrobes.[8] In the former case the cell is the organism. In the latter case all the cells derived from a fertilized egg, or zygote, constitute the one organism. We might summarize the view as 'one organism, one genome'. This concept of the organism

---

[7] This section summarizes ideas developed in considerably greater detail in the following section of this book, especially Chapters 11 and 12.

[8] It is bizarre that we should have a term covering the vast majority of organisms that have ever existed, but none for the small minority to which this term does not apply. 'Macrobe' seems the obvious candidate to fill this gap; and filling it should help to deflect the illusion that it is microbes that are the exceptional or unusual biological forms.

could be seen as the microlevel reflection of the macroscopic Tree of Life: both within and between organisms we find orderly and always divergent branching. But we might also want to approach the question of what constitutes an organism from a functional perspective: what are the systems of cells that interact with the surrounding environment as organized and generally cooperative wholes? From this starting point we would note that microbes do not typically function as isolated individuals but rather in complex associations often composed of highly diverse kinds of cells. Typical of such associations are biofilms, the generally slimy coatings that develop on practically any moist surface. Consider, for instance, one well-studied class of biofilms, those on the surfaces of our teeth known as dental plaque. Over 500 different bacterial taxa have been found living in the human mouth (Kolenbrander 2000) and, according to one authority, 'Oral bacteria in plaque do not exist as independent entities but function as a coordinated, spatially organized and fully metabolically integrated microbial community, the properties of which are greater than the sum of the component species' (Marsh 2004: 204). Why would we not consider this community, the organized functional whole, to constitute an organism?

If we concede that biofilms comprise a kind of multicellular organism, then the argument is also over as far as traditional monogenomic multicellular organisms are concerned. For all known such multicellular wholes exist in symbiotic relations to often enormous and diverse communities of microbes. In the human body, for instance, it is estimated that 90 per cent of the total number of cells are in fact microbial (Savage 1977), living mainly in our gut, but also on the surface of the skin and in all the bodily orifices. These microbes contain perhaps 100 times as many genes as those found in the more traditional human genome (Xu and Gordon 2003), which has led to the launch by the US National Institutes of Health of the Human Microbiome Project, which will explore this missing 99 per cent of the full human genome. The importance attached to this project reflects an increasing awareness that these symbiotic microbes have a fundamental influence on human health. They are known to be involved in digestive processes, and hypothesized to have a significant role in causing obesity. For model organisms it has been demonstrated that microbial symbionts are necessary for normal physiological development (Bates et al. 2006), that they affect gene expression in the 'host' cells (Hooper et al. 2001), and that they are involved in the maturing of the immune system (Umesaki and Setoyama 2000).[9] There is every reason to expect similar findings in humans.

I propose then that the typical organism is a collection of cells of different kinds, organized cooperatively to maintain its structure and reproduce similar structures. As Maureen O'Malley and I have put it (Chapter 12, this volume), an organism is a metabolically integrated community of lineage segments. It will immediately strike evolutionists that this conceptually separates the organism (functional whole) from the

---

[9] For a general review of this topic see McFall-Ngai (2002).

evolving entity (part of a lineage). But this, of course, is the point. The assimilation of these concepts obscures the empirical reality that evolution requires both (directly) reproducing lineages and the assembly of organisms from components of these lineages, and that these are in principle quite independent processes. While most of these lineage segments will have little chance of reproducing themselves except in so far as they are able to form parts of appropriate communities, this is nevertheless a contingent matter.[10] One consequence of this proposal is that what is an organism, and whether something is part of an organism or not, are not questions that necessarily admit of definitive answers. Whether a group of microbes is a closely connected ecological community or an organism may be a matter of biological judgement. The important point is that it, or most of it, will share an evolutionary fate. If its constituent cells are to send descendants off to participate in new biofilms it will be because the parental biofilm is thriving. What I have been calling organisms are units of selection, objects between which natural selection selects.

Cooperation again

I can imagine a frustrated reader complaining that I have yet to address the kind of cooperation that is of real interest to evolutionary biology, cooperation between conspecifics. Cooperation with other organisms is just adaptation to the environment, of which they are part. Some of them are to be eaten, most can be ignored, others are more useful as collaborators, and so on. Conspecifics, on the other hand, are always competitors for representation by their descendants in subsequent generations. So let me say something about this topic.

The orthodox neo-Darwinian view is that the only circumstance that brings about cooperation between conspecifics is kin selection. Here it is time to distinguish between some degrees of cooperation. If two lions can kill a wildebeest that neither could handle alone, and moreover it will provide plenty of food for both, they will do well to cooperate. Evolutionists tend rather to speak of 'altruism' in a technical sense according to which an act is altruistic only if it not only confers a benefit on the recipient but is also more costly to the donor than refraining from action. Any animal that acted in this way would lose out to natural selection in competition with others that avoided such acts of kindness. The only exception would be the case where the beneficiary is kin, perhaps one's offspring, as described by so-called inclusive fitness theory. Here the fundamental principle is said to be Hamilton's rule: $rB > C$. B is the benefit to the recipient, C the cost to the donor and r the so-called coefficient of relatedness. This coefficient is 1/2 for offspring or full siblings in sexual species, and is thought of as the proportion of genes that two organisms share by virtue of their

---

[10] This proposal is very much in the spirit of Developmental Systems Theory (Oyama 1985; Oyama et al. 2001), though developmental systems theorists have paid too little attention, to date, to the place of symbiotic microbes in developmental systems.

relations of descent.[11] If $rB > C$, for example if I make a sacrifice that provides more than double the benefit to my child, evolution will favour such behaviour. I don't want to deny that this is a powerful tool for analysing important aspects of evolutionary processes and their potential stability or instability. One very impressive example is its application to theoretical discussions of the evolution of eusociality, the often vast and complexly articulated social system characteristic of many ants, wasps, and bees (Hymenoptera), termites, and, alone among mammals, the naked mole rat. These arguments have shown that only under conditions of strict monogamy for an exclusive breeding couple is such a social arrangement likely to evolve. Recent work (Hughes et al. 2008) has confirmed that such strict monogamy was indeed the ancestral condition in a large number of Hymenoptera species studied, giving convincing support to inclusive fitness theory.

I want to make two somewhat more sceptical comments on this topic, however. First, it is often said that altruism outside the narrow confines of kin selection theory will be subverted by competition from less altruistic rivals. The assumption that there are indeed such rivals seems sometimes to be a matter of pure dogma. Consider, for instance, an example that seems to contradict standard kin selection theory, also from the Hymenoptera. The Argentine ant (*Linepithema humile*), while known for its inter-colony aggression in its native land, has now taken to behaving in a non-aggressive, cooperative way with relation to other colonies of conspecifics, in a range of newly colonized areas in Europe, North America, Japan, and Australia. Contrary to a speculation that this must be due to genetic relatedness between the recently landed colonists, colonies in the European case, at least, were found to be genetically diverse. There is considerable dispute about how to explain or even describe this phenomenon, though one thing that seems to be widely agreed is that the ants as a whole do very well out of the arrangement. As humans have also discovered, warfare may benefit a few, but it is hardly good for the species. Unsurprisingly, it is also speculated that the arrangement will be unstable. A mutant aggressive colony would perhaps do extremely well cutting a swathe through its amiable neighbours. But even if this could happen, it doesn't imply that it must. Perhaps eventually the system will collapse, and perhaps it is bound to do so in the very long run. But, to paraphrase Keynes, in the long run we are all extinct. The existence of cooperation between non-kin is sufficient to show that there are evolutionary processes capable of creating it. The most widely discussed such process is of course group selection, though this does remain controversial, if less so after the extremely influential work of Sober and Wilson (1998). Even if it is demonstrated that there are circumstances that would undermine these cooperative systems, this hardly shows that they could not, after all, have come into being in the first place. It is a contingent matter how long more or less cooperative, even altruistic, systems last.

---

[11] This is a trickier notion than is often noticed, as can somewhat amusingly be noticed when the same authors suggest that we share 50 per cent of our genes with our siblings and 98 per cent with a chimpanzee. Needless to say, different conceptions of 'the same genes' are at work here (see Barnes and Dupré 2008: 98ff.). This shouldn't cause any confusion in the present context, however.

This brings me conveniently to my second point, the one that has been the main focus of this chapter. It is that arguments about what entities can be expected to cooperate or compete with what others presuppose that we know what the individuals are that are cooperating or competing. Group selection is taken to be problematic because it is assumed that the members of the group are real, robust, indisputable individuals, whereas we see the group itself as a fragile coalition, a thoroughly dubious individual. But what I have been suggesting is that in fact there is no sharp line between the group of more or less cooperative individuals on the one hand, and the unified self-contained individual on the other. Indeed, it may well be that there is a tendency for the former to evolve into the latter, and that in the course of this process the individuals will act increasingly as parts subordinate to a larger whole. Presumably something like this must have happened in the evolution of multicellularity, and indeed is thought to have happened many times (Buss 1987).

There is even a bigger picture here. The idea that life is hierarchically structured is an ancient and obvious one. Molecules comprise cells; cells make organs and organ systems; organisms are composed of organs and the like; and organisms in turn make up larger social or ecological units. This is a useful picture in focusing the investigating mind on particular aspects of the biological world, but it can easily be taken too literally. Cells, organs, and even organisms are, in nature, embedded in larger systems, and their separate existence requires either a scalpel or a process of abstraction. Two further points reinforce both the significance and the plausibility of this observation. First, or so I would also argue, a full understanding of a biological entity at any of these intermediate levels is impossible without taking account both of its composition from smaller constituents, and of the influences exerted on it by the larger system of which it is part, though that is an argument beyond the scope of the present paper (see this volume, Chapter 8; Powell and Dupré 2009). Causal explanation runs both from smaller to larger and from larger to smaller. Second, we should recall that our hierarchy of entities is already itself an abstraction from a hierarchy of processes. It may be that many forms of scientific reasoning require descriptions of entities as if they had a set of static properties definitive of such entities. But the reality, as best we understand it, is of a series of nested processes at timescales ranging from nanoseconds for intercellular chemical reactions to hundreds of millions of years for some macroevolutionary processes (see chapters 4 and 5). The illusion of an objectively distinct and unique hierarchy of objects is much less compelling when this abstraction is borne in mind.

Lamarck redux

I turn now to the strictest taboo in neo-Darwinism, Lamarckism.[12] Lamarckism, here, must be understood in an even less historically grounded sense than Darwinism, and

---

[12] The only openly Lamarckian work that has been widely influential is Jablonka and Lamb (1995), who review in detail the Lamarckian implications of the epigenetic inheritance that will be briefly described in the next section.

has little to do with the great French naturalist. The taboo concerns the inheritance of characteristics acquired during the lifetime of the organism. According to strict neo-Darwinists only genetic mutations within the germline and the recombination of genetic resources brought about by sexual reproduction provide the resources on which selection acts. Curiously, however, though mention of Lamarckism can still bring a shudder to many evolutionary biologists, almost no one still believes in the strict form of the taboo. Or so, anyhow, I shall attempt to demonstrate.

The topic with which I began, lateral gene transfer, is one generally acknowledged qualification of strict anti-Lamarckism. Genes transferred laterally into the genome of an organism are certainly acquired, and may certainly be inherited. The reason that Lamarckism is such a profound potential challenge to traditional Darwinism is that somatic traits acquired during the lifetime of the organism may often be adaptive, constituting the organism's response to the environment. An animal may run as fast as it can to escape speedy predators or in pursuit of fleet-footed prey, for example, and in doing so it may develop stronger leg muscles. But the inheritance of such adaptive acquired characteristics would threaten the first principle of neo-Darwinism, the monopoly of natural selection in producing adaptation.[13] Here it may be thought that lateral gene transfer offers little threat of this kind. Perhaps we should see it as no more than the equivalent of a very big mutation. But first, there is a growing consensus that lateral gene transfer has been of fundamental importance, at least in microbial evolution. Boucher and colleagues (2003) review the evidence for its role in 'photosynthesis, aerobic respiration, nitrogen fixation, sulfate reduction, methylotrophy, isoprenoid biosynthesis, quorum sensing, flotation (gas vesicles), thermophily, and halophily'. Moreover, second, a large number of researchers suggest that lateral gene transfer is indeed often an adaptive response to the environment. According to Pal and colleagues (2005) 'bacterial metabolic networks evolve by direct uptake of peripheral reactions in response to changed environments'. And 'lateral gene transfer provides the bacterial genome with a new set of genes that help it to explore and adapt to new ecological niches' (Marri et al. 2007). Note the similarity with the kind of cooperative ventures I discussed earlier in this chapter. Whole microbial cells (or indeed macrobial cell systems) adapt to their environment by recruiting, or being recruited by, coalitions of cooperating cells. More complex organisms may recruit conspecifics or even members of other species to form social collectives that enhance their ability to cope with environmental challenges. And, finally, cells may sometimes recruit adaptively useful genetic fragments from their environments. All very Lamarckian.

One response to the issue of lateral gene transfer may be to downplay the importance of microbial evolution. Perhaps microbes are really rather insignificant little

---

[13] A very important book by Mary Jane West Eberhard (2003) explores a growing body of findings on the remarkable ways in which the development of organisms responds to environmental circumstances, and the implications of this for evolution. West Eberhard refuses to call any of her views Lamarckian, though this may possibly be for fear that this would lead to their marginalization.

beasts? To this, however, it is sufficient to respond that 80 per cent of evolutionary history is a history solely of the evolution of microbes; the vast majority of organisms alive today are microbes; and all known macrobes are dependent for their existence on symbiotic relations with microbes. As I have briefly mentioned above, the importance of lateral gene transfer in macrobial evolution is itself a matter of active debate. But anyhow, an account of evolution that doesn't apply to microbes is one that ignores the overwhelmingly dominant manifestation of life on Earth.

Varieties of inheritance

The Lamarckian aspects of the topic just considered at least do not violate the idea that the vast majority of inheritance passes through the nuclear genome. Lateral gene transfer may be very important in evolution, but it is very rare by comparison to the routine passage of genetic material from parents to offspring. However, there are other reasons to recognize that the neo-Darwinian restriction of inheritance to transmission of the nuclear genome provides a thoroughly impoverished picture.[14] The most widely discussed form of inheritance that is excluded is cultural inheritance. Much of this discussion is directed specifically to human evolution (e.g., Richerson and Boyd 2005). Although this work is very important in many ways, including in showing the inadequacy of the orthodox neo-Darwinian treatments of human evolution offered by Evolutionary Psychologists, in the present essay I shan't discuss the special problems of human evolution. There is still heated debate about whether human evolution raises unique issues, and every aspect of human evolution has been discussed and debated by numerous authors, including myself (Dupré 2001). In this chapter I shall avoid these very specific issues.

I mentioned in passing above the perspective of developmental systems theory (DST) (see note 10). DST abandons the myopic focus on the nuclear genome typical of much neo-Darwinism, and looks at the entire cycle of events by which the organism is reproduced. The fundamental unit of analysis is the life cycle of the organism and, given this unit of analysis, it should be clear from the preceding discussion that the requisite concept of an organism must also be the multigenomic, multilineage one advocated above. From a DST perspective a large body of work on the cultural transmission of behaviour can be seen as fitting fully into an evolutionary framework. Some fairly arbitrarily selected recent examples are the learning of frog calls by bats (Page and Ryan 2006), the use of sponges in foraging by bottle-nosed dolphins (Krutzen et al. 2005), or, perhaps the best-studied example, the transmission of bird songs (Slater 1986). The process of learning behaviour by immature individuals, and

---

[14] It is useful here to remember that inheritance systems of any kind are products of evolution. Perhaps some very primitive kind of inheritance must have emerged at the dawn of evolution, but whatever inheritance systems exist today were certainly products of evolution not, as some presentations of neo-Darwinism sometimes suggest, its prerequisite. In this light it will hardly be surprising if several have evolved.

the behaviour of mature individuals involved in mating and in rearing offspring, are clearly crucial parts of the developmental cycle, and potentially evolving aspects of the life cycle.

Less familiar, but perhaps even more important, is the fact that far from the idea occasionally suggested in popularizations of neo-Darwinism (e.g., Dawkins 1976) that the genome is the only significant material thing transmitted in reproduction, the minimal material contribution in any form of reproduction is an entire maternal cell. This is an extremely complex object with a great deal of internal structure and a bewildering variety of chemical constituents. For asexual organisms (most organisms, that is), it seems perverse to think of anything other than the cell as the basic unit of inheritance. For sexual organisms the issue is more complex, because each individual begins life with a new, generally unprecedented, inheritance, at least genetically. But of course there is a vast number of other materials that are passed on with the maternal cell (and a few even with the paternal sperm) that form a major part of the (inherited) developmental system.

It is sometimes supposed that all the non-DNA material passed on in reproduction is unimportant because it is the DNA that carries the inherited differences on which natural selection can act. But this seems to be a dogmatic assertion rather than anything for which there is empirical evidence. Why, for example, might not changes in the chemistry of the cell membrane be inherited in the process of cell division? But we do not need to speculate. There is a rapidly developing field of biological research, epigenetics, which may be seen as answering a fundamental question, but one that can seem mysterious from the radically DNA-centred perspective—why do different cells with the same genome do different things? Why do my liver cells differ so radically from my brain cells, for instance? Central to epigenetic research is the understanding of how other chemicals in the cell act on the genome to determine which parts of it are expressed (i.e., transcribed to RNA and (sometimes) translated to a protein).

Epigenetics is important in part for breaking the hold of the so-called 'Central Dogma' of molecular genetics, that causality, and hence information, runs only in one direction, from DNA to RNA to protein.[15] Epigenetics could be described, with a little hyperbole, as the study of the falsity of the Central Dogma. But, secondarily and consequently, it reveals the potential diversity of inheritance at the molecular level. In the first place, once it is seen that the surrounding cell acts on the genome, not merely the other way around, it is clear that the inter-generational transmission of any part of the cellular system may embody significant heredity. Second, one of the crucial ways in which epigenetic effects on the DNA occur is through actual modifications to

[15] In its original formulation by Francis Crick, the Central Dogma referred only to information about the sequence of bases or amino acids in macromolecules. Although even this is no longer a defensible position, in view of phenomena such as alternative splicing, the expression is nowadays typically used much more broadly, and in the present discussion I adopt this broader use. It is also plausible that Crick intended a degree of irony in using the term 'Dogma', something that has curiously dissipated in some contemporary references.

the structure of the DNA chain. The best-studied of these is methylation, in which a methyl group is attached to one of the bases, cytosine, that comprise the DNA sequence. This has the effect of inhibiting the transcription of the sequence of which the methylated cytosine molecule is part. It is an obvious possibility that these modifications could be inherited. The claim that they are indeed inherited has been highly controversial. In part this was because it had been understood that a process of demethylation took place during meiosis, the formation of sex cells. If this demethylation was total, then the epigenetic changes would not be transmitted. Recently, it has become increasingly widely agreed that demethylation is not complete, and hence that methylation is to some degree inherited (Chong and Whitelaw 2004). This has been a remarkably heated controversy, and it is impossible to avoid the suspicion that this is in significant part because if methylation patterns, something that can be acquired in the lifetime of an organism, can be inherited, this will raise the possibility of violating the taboo against Lamarckian inheritance.

It is very interesting to note that epigenetic changes might still be inherited even if they had proved to be entirely erased at meiosis. This is because when they are induced by external, environmental influences they may also contribute to the production of those same influences. The classic example substantiating this possibility derives from a series of experiments on maternal care in rats, carried out by Michael Meaney and colleagues. Grooming, especially licking, by mother rats appears to be very important for the proper development of rat pups, and rats that do not receive sufficient such maternal care grow up generally fearful and, most significantly, less disposed to provide high quality maternal care to their offspring (Weaver et al. 2004; Meaney et al. 2007). It has been demonstrated that these effects are mediated by maternal grooming causing changes of methylation within cells in the brain, which in turn affect the production of neurotransmitters. Thus, the trait of high quality maternal care appears to be transmitted through the induction of methylation patterns in young female rats through exposure to such maternal care. This might also be seen as an adaptive and heritable epigenetic switch: in a stressful and dangerous environment, perhaps, it is best to be fearful (even the paranoid can be right) and too risky to devote more than the minimum effort to caring for the young. It is, of course, possible, and a possibility that might be very widely significant, that this modestly Lamarckian mechanism could be an adaptation acquired by Darwinian means. As mentioned above, inheritance mechanisms are among the more interesting features of organisms that evolve.

Conclusion

I conclude very much as I began. With absolutely no disrespect to Darwin, biological insights gained over the last few decades have profoundly altered the way we can and should think about evolution. It appears that evolutionary processes may be more diverse than we had imagined, including Lamarckian mechanisms as well as neo-Darwinian, cooperative and symbiotic as well as competitive and individualistic.[16] The evolutionary histories of the entities that make up biological wholes may also be multiple. Genomes have different histories from the organisms in which they reside, both because they assimilate material from other sources, and because they have their own history within the organism—for example, of intragenomic duplications. And organisms, at least when understood as the functional wholes that interact with the rest of the world, are coalitions of entities with diverse evolutionary histories. Neo-Darwinism has much to say about the divergent processes that push biological entities apart, much less about the convergent processes in which the whole is constantly more than the sum of the parts.

None of this should be remotely shocking. But for some reason or reasons we have buffered an outdated view of evolution with a thicket of surrounding dogma and presumption that stands in the way of advancing the theory in line with the stunning insights that are being gained in other parts of biology. Part of this story surely is that this dogma has developed as an unintended response to competition with thoroughly anti-scientific perspectives (creationism, 'intelligent design') that have somehow positioned themselves as rivals to scientific evolutionism. And I suspect the links with creationist views may be more complex than that. Extreme neo-Darwinists sometimes share with creationists the yearning for an all-encompassing scheme, a single explanatory framework that makes sense of life.[17] One thinks, for instance of Daniel Dennett's (1995) paean of praise for natural selection, which he then deploys as the essential resource to explain everything from the breeding behaviour of bees to the deliberative processes of the human mind. But evolution is a mosaic of more or less related processes, producing a motley collection of outcomes. Just because one has a hammer, one should be careful not to suppose that everything is a nail.

If one of the things that needs to be done to remedy this partial paralysis of our evolutionary thinking is that we detach our view of evolution a little from our reverence for Charles Darwin, then I am sure he wouldn't mind.

---

[16] Here I should note that Darwin, as opposed to the later neo-Darwinians, was always a good pluralist and, incidentally, increasingly a Lamarckian.

[17] I am grateful to Staffan Müller-Wille for emphasizing this point to me in conversation.

PART III

Microbes

10

Size Doesn't Matter: Towards a More Inclusive Philosophy of Biology[1]

With Maureen A. O'Malley

Introduction: microbes and macrobes

The distinction between micro- and macro-organisms is one of the most widely assumed in thinking about life-forms. While we have two words for the first group—micro-organisms or microbes—there is none in common use for macro-organisms. We propose to fill this gap with the word 'macrobe'.[2] The contrast between microbes and macrobes is very close to that between multicelled and single-celled organisms. Microbes are also defined by features such as invisibility and a perceived lack of morphological and cellular sophistication; macrobes by a positive account of those features. But regardless of choice of defining features, neither of these categories would normally be attributed much biological coherence.

In general, any organism too small to be seen without a microscope is called a microbe or micro-organism, even though many of them are visible when clustered together (e.g., mould and algae filaments).[3] Microbes comprise two of the three superkingdoms,[4] Bacteria and Archaea, as well as single-celled eukaryotes (protists and yeasts) and viruses. Viruses, because they have no cells or metabolic function and require other organisms to replicate, tend to be placed in a grey zone between living

---

[1] Many thanks to our anonymous referee for very helpful advice and detailed comments; to Staffan Müller-Wille, Jane Calvert, and Jim Byrne for feedback; and also to the audiences at the first International Biohumanties Conference (Queensland, 2005) and the International Society for the History, Philosophy and Social Studies of Biology conference (Guelph, 2005).

[2] The word 'macrobe' has been used before (e.g., Postgate 1976; Dixon 1994), but the usage has not been widely adopted. We distance our use of it from any resonance with C. S. Lewis's in his book *That Hideous Strength* (1945), where 'macrobe' refers to a class of malign spirits.

[3] There are some bacteria visible as single cells, most notably *Thiomargarita nambiensis*, which is a recently discovered spherical sulphur bacterium with a diameter of 750 μm (Schulz and Jørgensen 2001).

[4] Superkingdoms or domains are the highest levels of taxa. The third superkingdom is Eukarya or Eukaryota, of which protists make up a substantial proportion (see the following note).

and non-living things (or organisms and chemicals), but their evolutionary history, involvement with prokaryotes and eukaryotes, and some surprising biological capacities (Luria et al. 1978; Raoult et al. 2004; Villarreal 2004a) make it difficult to dismiss them as non-living. We will focus on bacteria and archaea in this chapter, though many fascinating stories and philosophical complications could also be drawn from viruses and protists (e.g., Sapp 1987; Corliss 1999; Nanney 1999; Villarreal 2004b). Bacteria and archaea—until the 1970s considered under the single classification of bacteria—are now distinguished from each other by important differences in cell wall chemistry, metabolic pathways, and transcriptional and translational machinery (Woese and Fox 1977; Bell and Jackson 1998; Allers and Mevarech 2005).

Macrobes comprise the remainder of the Eukarya, the kingdoms Animalia (including the Metazoa), the Fungi, and the Plantae.[5] The distinction between macrobes and microbes is not entirely sharp: various social single-celled organisms, both prokaryotic and eukaryotic, such as the myxobacteria and cellular slime moulds, have long-recognized claims to multicellularity. We frame our argument round this distinction for two reasons, however. First, the macrobes are no more diverse a group than the microbes, so it is worth reflecting on why the latter seems so much more natural a concept than the former. But second, and this is the main thesis of this chapter, we believe that an indefensible focus on macrobes has distorted several basic aspects of our philosophical view of the biological world.

Micro-organisms dominate life on this planet, whether they are considered from an evolutionary or an ahistorical perspective. Evolutionarily, the first three billion years of life on the planet was primarily microbial, with the Cambrian explosion of modern multicellular metazoan body forms beginning only about 545 million years ago (Carroll 2001; Conway Morris 2003).[6] Microbes have far greater metabolic diversity than macrobes and can utilize a vast range of organic and inorganic energy sources via numerous metabolic pathways (Amend and Shock 2001). They are deeply implicated in the geochemical development of the planet, from the formation of ore deposits to the creation and maintenance of the oxic atmosphere on which macrobes depend (Kasting and Siefert 2002; Newman and Banfield 2002).[7] They can thrive in conditions that are intolerable for most plants and animals.[8] Prokaryotes flourish in temperatures

---

[5] A recent and less traditional division proposal for eukaryote kingdom divisions by Adl et al. (2005; Simpson and Rogers 2004) sets out six eukaryote kingdoms of which four are solely protists. Plants are part of Archaeplastida (which also contains single-celled algae) and animals merely a subset of Opisthokonta (which includes true fungi and several protist groups).

[6] Although there are numerous disputes about admissible data and interpretations, common dates for prokaryote origins are 3.8–3.5 billion years ago, followed by the first eukaryote micro-organisms 1.5–2.0 billion years later, with the first multicellular eukaryotes emerging around a billion years after that (see Carroll 2001; Nisbet and Sleep 2001; Waggoner 2001; Martin and Russell 2003; Kerr 2005).

[7] See Bryant (1991) and Lloyd (2004; also Biagini and Bernard 2000) for a discussion of whether there are any true obligate anaerobic eukaryotes.

[8] See http://www.nhm.ac.uk/research-curation/projects/euk-extreme/for an overview of eukaryote extremophiles (organisms that favour extreme environments), which are far fewer and more restricted than prokaryote extremophiles.

over $100°C$ and at least as low as $-20°C$. They colonize extremely acidic, alkaline, salty, metal-rich, radioactive, low-nutrient, and high-pressure environments. They can be found in high-altitude clouds and on human artefacts in space, several kilometres deep in the Earth's crust, as well as on and in every eukaryote organism alive or dead (Horikoshi and Grant 1998; Price 2000; Newman and Banfield 2002; Nee 2004). Just one gram of ordinary uncontaminated soil contains 10^{10} prokaryote cells which consist of as many as 8.3×10^6 species (Gans et al. 2005). Microbial species diversity in all of Earth's environments is only estimated but it exceeds all other life-forms, as do estimates of their global cell numbers.[9] The natural history of life on Earth was and always will be 'the age of bacteria' (Gould 1994).[10]

Even an exclusive interest in mammalian or human biology cannot justify ignoring microbes. There are estimated to be at least ten times as many microbial cells in our bodies as there are human somatic and germ cells (Savage 1977; Berg 1996),[11] as well as perhaps 100 times more genes (Xu and Gordon 2003). A full picture of the human organism sees it as a 'composite of many species and our genetic landscapes as an amalgam of genes embedded in our *Homo sapiens* genome and in the genomes of our affiliated microbial partners (the microbiome)' (Bäckhed et al. 2005: 1915; Lederberg, in Hooper and Gordon 2001).[12] Our microbiome functions as an additional 'multi-functional organ',[13] carrying out essential metabolic processes that we, in the narrow single-organism or single-genome sense, have never evolved for ourselves (Xu and Gordon 2003). Every eukaryote can, in fact, be seen as a superorganism,[14] composed of chromosomal and organellar genes and a multitude of prokaryote and viral symbionts (Lederberg 2000,[15] in Sapp 2003: 333). This multispecific interactionist perspective,

---

[9] A commonly accepted estimate is $4–6 \times 10^{30}$ prokaryote cells in all habitats (Whitman et al. 1998) and $\sim 4 \times 10^{30}$ viruses just in ocean waters (Suttle 2005). Even though microbial cells are usually much smaller than eukaryote cells, prokaryotes and viruses account for well over half the biomass on the planet (if the extracellular material of plants is excluded) and an even greater percentage (perhaps 90 per cent) if only the oceans are considered.

[10] Some important evolutionary biologists are entirely unconvinced by such arguments. Bacteria can claim only biochemical expertise and they occupy only leftover environments. Macrobes, particularly metazoans, are much more 'obviously' biologically interesting (e.g., Conway Morris 1998). Our discussion is trying to challenge all the assumptions in such arguments.

[11] Just the *E. coli* population in a single human is comparable to the entire human population (Staley 1997).

[12] The human genome (in the traditional, narrow, sense) appears to contain some microbial DNA (an initially exaggerated but still not clearly established amount) that was transferred directly into vertebrates rather than being inherited from non-vertebrates (Genereux and Logsdon 2003; Iyer et al. 2004), as well as an abundance of retroviral DNA (Griffiths 2001; Bromham 2002). A call for a research programme named 'the second human genome project' argues for an inventory and analysis of *all* the DNA in a human body in order to gain a better understanding of the system of interactions between humans and microbes (Relman and Falkow 2001).

[13] The metabolic activity of just the gastrointestinal bacteria in a human is believed to be equal to that of the liver—the most metabolically active organ in the human body (Berg 1996).

[14] Prokaryotes are similarly occupied by phages (bacterial viruses), which conduct a range of processes with the cellular machinery of their hosts.

[15] Lederberg's neologism for this community organism is 'symbiome'.

apart from fostering a far richer understanding of the biodiversity existing in the ecological niches provided by human bodies, should also lead to a better understanding of how human health, disease resistance, development,[16] and evolution have depended and continue to depend on interactions with microbes.

Despite the biological significance of microbes and the centrality of their study to some of the most exciting biology of recent decades (see below), the philosophy of biology has focused almost exclusively on multicellular life.[17] Decades of heated philosophical discussion about systematics and concepts of species have either not noticed the microbial world or found it convenient to dismiss it. It is rare, even in classification and species discussions, for philosophers to invoke microbial phenomena. Philosophical discussions of biodiversity produce only apologies for ignoring microbial biodiversity (e.g., Lee 2004). Even in philosophical debates about evolutionary processes, little notice is taken of microbes except when they are placed as backdrops to what is in truth merely 'the sideshow of metazoan evolution' (Sterelny and Griffiths 1999: 307).

In our conclusion we speculate briefly on why this has happened. Our main aim in this chapter, however, is to argue for an end to this myopia. We aim to show the radical revisions new understandings of microbes force upon some long-established ways of thinking in the philosophy of biology, specifically with respect to ontology, evolution, and taxonomy (including biodiversity). We will start with outlines of some recent developments in microbiological understandings of sensory capacities, communication processes, and gene transfer, and show how these present fundamental challenges to traditional ways of thinking about microbes as primitive individual cells.

Microbiology: a brief history

Early microbiology and the pure culture approach

The history of microbiology begins with the invention and development of the microscope in the late sixteenth and early seventeenth century, but it took a considerable time for any deep understanding of microbes to develop. Their long-hypothesized association with illness, fermentation, and food spoilage became an important topic of investigation in the late 1700s. In the early 1800s the stage was set for the first 'golden age' of microbiology with experimental tests of the spontaneous generation hypothesis, followed some decades later by the rejection of bacterial pleiomorphism (the thesis that all microbes could shift from their present form to any other and thus did not have

---

[16] See McFall-Ngai (2002) for a discussion of the influence of bacteria on animal development.

[17] There are, of course, exceptions to this tendency. Amongst them are Jan Sapp (1987, 2003), whose historical work on microbiology delves deeply into the philosophical issues of the discipline; Carol Cleland (Cleland and Copley 2005), who has written about alternative definitions of life with particular reference to prokaryotes; and Kim Sterelny (2004), who proposes the transmission of bacterial symbionts as an inheritance system. We are sure there must be others, but our general point—that detailed philosophical attention to microbes is rare—still stands.

constant effects or species characteristics) and the development of methods for the identification of numerous pathogens involved in disease and putrefaction (Drews 2000). The key method for such rapid success was formalized by Robert Koch,[18] whose 'postulates' of removing organisms from their complex communities and experimentally isolating the disease-causing process dominated microbiology for more than a century (despite the fact that alternative 'mixed culture' and ecological approaches were available).

Koch's postulates emphasized two things: microbes as static individuals of single-cell types from which pure cultures could be developed, and tightly controlled uniform environments that were laboratory creations (Penn and Dworkin 1976; Bull and Slater 1982a; Caldwell et al. 1997;[19] Shapiro and Dworkin 1997). Both these emphases have skewed microbiology, and only in very recent decades has alternative work on bacteria as dynamically interacting components of multicellular systems in a diverse range of non-laboratory environments taken hold.

Microbial biochemistry, genetics and molecular microbiology[20]

As bacteriology matured from medical and industrial applications into a biological discipline at the end of the nineteenth century, it increasingly used biochemical tools and analyses to understand the biological processes of bacteria and other microbes (Brown 1932; Summers 1991). The origins of modern biochemistry are, in fact, attributed to the isolation of fermentation enzymes from the microbe yeast in the late 1890s[21] (Kohler 1973; Manchester 2000). Biochemical investigation generated rapid growth of understanding of intracellular processes in bacteria and other microbes, but these insights were retained within the specialized domain of bacteriology and were of little interest to mainstream biology and genetics.

The transition from microbial biochemistry to molecular microbiology and microbial genetics took microbiology right into the centre of modern biology (Magasanik 1999). It was not until the 1940s that bacterial genetics was founded on the basis of the realization that bacteria have genetic material and that their study would enhance investigations of genotype–phenotype relations. This merger of biochemistry and genetics to study bacteria, viruses, and unicellular eukaryotes was responsible for the

---

[18] Better microscopes and microscopy, chemical studies of metabolism, developmental investigations of eukaryotic microbes, and better classification systems all contributed to this period of success. See Drews (2000) for a comprehensive and succinct overview.

[19] Penn and Dworkin (1976: 279–80) categorize these approaches as 'essentialist' (microbes as independent entities possessing intrinsic unchanging characteristics) in contrast to an 'interactive' or dynamic developmental understanding of microbes and microbial processes—an understanding available even in Koch's time.

[20] Our description in this subsection of the period from the 1900s to the 1970s passes over the development of several other techniques and technologies in microbiology, perhaps most notably the electron microscope.

[21] For an alternative history, see Wainwright (2003).

greatest triumphs of molecular genetics in the second half of the twentieth century and had a profound impact on a range of other disciplines from evolutionary biology to epidemiology (Luria 1947; Brock 1990). Major breakthroughs gained via microbial analysis included many of the most famous insights into DNA, RNA, and protein synthesis (e.g., Beadle and Tatum 1941; Luria and Delbrück 1943; Avery et al. 1944; Lederberg and Tatum 1946). In addition, the subsequent (1970s) development of recombinant DNA technology on the basis of knowledge of bacterial genetic systems generated a huge body of biological insight and biotechnological applications (Brock 1990).

Microbial sequencing and genomics

The experimental focus of molecular microbiology achieved enormous advances in microbiology and genetics, but it was painstaking work that continued to revolve around lab-cultured microbes. These approaches were still unable to produce data sufficient for a 'natural' classification system that would surpass the purely pragmatic one often considered unsatisfactory for a true microbial science (Stanier and Van Niel 1941; Stanier et al. 1957).

The advent of sequencing technology transformed microbiology's datasets and breadth of knowledge. The early sequencing revolution in microbiology was initiated by Carl Woese and his colleagues as an implementation of Zuckerkandl and Pauling's methodological outline of how to use molecules as fossils or documents of the evolutionary history of organisms.[22] Zuckerkandl and Pauling had proposed that the evolutionary trees inferred from the comparison of genetic or protein sequence data from different organisms would map onto those inferred from traditional phenotypic characters and thus converge upon real macroevolutionary patterns (Pauling and Zuckerkandl 1963; Zuckerkandl and Pauling 1965). They posited that a molecular clock was ticking in these sequences in the form of accumulated mutations, and because of its regularity, the time of evolutionary divergence in sequences could be calculated (within a margin of error) and ancestral relationships much more firmly established. Early molecular work on the phylogenetic relationships between microbes used a variety of amino acid and nucleotide sequences, but Woese settled on small subunit ribosomal RNA (SSU rRNA) and rDNA sequences, particularly the 16S gene, as the best 'molecular chronometers' because of their ubiquity, highly conserved structure, functional constancy, predictable rates of variation in different regions, and practical ease of sequencing (Woese and Fox 1977; Fox et al. 1980).

Woese's discovery of the archaea dramatically transformed biology's basic classificatory framework of life from two fundamental domains or superkingdoms (prokaryotes

---

[22] Molecular sequences had been used to infer evolutionary relationships since the 1950s (Olsen et al. 1994), but Zuckerkandl and Pauling gave such efforts a much-needed theoretical and analytical boost.

and eukaryotes[23]) to three, and cast new light on the origins and subsequent differentiation of biological lineages. Although disputed by many taxonomists, especially those outside microbiology (e.g., Mayr 1998[24]), Woese's work made more sense of molecular data and appeared finally to enable a 'natural' phylogenetic classification of bacteria instead of the prevailing phenetic approaches used—however reluctantly—as defaults (Olsen et al. 1986; Woese 1987; Woese et al. 1990).

The cumbersome methods and limited data of early microbial sequencing were rapidly overwhelmed by high-throughput whole-genome sequencing methods. The first microbial genome sequenced was that of *Haemophilus influenzae* in 1995 (Fleischmann et al. 1995), followed quickly by the smallest bacterial genome then known— *Mycoplasma genitalium* (Fraser et al. 1995)—and then the archaeal genome of *Methanococcus jannaschii* (Bult et al. 1996).[25] There are now more than 230 whole prokaryote genomes sequenced (with 370 in the pipeline, and over 1500 virus genome sequences)—more than twelve times the number of eukaryote genomes available (http://www.ncbi.nlm.nih.gov/genomes). The comparative work done with these sequences has been enormous and has enabled an increasingly complex understanding of gene function and evolution (Brown 2001; Ward and Fraser 2005). Genomic insights have illuminated enquiries into the transition from prokaryotes to eukaryotes, indicated the minimal genome required to support cellular life, and tracked pathogenic diversity over the course of a disease and virulence mechanisms across a range of species (Schoolnik 2001; Ward and Fraser 2005). Simultaneously, however, genomic data pointed to phylogenetic contradictions between the 16S and other genes used as markers of evolutionary history. The inconsistent stories such markers tell challenge the practice of equating the evolutionary history of organisms with the history of molecules—a challenge we will outline and explore in the section below on lateral gene transfer.

Microbial ecology and environmental microbiology[26]

Ecological studies of microbes (historically not part of general ecology, but a subfield of microbiology) have been marginalized thoughout most of the history of microbiology by the pure culture paradigm and the lack of effective alternative methods (Brock 1966; Atlas and Bartha 1998; Costerton 2004). Early articulations of microbial ecology are attributed to Russian soil microbiologist, Sergei Winogradsky, and the founder of the famous Delft school of microbiology, M. W. Beijerinck, at the end of the

---

[23] We continue using the convenient label of prokaryote throughout this chapter because it does usefully describe both archaea and bacteria in terms of cellular and genomic size and organization. See Walsh and Doolittle (2005) for a better argument along these lines.

[24] 'It must be remembered,' sniffs Mayr (1998: 9721), 'that Woese was not trained as a biologist and quite naturally does not have an extensive familiarity with the principles of classification.'

[25] Since renamed *Methanocaldococcus jannaschii*.

[26] Microbial ecology is sometimes described as the 'basic' study of microbial interactions in environments, and environmental microbiology as their 'applied' study especially in relation to their effects on humans (Maier et al. 2000).

nineteenth century. It was not until the late 1960s, however, with the availability of a range of new molecular methods and a revived ecological sensibility that microbial ecology began to flourish as a subfield that proclaimed the limitations of studying bacteria as isolated individuals in artificial environments (Brock 1987; Caldwell and Costerton 1996). These limitations were highlighted by the 'great plate count anomaly', which drew attention to the several orders of magnitude of discrepancy between microscopic cell counts of environmental samples and plate counts of bacteria cultured from those samples (Cutler and Crump 1935; Jannasch and Jones 1959; Staley and Konopka 1985). Once these discrepancies were no longer attributed to observed cells being 'non-viable', they led to estimates that as many as 99 per cent of prokaryotes could not be observed or studied further because their culture evaded all available techniques (Amann et al. 1995).[27] Molecular microbial ecology is increasingly integrated with biogeochemical approaches that study microbial interactions with the chemistry and geology of ecosystems (Newman and Banfield 2002; Croal et al. 2004; Doney et al. 2004) and has been further enhanced by the development of imaging technologies that enable *in situ* observation at the cellular and subcellular level (Brehm-Stecher and Johnson 2004; Daims et al. 2006).

This environmental turn has also occurred within microbial genomics itself, which has extended its approach beyond laboratory cultures of micro-organisms to DNA extracted directly from natural environments (Stahl et al. 1985; Olsen et al. 1986; Amann et al. 1995). While this move out of the laboratory vastly expanded the scope of the data collected as well as understandings of biodiversity and evolution (Pace 1997), the continued focus on particular genes as phylogenetic markers still gave limited assessments of diversity (Dykhuizen 1998; Schloss and Handelsman 2004) and did not provide much information about the physiological or ecological characteristics of the organisms (Staley and Gosink 1999; Brune and Friedrich 2000; DeLong and Pace 2001; Rodríguez-Valera 2002).

A potential remedy to these shortfalls lies in the development of metagenomics, an approach in which the DNA of entire microbial communities in their natural environments (the metagenome) is sequenced and screened and then further analysed in attempts to understand functional interactions and evolutionary relationships (Handelsman et al. 1998; DeLong 2002a; Handelsman 2004; Riesenfeld et al. 2004; Rodríguez-Valera 2004).[28] These studies are not only discovering new genes and

---

[27] These observations do not mean the abandonment of culturing, and many new culturing techniques are addressing microbes previously thought to be unculturable in order to supplement molecular and other ecological investigations (Joseph et al. 2003; Leadbetter 2003).

[28] Sampled environments include ocean sediments (Breitbart et al. 2004), the human gut (Breitbart et al. 2003), the human oral cavity (Diaz-Torres et al. 2003) and drinking-water valves (Schmeisser et al. 2003). The most comprehensive metagenomic studies have shotgun-sequenced all the DNA in an environmental sample—both from environments with low species densities (Tyson et al. 2004) as well as from considerably more complex oceanic communities (Venter et al. 2004). However, the full metagenome sequence of the most complex and diverse communities (especially in soils) is still beyond the reach of current technologies because of the size and complexity of the communal genome, which requires formidably high numbers of

strains of prokaryotes and viruses, but are also revealing wholly unanticipated functions and mechanisms such as photobiology in oceanic bacteria (Béja et al. 2000; DeLong 2005) and the molecular complexities of symbiotic relationships (Kitano and Oda 2006a). Metagenomics is still at a very early stage of constructing inventories of microbiodiversity, however, and it will need to integrate many other approaches in order to understand the complexity of microbial interactions in their diverse environments.

Prokaryotes as multicellular organisms

The tendency for other disciplines to ignore or marginalize microbes and microbiology may be because of assumptions that prokaryotes are simple separate cells that are behaviourally limited and the equivalent of evolutionary fossils of life's primitive beginnings. A great deal of recent and older evidence can be marshalled in support of the very opposite conclusion: that bacteria are complexly organized multicellular entities with sophisticated and efficient behavioural repertoires (many elements of which are not available to multicellular eukaryotes) and that microbes are, in fact, the evolutionary sophisticates who exhibit far more capacity to adapt to dramatic environmental change than does multicellular eukaryotic life.

A growing group of microbiologists now argue that to study prokaryotes exclusively as unicellular organisms is highly misleading (Slater and Bull 1978; Caldwell and Costerton 1996; Shapiro 1998; Davey and O'Toole 2000; Kolenbrander 2000). Prokaryotes rarely live in isolation but in a variety of communal organizations that often include macrobes. Microbes engage in a range of associations with other organisms, some of which are competitive or parasitic, and others of which are commensalisms (benefiting one partner) or mutualisms that benefit all involved (Bull and Slater 1982b; Wimpenny 2000). Many of these may be loose or temporary, whereas others are more stable and obligate (e.g., endosymbiont or intracellular symbiotic relationships[29]).

Everyone may agree that there are intercellular relationships and loose communities, but the argument is about whether such interactions justify the postulation of multicellularity (e.g., Jefferson 2004). Traditional definitions of multicellularity emphasize task sharing by tissue differentiation and the permanent alteration of gene expression

clones and sequence coverage to accurately represent the genetic composition of the community (Riesenfeld et al. 2004). In addition, the harsh process of extracting DNA from the soil sample breaks the DNA into very small fragments which may be unsuitable for studies that are interested in networks of genes rather than single genes (Handelsman et al. 1998; Daniel 2004).

[29] Endosymbionts such as *Buchnera* in aphids and *Wolbachia* in numerous insects and other invertebrates are so integrated into their partner's cells that their genomes are greatly reduced, partly by loss and partly as genes are transferred from the symbiont's genome to the host's nucleus and the gene products are transported back to the endosymbiont (Andersson 2000; Douglas and Raven 2003). They may eventually become organelles of the host cell as did the proteobacteria that are now the mitochondria and the cyanobacteria that became the chloroplasts.

patterns, thereby excluding non-macrobial forms of cellular organization. However, a more encompassing definition is suggested by the molecular and cellular study of microbial communities. These communities exhibit well-defined cell organization that includes specialized cell-to-cell interactions, the suppression of cellular autonomy and competition, and cooperative behaviour that encompasses reproduction (Carlile 1980; Kaiser 2001; Keim et al. 2004a).

By working together as functional units, microbes can effect a coordinated division of labour into zones of differentiated cell types that enables them to access a greater variety of energy sources, habitats, protection, and other collective survival strategies (Gray 1997; Shapiro and Dworkin 1997; Crespi 2001; Webb et al. 2003). Many of these are activities that individual microbes are unable to accomplish and which are, in fact, often achieved at the expense of 'altruistic' individual micro-organisms.[30] In the most common community structure of biofilms, individual cells usually show lower growth rates than do free-living individuals (Kreft 2004). The 'suicidal' programmed cell death or autolysis (self-disintegration) of individual cells appears to directly benefit the group (Dworkin 1996; Lewis 2000; Ameison 2002; Rice and Bayles 2003; Velicer 2003).[31] A great variety of communal strategies has been observed and experimented on in single-taxon populations, but the most common forms of complex cooperation are found in mixed (multi-taxa) consortia of prokaryotes and other microbes.[32] Their communal activities range from carrying out coordinated cascades of metabolic processes to the regulation of host–parasite interaction and environmental modification (Dworkin 1997; Shapiro 1998; Hooper et al. 1998, 2001; Kolenbrander 2000; Crespi 2001). Recent decades of studies of the collective behaviours involved in biofilm formation, chemotaxis, quorum sensing, and genetic transfer give a great deal of support to the multicellular description of microbial communities.

Biofilms

Biofilms are the favoured lifestyle of most prokaryotes and are found in all microbial environments with surfaces, nutrients, and water, from fast-flowing hot springs to catheters. They are often visible and may contain many millions of cells. Biofilms are constructed by micro-organisms exuding and surrounding themselves with slimy biosynthetic polymers. Formation occurs in clear stages of adhesion, attachment, maturation, and detachment (Costerton et al. 1995; Stoodley et al. 2002). Different environmental conditions influence a variety of biofilm architectures, and other

---

[30] Cheater controls are obvious objects of investigation to understand the fine-tuning of cooperation in prokaryote communities and there is some evidence to indicate they exist (Velicer 2003; Travisano and Velicer 2004), although this interpretation of the data is still somewhat controversial.

[31] Even apparently non-cooperative acts of cannibalism appear to be beneficial for the group, because some components of the group are digesting other components in order to keep the whole alive (Engelberg-Kulka and Hazan 2003).

[32] One reason so few prokaryotes have been cultured may be because laboratory environments provide only nutrients and not signals from community members (Kaeberlein et al. 2002).

materials and new species are incorporated into (or break away from) the biofilm as it develops. The prokaryotes in biofilms express genes in patterns that are very different from free-floating (planktonic) microbes, and gene expression in a biofilm changes at each stage of its development (Stoodley et al. 2002).

Living in a biofilm prevents the annihilation of bacterial communities in adverse conditions, even those of heavy and repeated antibiotic therapy (Davey and O'Toole 2000; Wimpenny, 2000; Stewart and Costerton 2001).[33] Biofilms enable close inter-cellular contact that involves the exchange of many different molecules and allows greater metabolic diversity, as in the multistage digestive processes carried out by prokaryotes in the bovine rumen, as well as genetic transmission between cells and the rapid acquisition of antibiotic-resistance or virulence genes (Watnick and Kolter 2000). Although biofilms have been studied intensively since the late 1970s, it is only in recent years that researchers have emphasized their biological aspects (over their physico-chemical) and begun to conceptualize biofilm formation as a multicellular developmental process (Davies 2000; O'Toole et al. 2000; Stewart and Costerton 2001). It is a more flexible form of development than metazoan development because although biofilm formation is directional, it is strongly influenced by environmental conditions, and is reversible and not locked into a rigid sequential process as is metazoan development (Parsek and Fuqua 2004; see note 36).

Chemotaxis

Chemotaxis is the directed movement of cells to or away from chemical stimuli. First studied in the late nineteenth century, its molecular mechanisms were not understood until the late 1960s (Adler 1969; Eisenbach 2005). 'Bacterial' (including archaeal)[34] chemotaxis is achieved by a two-component signal transduction system that involves transmembrane receptors on the prokaryote cell. These respond to subtle changes in environmental chemicals and regulate the motor activity and type of movement, thereby altering the cell's direction (Falke et al. 1997). Moreover, chemotaxis is a social process in which prokaryotes are attracted by the chemicals secreted by neighbours. The assemblies they then form enable and enhance further social interactions associated with biofilm formation, communication, and genetic exchange (Park et al. 2003).

A feedback methylation system (in which the methylation states of the receptors are modulated by enzymes affected by stimulus response) allows the cells to adapt to the initial stimulus. This process is frequently analogized to memory because it allows cells

---

[33] Some researchers estimate that prokaryotes in biofilms have 1000 times more resistance to antibiotics than do planktonic prokaryotes (Davey and O'Toole 2000).

[34] Different chemotaxis systems operate in a great variety of prokaryote and eukaryote cells. The most well-studied prokaryote system is that of *E. coli*, but *Bacillus subtilis* and *Rhodobacter sphaeroides* systems are also important as models (Wadhams and Armitage 2004). Eukaryote chemotaxis is often investigated in *Dictyostelium discoides* (cellular slime mould) and neutrophils (mammalian cells that track down infections) (van Haastert and Devreotes 2004).

to compare their present situation with the past and respond accordingly (Koshland 1979; Falke et al. 1997; Grebe and Stock 1998).[35] The sophistication of these chemotaxis receptor systems has led some researchers to argue that they are 'nanobrains'—tiny organs with enormous computational power that use sensory information to control motor activity (Webre et al. 2003; Baker et al. 2005).

Quorum sensing

Quorum sensing is a form of communication-based cooperation that is often called 'chemical language' and analogized to hormonal communication between metazoan cells (Bassler 2002; Shiner et al. 2005). Quorum sensing can only be carried out in communities because it is population-density dependent. It involves the release of small signalling molecules (called 'autoinducers'), through which cells are able to assess population density.[36] When it is high and the molecules reach a threshold of concentration, they interact with proteins that regulate gene expression thereby activating collective behaviours from biofilm formation to the production of virulence or bioluminescence (Dunny and Winans 1999; Miller and Bassler 2001; Henke and Bassler 2004). The behaviour of individual cells thus reflects regulation at a multicellular level (Gray 1997) and indicates 'primordial social intelligence' (Ben-Jacob et al. 2000). The communities in which quorum sensing operates include not only prokaryote species but also eukaryote hosts, where interactions may involve the bi-directional modulation of gene expression in host and commensals (Brown and Johnstone 2001; Federle and Bassler 2003; Shiner et al. 2005; Visick and Fuqua 2005).

Lateral gene transfer

The genome itself participates in the multicellular life of prokaryote communities through processes of genetic transfer between cells—perhaps the 'ultimate interaction' between organisms in communities (Dworkin 1997: 10; Shapiro 1997). Lateral or horizontal gene transfer (LGT or HGT) involves the transfer of diversely packaged genetic material from one organism to another most commonly by conjugation, transduction, or transformation. Conjugation is the transfer of DNA that involves cell-to-cell contact between organisms and the transfer of a mobile genetic element (a conjugative plasmid or transposon); transduction is the transport of DNA from one organism to another by bacteriophages; transformation is the direct uptake of free environmental DNA by a 'competent' organism into its genome (Ehlers 2000;

---

[35] For other instances of memory in prokaryotes and phage, see Casadesús and D'Ari (2002).

[36] There are three canonical quorum sensing systems or circuits, which are discussed in detail in Miller and Bassler (2001). Two are used for intraspecies communication; the other for a wide range of interspecies communication (Federle and Bassler 2003). Many prokaryotes possess versions of more than one system (Henke and Bassler 2004). There is a little scepticism about whether quorum sensing is group communication or merely individual sensing of chemical diffusion (e.g., Redfield 2002) but the latter is a minority interpretation.

Bushman 2002; Thomas and Nielsen 2005). Competence is an induced state of ability to bind, import and recombine free DNA (Solomon and Grossman 1996)—an ability that is at least partly regulated by extracellular chemical signals between organisms in communities (Dunny and Leonard 1997; Lee and Morrison 1999; Peterson et al. 2004).

The transfer of genetic material enables communities to adapt rapidly to changing environments (Reanney et al. 1982). Laterally acquired advantages include novel capacities with which to take over new environments, new metabolic functions, resistance to antibiotics, and increased pathogenic virulence (Levin and Bergstrom 2000; Ochman et al. 2000; Feil and Spratt 2001; Sonea and Mathieu 2001). The genes for the entire chemotaxis system, for example, were probably transferred as one unit between bacteria and archaea (Faguy and Jarrell 1999; Aravind et al. 2003). Current research indicates that genetic transfer by conjugation and transformation is much more frequent and efficient in biofilms than amongst planktonic bacteria (Hausner and Wuertz 1999; Molin and Tolker-Nielsen 2003; see Ehlers 2000 for methodological limitations of these studies). Genetic transfer and its mechanisms also appear to have positive effects on the development and stability of biofilms, meaning it is a communal activity that has both short-term lifestyle benefits as well as longer-term evolutionary benefits (Molin and Tolker-Nielsen 2003).

The capacity for lateral gene transfer in communities has many implications for evolutionary theory and taxonomic practice (discussed below), but the main point we are making here is that the 'one-organism, one-genome' equation is insufficient to describe the genetic constitution of microbial communities. The concept of the metagenome is based on this extended understanding of a community genome as a resource that can be drawn on by the community organism—the metaorganism or superorganism. This genomic perspective backs up the notion of microbial communities as multicellular organisms.

The body of evidence above not only challenges the unicellular perspective in microbiology itself but also raises important issues for the philosophy of biology, especially in relation to how philosophers understand biological individuality, evolutionary transitions and processes, and the concept of species. We will examine each of these areas from the microbiological platform we built above, and outline some issues of major relevance to philosophers of biology.

Ontology

The central ontological categories for traditional philosophy of biology have been the individual organism and the lineage, the latter sometimes extended to include the more controversial notion of species as individuals (Hull 1987b). Populations, whether sexually or asexually reproducing, have been conceived of as constructed out of individuals. Individual microbes have an unproblematic status in microbiology as well but, as explained above, the notion of community in its various forms has also deeply informed the discipline's theory and research.

If communities are self-organizing entities that operate as functional units and are more than simple aggregations of individuals (Andrews 1998; Ben-Jacob et al. 2000; Kolenbrander 2000), they can only be excluded from multicellular status if the definition of multicellularity is closely based on knowledge of multicellular eukaryotes. Broader definitions (mentioned above) are able to include groups of interacting microbes, of one or many taxa, including sometimes eukaryote hosts (Dworkin 1997). This, in turn, suggests that rather than see macrobes as a 'higher' level of biological organization, we should view macrobes and microbial communities as constituting alternative strategies for coordinating the activities of multiple differentiated cells.

Philosophers may want to ask some basic questions about the ontological status of microbial communities, particularly whether the community organism is more fundamental than the individual organism. Macrobial ecologists have tended to shy away from any notion of communities having functional properties analogous to organisms because clear spatial and temporal boundaries appear to exist only at the level of the individual organism (Looijen 1998; Parker 2004). Communities of plants, for example, do not typically appear to have firm boundaries or discrete forms due to the continuous nature of the environmental conditions that shape them. Consequently, communities are defined very loosely, usually as groups of populations in a place the ecologist happens to be studying rather than as biological individuals (Underwood 1996; Collins 2003). The notion that communities might have emergent properties that individuals do not is explicitly rejected by many ecologists (e.g., Underwood 1996). This 'boundary problem' for communities of plants and animals is presumed to be even worse for microbes, which are generally considered to be globally distributed and environmental will-o'-the-wisps (Finlay and Clarke 1999).

A first response to these doubts might be that clear boundaries are not necessarily connected to ontological fundamentality. Philosophers of biology willing to accept the thesis of species as individuals in conjunction with even limited hybridity should have no difficulty acknowledging this point. Second, the biofilms that are the preferred lifestyle of prokaryotes make possible their study as bounded multicellular entities as well as contradicting common conceptions of bacteria as free-floating individuals in occasional and highly impermanent contact. Finally, there is a large body of empirical work which challenges standard views of boundaries because it reverses expectations about organismal integrity and microbial ubiquity. In regard to the former, the omnipresence of genetic exchange in microbial communities shows organism boundaries to be much more permeable than might have been thought. For the latter, although it has long been presumed that 'everything is everywhere' in relation to microbial distribution, meaning that microbes have *no* biogeography (Finlay and Clarke 1999), recent studies taking a more extensive and finely resolved genomic perspective have found that communities of bacteria and archaea in hot springs and soils, for example, do actually have geographic limits at the strain level (Cho and Tiedje 2000; Whitaker et al. 2003; Papke and Ward 2004).

Communities may not possess the level of physiological integrity that individual (monogenomic) organisms do, but the recent research that we have outlined clearly indicates that they are much more than just individuals who happen to have blundered together. It seems more promising to conceptualize microbial communities as individuals with somewhat indeterminate boundaries that have some 'un-organism-like properties' (McShea 2004) while still possessing many organismal (or proto-organismal) character-istics. If the community system is posited as more ontologically fundamental than the individual components, then its causal properties will have detectable and important influences on the constituents. The avenues of research mentioned above concerned with understanding the multicellularity of bacterial communities appear to demonstrate such 'downward' causation, and at the least provide strong reasons for pursuing this issue further.

Evolution

Evolution has, for the most part, been about microbes, and many of the most fundamental evolutionary questions revolve around unicellular life: how life began, how prokaryotes evolved to eukaryotes, and how transitions from unicellular to multicellular life were accomplished. The philosophy of biology is, of course, interest-ed in these issues but primarily as a background to its evolutionary focus on multicel-lular organisms. The neglect of microbes can be particularly striking in one of the most exciting topics in philosophy of evolution, evolutionary developmental biology or 'evo-devo'. For example, Robert (2004: 34), in a pioneering philosophical treatment of ontogeny, writes: 'Development is what distinguishes biological systems from other sorts of systems, and it is the material source of evolutionary change.' Since microbes, though they go through cycles of internal reorganization do not, in the macrobial sense, develop at all,[37] it would appear that on this view they are not biological systems

---

[37] Prokaryote development has been intensively researched for over two decades (Figge and Gober 2003; Kroos and Maddock 2003) but it is about something very different from eukaryote multicellular develop-ment, which is how development is almost invariably conceived outside microbiology. Eukaryote develop-ment involves the differentiation of cell lineages leading to tissues with specialized physiological functions, morphological complexity and growth, with sexual reproduction as the main source of genetic diversity. Prokaryote development is primarily environmentally initiated (although it can also be an internally cued stage in a cell division cycle, such as in *Caulobacter*), and is usually uncoupled from single-cell growth. Genetic diversity is obtained via a number of other strategies (see above). A commonly used definition of prokaryote development is 'a substantial change in form as well as function in the life cycle of the cell' (Dworkin 1985: 3), which may take either unicellular or multicellular forms (as in myxobacteria aggregations). There are four main categories or cycles of prokaryote development: resting cells, complementary cell types, dispersal cells, and symbiotic development (Shimkets and Brun 2000). Individual cells can still leave developing multicellu-lar units and enjoy their own singular fate rather than the developmental fate of the multicellular group (Shimkets 1999). Prokaryote development therefore involves different organizational strategies, different selective pressures, and much more genetic and biological diversity than does eukaryotic multicellular

and apparently could not have evolved. Of course, as we have been arguing, it might turn out that individual microbes are not the best way to understand microbial organization and development, and it may be that only as communities could they have evolved. But it is doubtful whether communities have exactly the kind of developmental properties that the eukaryotic multicellular vision requires, and it is certain that Robert did not intend to describe the development of prokaryote communities. Surely it reflects an oversight, but one we think is very telling of the tendency for philosophy of biology to focus exclusively on macrobes. It also nicely illustrates how evolutionary microbiology can enrich and challenge standard evolutionary theory.

Units of selection and evolutionary transitions

A long-standing debate in the philosophy of biology has been about the units and levels of biological organization on which selection acts. A key divide has been whether selection operates in a privileged way on genes and organisms, or whether it also operates at group and other levels (Brandon and Burian 1984; Sober and Wilson 1994; Wilson 1997). Although considerable conceptual progress has been made over the last two decades (Brandon 1999; Lloyd 2000; Okasha 2003), prokaryote communities have hardly ever been used as illustrations or objects of analysis in the debate.[38] One of the obvious questions the discussion of community function raises is whether these apparently co-evolved relationships and community-level properties are selected for, or whether their existence can be fully accounted for by selection at the individual gene/organism level (Collins 2003; Whitham et al. 2003). Can such entities as prokaryote communities be conceived of as units of selection? There is experimental evidence that supports group selection in prokaryote communities (e.g., Queller 2004).[39] Is there competing selection of individual cells and genes that threatens the cooperation achieved at the community level? If we accept the arguments for microbial communities as biological individuals, then it is a plausible speculation that systems involving commensal microbes and sometimes macrobes could be considered to be *the* standard unit of selection. Community-level accounts of selection may even provide the key to identifying the mechanisms that allowed a hierarchy of biological organization to evolve in the first place (Okasha 2003, 2004).

---

development (Shimkets and Brun 2000). There are also some phenomena common to both, however, and these include self-recognition, spatially directed growth, specialized cell differentiation, intercellular signalling, and programmed cell death (Shimkets 1999).

[38] See, for example, the table in Goodnight and Stevens (1997). Parasite populations are popular illustrative examples, but they are usually metazoan parasites (e.g., Sober and Wilson 1994). The myxoma virus infection of rabbits used in the earlier stages of the debate (e.g., Lloyd 1989) is an exception to the focus on multicellular organisms.

[39] Queller (2004) reports on the experimental results of Griffin et al. (2004), who find that the best interpretation of social behaviour in *Pseudomonas aeruginosa* is group selection, not kin selection.

One of the great benefits of attention to microbes is that it draws attention to the problem, easily overlooked when the transition to multicellularity is interpreted as self-evident progress, of *why* multicellularity evolved at all. Explanations of the evolution of multicellularity tend to take it for granted that eukaryotic multicellularity is obviously superior, so the discussion tends to be about *how* it evolved. For the multicellular organism to have become an individual in its own right (as opposed to an aggregation of cells), selfish tendencies of single cells would have had to have been regulated and cooperative interactions promoted (Buss 1987; Michod 1997a, 1997b; Okasha 2004). Maynard Smith and Szathmáry's (1995) account of major evolutionary transitions specifies that entities that replicated independently before the transition can replicate only as part of the larger whole (or next level of organization) afterwards. Okasha (2003) and Michod (1997a, 1997b) make this point more subtly and argue that the transition to multicellularity would begin on the basis of group fitness equalling average (lower-level) individual fitness, but that higher-level fitness would eventually decouple from component fitness as the transition proceeded.

It may be that while this point is basically correct, its formulation still suffers from a residually macrobial perspective. The components of an integrated community would not be capable of independent replication, not because replication had become a specialized function but because the various components could only function cooperatively. Sequestered reproduction or the specialization of reproductive cells grounds one very interesting form of cellular cooperation, but perhaps we should avoid thinking of it as the only possible form. If there is something incoherent about the idea of an organism reproducing through the independent reproduction and subsequent reintegration of its parts, it is an incoherence that needs to be demonstrated.

The preceding point can be seen as part of the broader project of rethinking much more generally the possibility for aggregation of cells into more complex structures. We are inclined to speculate that macrobial multicellularity (like organelles in eukaryote microbes) is just a frozen, less flexible, obligate analogue of bacterial multicellularity. Prokaryote cell differentiations can de-differentiate whereas metazoan multicellularity is irreversible. For example, in eukaryote multicellularity aerobic metabolism is essential because this form of multicellularity has high energy demands that cannot be met by anaerobic means (Fenchel 1996). Prokaryote multicellularity, however, is an energy-efficient form and metabolic diversity is not sacrificed. The eukaryote multicellularity we commonly think about had to be selected for, to be sure, but in the long run of evolution it is likely to be much less well able to adapt to major changes in environmental conditions, such as atmosphere. Or, if it does adapt, this may be very much dependent on the more diverse capacities of microbial commensals. Microbes have a proven track record of living in a world devoid of eukaryotes, but multicellular eukaryotes are unlikely to be able to manage in a microbeless ecosphere.

In many ways, microbial communities have experienced a great deal more evolutionary and ecological success than macrobes. No doubt the key to understanding how macrobes evolved at all is to locate more clearly what it is that they do better than microbial communities (unless, indeed, we should see macrobes in a neo-Dawkinsian way, as primarily vehicles for the billions of microbes that live in the many niches macrobes provide, designed to transport them to especially large and attractive energy resources).[40]

At any rate, we need to resist the temptation to see microbes as primitive precursors of macrobes and the transition to multicellularity as representing unambiguous progress. Rather, we must face the fact that much of our evolutionary theory is grounded in features peculiar to macrobes and has questionable relevance to microbial evolution—which is to say, by far the largest part of all evolution. It is also, in a real sense, the most important part of evolutionary history. For it is clear that the basic machinery of life evolved in microbes prior to what might, in relative terms, be seen as no more than a severe narrowing and slight diversification of the applications of that chemistry in macrobes. And, of course, it is only due to ancient prokaryotic mergers that there are eukaryotes at all (Margulis 1970).[41]

Evolutionary process and pattern

As important as these questions about major evolutionary transitions is the need to reflect on the mechanisms by which microbial communities adapt and evolve. The philosophy of evolutionary biology must take account of the rapidly growing body of work in microbial phylogeny on horizontal or lateral gene transfer. The capacity for resource exchange that LGT allows has been described as a distributed genome or a genetic free market (Sonea and Mathieu 2001)—a global resource too big for single cells but accessible when populations find ecological reasons to acquire DNA for new functions. A strong interpretation of gene transfer means that individual genomes are ephemeral entities fleetingly maintained 'by the vagaries of selection and chance', and taxa are only an 'epiphenomenon of differential barriers' (environmental, geographical, and biological) to lateral gene transfer (Charlebois et al. 2003).

The findings of comparative evolutionary genomics have raised enormous problems for the dominant eukaryo-centric paradigm of vertical inheritance and mutation-driven species divisions that give rise to a single Tree of Life (Doolittle 1999, 2002; Stahl and Tiedje 2002; Gogarten and Townsend 2005; O'Malley and Boucher 2005). While comparative genomic studies confirmed the distinctiveness of the archaea, they also complicated the simpler stories told by popular single-gene phylogenetic markers (such as the 16S ribosomal gene) by revealing huge amounts of atypical DNA in numerous genomes. Many genomic sequences do not match organismal or species patterns due to the complex histories of gene exchange. Frequent transfers result in

---

[40] Bonner (1998) points out that it is likely early multicellular clusters may have had no adaptive advantages.

[41] See Martin and Russell (2003) for an evaluation of competing hypotheses on eukaryote origins, and McFall-Ngai (2001) for an argument that symbiosis with microbial communities has been a key factor throughout the evolution of multicellular organisms.

mosaic genomes which consist of genetic contributions from many sources, even phylogenetically distant ones (Koonin et al. 2001; Doolittle et al. 2003; Lawrence and Hendrickson 2003). This lack of a unilinear history to genomes has inspired a number of methods that attempt to capture not only vertical lines of descent (as bifurcating tree branches) but also the web-like complexity of lateral movement between lineages (e.g., Huson 1998; Bryant and Moulton 2004).[42]

Microbial populations exhibit much more rapid rates of evolutionary change than do their macrobial equivalents, the variety of dynamics and mechanisms of evolution is more diverse, and extinction means something quite different if indeed it has any relevance at all to microbes (Staley 1997; Lawrence 2002; Stahl and Tiedje 2002; Weinbauer and Rassoulzadegan 2004; Myers et al. 2006). It seems likely that the biologically significant loss in a microbial context would be something like a metabolic capacity rather than a particular microbial strain. But given the possibility of a wide distribution of genomic resources underlying these capacities, such extinction may be an improbable event. If so, then extinction, which plays a major role in standard models of macroevolution, is irrelevant for theorizing the evolution of microbes.

Most importantly, the genetically isolated lineage, often conceived of as the funda-mental unit of evolutionary theory, may have no real analogue in the microbial world. It might be possible in principle to construct evolutionary models in which microbial clones play a similar role to the familiar macrobial lineages. But even apart from the great diversity of clonal structure exhibited by different microbial taxa, there are some serious difficulties with such models. The most obvious is timescale. Microbial clones have lifespans of hours or days rather than the thousands of years typical of macrobial lineages. This suggests a need for higher level models if any sense is to be made of long-term evolutionary change. It further needs to be decided how the beginning and end of a clone are to be defined for this purpose, especially in light of a large body of evidence that shows little true or enduring clonality in most bacterial populations (Maynard Smith et al. 1993, 2000). The prevalence of mobile genetic elements moving between microbial units again points to a focus on larger units within which these movements take place.

This point suggests a slightly different formulation of the question raised earlier about the boundaries of communities. If it turns out that the lateral circulation of genetic material takes place within reasonably clearly delineated microbial commu-nities, it may be useful to consider these as units of selection. Surely such relative isolation will apply to communities defined by their residence in, for example, a particular human gut. Whether the same applies to aquatic bacteria, say, is another matter. If not, either microbial evolution is limited to more peripheral, isolated

---

[42] The debate continues about whether the vertical lines in molecular phylogenies of prokaryotes are overwhelmed by lateral lines. Some recent studies have managed to recover an approximate 16S-defined tree structure from very large datasets (e.g., Beiko et al. 2005).

environments or, more likely, we will need to expand on traditional macrobial models in search of an adequate understanding of microbial evolution.

Microbial genomics and metagenomics have evolutionary implications that reach into the most basic representations of evolution since they make clear that most of life and its history cannot be simply configured as a tree-like pattern of evolutionary outcomes (Doolittle 2005). This realization makes yet further deep inroads into the philosophy of biology because of its extensive implications for microbial taxonomy, the units of taxonomy, and the philosophical appreciation of biodiversity.

Taxonomy and biodiversity

Taxonomy

Identifying categories of organisms is central to the task of understanding the diversity of past and present forms of life and the evolutionary relationships between them. While the philosophy of biology has often recognized prokaryote classification as a special case (e.g., Hull 1987a; Sterelny 1999; Wilkins 2003), it has paid the issues involved hardly any attention and continues to believe that evolutionarily defined categories of organisms can be represented as bifurcating lineages that compose a Tree of Life. A variety of concepts have been proposed to define the species that make up this tree, but all of them prove unsatisfactory when gene exchange and genomic heterogeneity are brought into the picture. Prokaryote taxa simply refuse to show the clear, consistently definable characteristics often associated with eukaryotic species and classification schemes (Roselló-Mora and Amann 2001). There is, of course, controversy over how sharp the species boundary is even in eukaryotes but to whatever extent it is a problem there, it is considerably worse in prokaryotes (Dupré 2002).

The early history of microbial classification is a struggle for the specificity of bacteria and the recognition that groups have inherent characteristics that distinguish them from other putative species groups (Cohn 1875, in Drews 2000). The key issue from a microbial genomics perspective is whether to think of prokaryote taxa as continua or as discrete clusters of species-specific genetic diversity (Lan and Reeves 2000; Doolittle 2002; Konstantinidis and Tiedje 2005). Although the biological species concept (BSC) has never found much purchase in microbial systematics because of its exclusion of asexual reproduction and difficulties in coping with gene transfer between evolutionarily distant lineages (Maynard Smith 1995; Cohan 2002; Dupré 2002), there is an active debate between microbiologists about what constitutes an appropriate evolutionary or phylogenetic definition (Roselló-Mora and Amann 2001). In its simplest form, this simply means species are defined by common ancestry. Usually, however, this basic concept is accompanied by assumptions about which molecules are more reliable bases of such phylogenetic inference, and ribosomal DNA sequence is generally considered to be the prime candidate for divulging 'natural relatedness groups, the phylogenetic divisions' (Hugenholtz et al. 1998; Ward 2002).

As we outlined above, the role of 16S rRNA gene sequence as the ideal phylogenetic marker has been undermined by conflicting genomic evidence, which has also damaged more generally the idea of a single true marker for micro-organismal evolutionary history. Other microbiologists emphasize the importance of ecological forces on populations, with 'ecotypes' (equivalent to strains) being the product of ecological (but not reproductive) divergence (Palys et al. 1997; Cohan 2002; Gevers et al. 2005). Pragmatists, generally more convinced of the extent and implications of gene exchange, use the word 'species' as a purely practical term that means 'assemblages of related organisms for which microbiologists have attached specific names rather than natural kinds' (Gogarten et al. 2002: 2226). These are 'species-like' entities (Rodríguez-Valera 2002) whose classifications are created by classifiers, not nature, and these must be constantly revised in light of new evidence and emerging inconsistencies.

Popular operational measures reflect the mixture of concepts and conceptual problems at work in microbial systematics. The currently predominant measure of where the boundary falls between prokaryote species is below a 70 per cent rate of DNA– DNA reassociation in hybridization tests of the total genomic DNA of two organisms (Dijkshoorn et al. 2000; Roselló-Mora and Amann 2001). This crude measure of genomic distance is commonly considered equivalent to 97 per cent rDNA identity. The first value was chosen because it appeared to map onto phenotypic clusters for no known evolutionary reasons; the second because it conveniently mapped onto the 70 per cent measure (Lan and Reeves 2000; Cohan 2002). Apart from the fact that both measures ignore apparently important genomic differences, there is no evolutionary reason why 70 per cent DNA–DNA similarity values should be a species boundary, nor for 16S genes to be considered adequate representatives of a species history (Palys et al. 1997; Boucher et al. 2001; Lan and Reeves 2001). Moreover, the correlation between DNA–DNA reassociation and 16S sequence varies in different genera, and it is well known that the 16S gene lumps together physiologically diverse strains (Staley and Gosink 1999; Kämpfer and Roselló-Mora 2004).

An influential proposal designed to overcome these problems is the quasi-official (American Society of Microbiology) species definition (Vandamme et al. 1996; Stackebrandt et al. 2002). It combines genomic, phylogenetic, and phenotypic approaches into a pragmatic and 'phylophenetic' (or 'polyphasic') taxonomic framework in which a species is 'a monophyletic and genomically coherent cluster of individual organisms that show a high degree of overall similarity with respect to many independent characteristics, and is diagnosable by a discriminating phenotypic property' (Roselló-Mora and Amann 2001: 59). In practice, however, any such practical species measure is still anchored phylogenetically by the 16S rRNA gene (Dijkshoorn et al. 2000; Young 2001) which is seen as a proxy for natural units and their boundaries, and helps overcome the discomfort of many microbial systematists with 'non-natural' classification concepts and methods (e.g., Ward 1998; Coenye et al. 2005).

Another operational measure with the aim of natural classification uses the concept of a 'core' genome. Although there were earlier hopes of finding a phylogenetically definitive universal core of genes common to all prokaryotes, current measures focus on pools of genes that determine 'properties characteristic of all members of a species' (metabolic, regulatory and cell-division genes) and are seldom transferred (Lan and Reeves 2000). Because there is presumed to be a barrier to the interspecific recombination of core genes, they reveal the evolutionary history of the species (Wertz et al. 2003). Core genes are contrasted to more variable 'auxiliary' genes which often enable niche adaptation but are unreliable as species indicators.[43] There is still, however, great difficulty in finding genes that provide core conserved functions but are not transferred (Boucher et al. 2001; Doolittle 2005; Saunders et al. 2005) and different patterns of variability and stability in genomes of different species may require a range of species-genomes concepts. The idea of a core genome may be capable of providing a definition of species, but is unlikely to ground a fully phylogenetic taxonomy given the prevalence of lateral gene transfer over deep time.

If, as is strongly suggested by the several lines of research outlined above, the individual microbe is not the fundamental ontological unit in microbiology, then it should be no surprise that attempts to find a division of individual microbes into natural kinds are doomed to failure. Microbiologists should be well prepared for the discovery that species-genomes or phylotypes (a taxon defined by a particular gene marker) fail to capture the way microbial life has organized itself or, indeed, that microbial life and evolution does not lend itself to a monistic, consistently applicable species concept that allows evolutionary history to be represented as one true Tree of Life.[44]

Many further questions remain in this area. Is there potential for a taxonomy of communities or community lineages, or do these entities have limited taxonomic significance because of their weak boundaries and evolutionary lability? Should genomic identity or functional role guide the classification of participants in community systems? Finally, if we let the idea of the communal genome as a dynamic community resource further undermine the notion of stable species boundaries, what are the implications for how we understand biodiversity?

Biodiversity

Microbial diversity is generally given short shrift by biodiversity studies and philosophers of biodiversity (Ehrlich and Wilson 1991; Loreau et al. 2001; Sarkar 2002; Oksanen and Pietarinen 2004; Nee 2005), mostly because of methodological and technical limitations. Microbiologists have long known that their understanding of microbial diversity has been restricted both by technology and by a health- or

---

[43] Together, these categories of genes make up the 'pan-genome' of a species, sometimes called the 'clade-specific metagenome' (Lawrence and Hendrickson 2005; Medini et al. 2005).

[44] As noted above, there is a question of how true this is for eukaryotes, but the problems for prokaryotes are surely more extreme.

agriculture-based bias towards pathogens. Microbes' enormous diversity of habitats, metabolic versatility, and physiological adaptability are still only beginning to be understood. Genomics-driven estimates have risen to as many as 10^7-10^{12} prokaryote 'species' (Dykhuizen 1998),[45] of which fewer than 36,000 are indicated by rRNA sequence analysis (Schloss and Handelsman 2004) and only 7,800 of those are named and described (Kämpfer and Rosselló-Moro 2004).[46]

Simple numerical comparisons of eukaryotic and prokaryotic diversity by species counts or estimates are inadequate for several reasons. As we have just seen, there are deep conceptual problems in defining the microbial species. If eukaryote species were designated by the same broad genomic hybridization criteria that prokaryote species are, then groups such as humans, chimpanzees, orang-utans, gibbons, baboons, and lemurs would all belong to the same species (Staley 1997). Environmental genomics is centrally concerned with escaping these limitations, although it still relies heavily on ribosomal gene sequence to do so. One of the early benefits anticipated for metagenomics is the contribution to a broader and deeper understanding of microbial diversity.

At present, broad studies of microbiodiversity are largely occupied by cataloguing exercises, but as the research deepens to include multilevel interactions and processes rather than things, the object of study could become biodiversity in the extended functional sense of how micro-organisms are involved in ecosystem processes such as resource use, decomposition, and nutrient cycling (Finlay et al. 1997; Loreau et al. 2001). Appropriate ecological assessments of biodiversity need to be able to take into account the variability of microbial populations as well as the relationship between community structure, biogeochemistry, and ecosystem function (O'Donnell et al. 1994; Stahl and Tiedje 2002; Ward 2002; Buckley 2004a). They also need to incorporate explanations of 'the tempo, mode and mechanisms of genome evolution and diversification' in relation to higher-order biological and ecological processes (DeLong 2004b: 26; Falkowski and de Vargas 2004) and obviously the findings of biogeographic patterns in the distribution of prokaryotes and other microbes (Martiny et al. 2006; see above) will be part of this analysis.

Clearly, these are not straightforward research programmes that will give simple answers about biodiversity, but they are aspirations towards understanding complex phenomena for which technology and tools of analysis are beginning to develop. As understanding of the role of microbial communities in ecosystem function grows, and microecological studies are integrated with macroecological, it is likely that philosophical and practical arguments for microbial conservation—not recognized at all in the philosophy of conservation—will also develop (Colwell 1997; Staley 1997). It remains to be seen whether we *should* be much concerned about microbial conservation. Our remarks above about extinction raise the question of whether there is any serious risk to

---

[45] Rough estimates of virus species posit ten times more of them than prokaryote species (Rohwer 2003).

[46] Versus over a million named plants and animals (Staley and Gosink 1999).

be evaluated. However, given the fundamental role of microbes in all life, it would be good to know how microbial diversity is affected by environmental changes already profoundly affecting macrobial biodiversity. Philosophical analysis could make important contributions to framing the questions that need to be asked.

Towards a more inclusive philosophy of biology

Even prior to recent developments stemming from the growth of genomic technology, philosophy of biology has been culpable in its failure to take serious account of the microbiological realm. Today this omission is inexcusable. The range of diverse and interconnected microbiological perspectives that we have outlined above have fundamental importance for how we understand life. These reconceptualizations are not just a background development but a major transformation in understanding that needs to be reflected in the philosophy of biology.

Finally, it might be worthwhile hazarding a guess as to why the philosophy of biology has been so willing to ignore microbes and microbiology. Candidate reasons could be the intractability of microbial analysis, ignorance, authority, invisibility, and a progressive view of evolutionary history. Intractability of analysis (difficulties in coming up with a natural classification system and measures of diversity) is an implausible answer, as it might just as easily have stimulated philosophical scrutiny. It is not a simple matter of ignorance either, because many philosophers of biology are at least aware enough to sweep microbes aside. Does philosophy of biology focus on metazoans simply because of some old and still unchallenged attributions of status to zoology and animals (over botany and plants as well)? An even more basic explanation could be a cognitive bias towards larger, more visible phenomena—the same reason Sean Nee (2004) gives for the public indifference to microbes. But philosophers have shown no reluctance to get involved in debates about the molecular minutiae of other biological findings, so this explanation is not compelling either. Similarly one might point to the rapid development of techniques and theoretical frameworks in microbiology as inhibiting factors, but this rapidity would not distinguish it from various other biological subfields, especially in molecular biology, with which philosophers have been quite willing to keep up to date.

Some scientists perceive 'an unspoken philosophy of "genomic supremacy"' (Relman and Falkow 2001: 206) that is accorded to more complex animals because of genome size and number of predicted genes. If this were strictly true, then cereals, amphibians, and some amoeba—whose genomes are up to 200 times larger than those of humans (Gregory 2001)—would be ranked higher and receive more philosophical attention than mammals, which is patently not the case. Any unspoken philosophical ranking of life-forms and their study would need to propose a broader view of human supremacy (Paabo 2001) and comparative genomics is more likely to challenge such a notion than to support it.

Taking this explanation in terms of human supremacy further, Stephen Jay Gould (1994: 87) sees general indifference to microbes as part of the 'conventional desire to view history as progressive, and to see humans as predictably dominant' thus leading to overattention to 'complexifying creatures'. This view places at the centre of life a 'relatively minor phenomenon' instead of the most salient and enduring mode of life known to this planet. Is it possible that philosophers, usually amongst the first to condemn notions of progressive evolution, are under the influence of this view of the history of life when they ignore microbes? Perhaps a more charitable interpretation is that the discontinuity of life-forms implied by the prokaryote–eukaryote division (Stanier and Van Niel 1962; Olsen et al. 1994; Sapp 2005b; Woese 2005) and the emphasis of negative characteristics of prokaryotes (no nucleus, no internal membranes, small size) gave rise decades ago to a generally unchallenged notion amongst philosophers that microbes were less interesting than their (assumed-to-be) categorically different multicellular descendants. That this notion is maintained despite the growth of knowledge and theory in microbiology means that adherence to a bad habit is the only reasonable explanation for the reluctance of philosophers of biology to deal with microbes. In that case, delving even briefly into the recent microbiological literature might provide just enough of a conceptual kick to initiate a wider range of thinking in the philosophy of biology and perhaps even stimulate a philosophy of microbiology.

11

Metagenomics and Biological Ontology[1]

With Maureen A. O'Malley

Introduction

Life is commonly considered to be organized around the pivotal unit of the individual organism, which is traditionally conceived of as an autonomous cell or a group of coordinated cells with the same genome.[2] Hierarchies of other biological entities constitute and are constituted by organisms. Macromolecules are often placed at the bottom of the organism-constituting hierarchy, and they are succeeded by various subcellular and cellular levels of organization, including the tissues and organs of multicellular organisms. Above the level of organism rises the organism-constituted hierarchy, in which groups of organisms form ecological communities across space and lineages or species across time. Implicit in this hierarchical approach to the organization of life is an evolutionary timeline that runs from most primitive to most complex. Unicellular organisms are generally regarded as inhabiting the lower end of the complexity spectrum and large mammals are placed at the upper end.

All biologists and philosophers of biology know the difficulties of uniquely dividing different groups of organisms into species or genomes into genes, and both communities of investigators are divided about the inevitably of pluralism or the possibility of defining natural kinds. Our view is that these problems reflect a more fundamental difficulty, that life is in fact a hierarchy of processes (e.g., metabolic, developmental, ecological, evolutionary) and that any abstraction of an ontology of fixed entities must

[1] We thank the participants at the Philosophical and Social Dimensions of Microbiology Workshop (University of Exeter, July 2007) for discussion and ideas, and our two referees for helpful comments on the chapter. We gratefully acknowledge research funding from the UK Arts and Humanities Research Council, and workshop funding from the Wellcome Trust.
[2] If viruses (which are included in the usual definitions of microbes) are given organismal status, then an appropriate definition of organism could not insist on cellularity. Although there is a strong case for defining life as fundamentally cellular (most biologists would agree on this), the exclusion of viruses from the category of organisms raises deep conceptual problems due to the increasing recognition of the centrality of viruses to life processes at all scales.

do some violence to this dynamic reality. Moreover, while the mechanistic models that are constructed on the basis of these abstracted entities have been extraordinarily valuable in enhancing our understanding of life processes, we must remain aware of the idealized nature of such entities, and the limitations of analogies between biological process and mechanism (we shall say a bit more about this last point in the concluding section of the chapter). Despite the philosophical challenges just mentioned to species and gene concepts, there are few similar doubts about the notions of unicellularity and multicellularity or, indeed, about the prima facie most unproblematic concept of all, the individual organism. We will raise these questions through a discussion of an emerging scientific perspective, metagenomics, and conclude with some further reflections on biological ontology.

Multicellularity

One of the most fundamental divisions in conceptions of the way life is organized is that between unicellular and multicellular life-forms. The distinction between these two modes of life is usually seen as a major evolutionary transition (Maynard Smith and Szathmáry 1995). It is standard to understand multicellularity as the differentiation of 'monogenomic' cells (differentiated cells possessing the same genome) and to think it pertains only to plants, animals, and fungi (excluding yeasts). The most sophisticated forms of multicellularity are usually attributed to vertebrates because of the complex cell differentiation and coordination necessary throughout the lifespan of these animals. Non-vertebrates and plants are often considered to be less sophisticated multicellular organisms, perhaps because many of them can and do 'revert' to reproducing themselves by non-specialized cells. The complexities of incorporating well-known symbiotic forms such as lichens and 'Thiodendron' mats cannot even be addressed within this scheme. This common way of ranking organisms causes problems, however, for an adequate understanding of multicellularity.

Many of the characteristics that are used to define multicellularity do not exclude unicellular life when proper attention is paid to bodies of research that illuminate cellular cooperation, developmental processes, competition, and communication strategies amongst unicellular organisms (Shapiro and Dworkin 1997; Shapiro 1998; Wimpenny 2000). Less formal criteria for designating unicellularity, such as visibility, are also problematic, because many supposedly unicellular forms of life live in groups visible to the naked eye, such as filaments, stromatolites (microbial mats that bind sediments and are best known in their fossilized forms), and biofilms, as well as some very large unicellular organisms. The research we refer to above reveals that microbial communities do indeed exhibit a multitude of multicellular characteristics. We have discussed several philosophical aspects of conceiving of microbial communities as multicellular organisms in earlier work (Chapter 10). Here, we will focus on a new genomics-based strategy called metagenomics that looks at microbial communities with fewer implicit preconceptions as to what constitutes a single organism.

Metagenomics

Metagenomics—also called environmental genomics, community genomics, ecogenomics, or microbial population genomics—consists of the genome-based analysis of entire communities of complexly interacting organisms in diverse ecological contexts. The term 'metagenome' was first defined as the collective genome of the total microbiota of a specific environment by Jo Handelsman and colleagues in 1998 (Handelsman et al. 1998) and 'metagenomics' is now applied retrospectively to some earlier studies that preceded the coining of the label (e.g., Stein et al. 1996). A brief history of where metagenomics fits in both recent molecular biology and microbiology in general will help clarify its scientific and philosophical implications.

Historical overview

Microbiology has traditionally been a discipline marginal to the mainstream biology of macro-organisms.[3] Its research interests entered the mainstream of biological thinking through genetics and molecular biology in the 1940s, when most molecular and genetic analysis was tied to micro-organisms for technical reasons (Brock 1990). The use of microbes, especially viruses and bacteria, as tools to understand genetic inheritance was based on the conviction and eventual demonstration that microbes possessed not only the same genetic material that multicellular organisms did, but also engaged in many similar biological processes, including reproductive ones. As single-gene studies gave way to whole-genome studies in the 1990s, the centrality of microbes and microbial knowledge to molecular biology was further reinforced, especially as the sequencing of prokaryotes and viruses rapidly advanced beyond that of macrobial organisms. Genomic studies of individual isolated microbes, while adding vast amounts of new information and understanding to comparative evolutionary biology in particular (Ward and Fraser 2005), has so far provided only limited knowledge about function and about the many millions of uncultured microbial taxa (Nelson 2003).

Attempts to remedy these shortfalls have resulted in the broadening of molecular microbiology's environmental scope. Ecological studies of microbes have historically been peripheral to both general ecology and mainstream microbiology, the former because of more general tendencies in biology and the latter because of the predominance of the pure culture paradigm (Brock 1966; Costerton 2004). Although microbial ecology had late nineteenth-century roots in the work of Russian soil microbiologist, S. Winogradsky, and the founder of the famous Delft school of microbiology, M. W. Beijerinck, it took until the late 1960s for microbial ecology to gain real disciplinary recognition. Its unifying theme, whatever the methods used, is that micro-organisms have to be understood in their ecological contexts (which include the context of other organisms), rather than as isolated individuals in artificial

---

[3] In Chapter 10 we suggest the word 'macrobe', in contrast to 'microbe', to refer to the diverse group of eukaryotic organisms generally thought of as multicellular.

environments (Brock 1987). Most microbes live preferentially in complex, often multi-species, communities such as biofilms, and as many as 99 per cent of prokaryote taxa are not currently culturable in 'unnatural' laboratory conditions (Amann et al. 1995).

Taking an environmental approach enabled microbial genomics to extend beyond the sequencing of laboratory cultures of isolated microorganisms to the sequencing of DNA extracted directly from natural environments (Pace et al. 1985; Olsen et al. 1986; Amann et al. 1995). This move out of the laboratory vastly expanded the scope of the data collected as well as understandings of biodiversity and evolutionary relationships (Pace 1997; Xu 2006). It took the rest of the 1990s, however, to overcome another associated narrowness of approach, in which very limited DNA sequences (often non-protein-coding) were used as markers of species and indicators of biodiversity. A focus on particular genes and an assumption they would be the same in each species—whatever the environment—limited the information gained about the physiological or ecological characteristics of the organisms (Ward 2006; Tyson and Banfield 2005; Rodríguez-Valera 2002).

Metagenomics is conceived of as an approach through which such limitations can be transcended. Instead of individual genomes ('monogenomes') or single gene markers, metagenomics starts with large amounts of the DNA collected from microbial communities in their natural environments in order to explore biodiversity, functional interactions, and evolutionary relationships. To understand the full range of metagenomic approaches and the ambitious agenda those practising it have set for themselves, we will briefly cover the various kinds of insight generated by the field so far. These range from the generation of catalogues of biodiversity to projects in evolutionary and functional metagenomics, and culminate in what we might call 'metaorganismal' metagenomics.

Biodiversity metagenomics

The primary focus of early metagenomics has been a better appreciation of biodiversity through the analysis of metagenomes or environmental DNA. Currently, metagenomicists take two approaches to the study of metagenomes. The most common practice consists of extracting DNA from environmental samples and cloning it in large-insert libraries.[4] These are screened for clone activity (particular functions expressed in the host cell) or specific gene sequences (Riesenfeld et al. 2004). Genes of interest continue to include ribosomal RNA genes, genes that have been favoured for phylogenetic analysis since the early days of molecular microbiology, and which continue to be used as phylogenetic 'anchors' for further analysis of diversity and function (Tringe and Rubin 2005). Sampled environments include ocean waters of various depths and temperatures (Béjà, Suzuki et al. 2000; DeLong et al. 2006;

---

[4] DNA fragments of 100 kb and more can be propagated in bacterial artificial chromosomes (BACs) and up to 40 kb in fosmids (modified plasmids). These cloning vectors were originally used primarily for plant and animal genomics, and BACs were essential to the sequencing of the human genome.

Grzymski et al. 2006), marine sediments (Hallam et al. 2004), agricultural soils (Rondon et al. 2000), the human gut (Manichanh et al. 2006) and mouth (Diaz-Torres et al. 2003), as well as human-made environments such as drinking-water valves (Schmeisser et al., 2003). These approaches deliver a mix of targeted and 'undirected' biological information about all levels of biological organization, from single genes and metabolic pathways to ecosystem activities (Riesenfeld et al. 2004).

The second approach is even more comprehensive and involves the random shotgun sequencing of small-insert libraries of all the DNA in an environmental sample.[5] The two 'classic' examples include, first, a study that sequenced all the DNA from an environment with low species density, specifically a biofilm in the highly acidic, metal-rich runoff in a mine that exposed pyrite ore to air and water (Tyson et al. 2004). The genomes of only five taxa were identified in this community's DNA, the relative simplicity of which allowed two individual genomes to be reconstructed from the sequence data. Analysis of genes for metabolic pathways gave insights into the roles of community members in metabolic processes and changed the authors' understanding of how the community functioned (Tringe and Rubin 2005; Tringe et al. 2005).

The second example involved the DNA from a considerably more complex oceanic community in the low-nutrient Sargasso Sea (Venter et al. 2004). This catalogue of subsurface prokaryotic DNA is still the largest genomic dataset for any community and has been described as a 'megagenome' (Handelsman 2004). Almost 70,000 divergent genes and 148 previously unknown phylotypes (taxa with divergent sequence in commonly used phylogenetic markers) were amongst this study's findings. Special database (GenBank) provisions had to be made so that the megagenomic data did not overwhelm all the ordinary monogenomic data contained in the databank and skew subsequent comparative analyses (Galperin 2004). Venter's team continues to gather environmental DNA from oceans around the world (following the route of *The Beagle*) and has also commenced the 'Air Metagenome Project', which will sample and sequence the midtown Manhattan microbial communities in the air (Holden 2005).

Viral metagenomics, which also focuses on shotgun sequencing of metagenomes, gives insight into the vast and previously untapped diversity of viral communities in, for example, near-shore marine environments (sediments and water column), and human and horse faeces (Breitbart et al. 2002, 2003, 2004; Edwards and Rohwer 2005). These data, which indicate an even greater amount of biodiversity than that attributed to prokaryote communities, allow further hypotheses to be developed about the role of viral communities in the evolution and ecology of not only microbial communities but the entire natural world (Hamilton 2006).

An epistemologically paradoxical but methodologically intriguing use of metage-nomics was to overcome the technical challenges of sequencing ancient DNA—in the

---

[5] These libraries consist of very small fragments of DNA cloned in plasmids and amenable to high-throughput sequencing.

most well-known case from a Pleistocene cave bear, *Ursus spelaeus*. By sequencing all the DNA in the sample, which included microbial, fungi, plant, and animal contaminants, then comparing the metasequence against modern dog and bear sequences, it became possible to distinguish the small amount (<6%) of cave bear DNA in the sample (Noonan et al. 2005). With similar techniques applied to mammoth DNA (Poinar et al. 2006), the viability of ancient DNA sequencing via 'palaeometagenomics' now seems established and may result in the success of the 'Neanderthal Metagenome Project' (Rubin, in Pennisi 2005).[6] In these cases, the interest is obviously not the metagenome itself but an individual genome, and 'metagenomics' is clearly conceived of as no more than a bioinformatics technique.

There is much more to metagenomics than shallow catalogues of sequence and biodiversity, however. These inventories also give unique insights into microbial community structure and biogeography. They enable subtle understandings of eco-physiological characteristics of communities, in which adaptations to different environmental gradients result in different metabolic and morphological strategies (e.g., capacities for movement) that spread vertically and horizontally through community members (DeLong et al. 2006). The genomic heterogeneity in environmental samples shows that the genomes of single isolated organisms can no longer be considered as typical of whole populations or species (Allen and Banfield 2005). Rich as such understanding is, it is only the beginning of metagenomic analysis because community genome data are also the basis for more complex evolutionary understanding as well as for the investigation of community function and ecological dynamics.

Evolutionary metagenomics

Metagenomics has amplified insights into and questions about the genetic heterogeneity of populations and the genomic mosaicism of individuals. Understanding this genetic variability requires a deeper understanding of evolutionary processes and the mechanisms of genetic exchange and recombination (Falkowski and de Vargas 2004). Metagenomics naturally aligns with an area of investigation that is sometimes called 'horizontal genomics' because both are concerned with the plethora of mobile genetic elements available to microbial communities and with the ways in which the metagenomic resources they inherit are shared and utilized (DeLong 2004a).

The metagenomic role of gene cassettes provides an interesting example of how such study is being pursued. Gene cassettes are intergenomically mobile genes that are integrated into genomes in units called integrons. Such cassettes usually carry genes for environmental emergencies, such as antibiotic assaults on prokaryote communities, rather than genes for everyday function. Cassettes, and the integron elements in the host genome that allow the cassettes to be inserted and expressed (or excised), are

---

[6] (Note added to this publication) A draft Neanderthal sequence was published in the journal *Science* in 2010 (Green et al. 2010).

efficient mechanisms for the movement and expression of genes within and between species, and are implicated heavily in antibiotic resistance (Rowe-Magnus et al. 2002; Holmes et al. 2003; Michael et al. 2004). Gene cassettes were originally studied individually but a metagenomic perspective[7] allows them to be treated as a 'floating' evolutionary resource of high diversity and widespread activity that exists independently of individuals and is likely to have a high impact on bacterial genome evolution (Holmes et al. 2003; Michael et al. 2004).

Although the extent, types, and precise effects on the metagenome of mobile resources (cassettes, as well as all the genetic material available for exchange to greater and lesser degrees by conjugation, transformation, and transduction) still require much more research, the conceptual implications for evolutionary understanding are already powerful, particularly because such studies back up extensive work done on lateral gene transfer and recombination processes. Metagenomic analysis supports and extends the earlier unexpected findings of comparative microbial genomics, which contradicted the dominant eukaryo-centric paradigm of vertical inheritance and mutation-driven species divisions that give rise to a single Tree of Life (DeLong 2002b; Rodríguez-Valera 2004; Allen and Banfield 2005; Doolittle 2005). Rather than focusing on individual organismal lineages, such metagenomic studies enable a shift in scientific and philosophical attention to an overall evolutionary process in which diverse and diversifying metagenomes underlie the differentiation of interactions within evolving and diverging ecosystems. Conceptually, metagenomics implies that the communal gene pool is evolutionarily important and that genetic material can fruitfully be thought of as the community resource for a superorganism or metaorganism, rather than the exclusive property of individual organisms (Sonea and Mathieu 2001).

Functional metagenomics

Most functional metagenomics involves screening library clones created from environmental DNA for functional activity and specific genes, particularly those for which function has already been established. Many metagenomic studies reconstruct metabolic pathways based on genome sequence by assigning functional roles to different taxa in the community based on those genes (Allen and Banfield 2005; Tringe et al. 2005). These analyses are not only discovering new genes, but also revealing wholly unanticipated functions and mechanisms such as photobiology in oceanic bacteria. Genes for light-driven energy production (proteorhodopsin genes) were well known in halophilic or salt-loving archaea but never before suspected in oceanic bacteria until

---

[7]   The metagenomic technique is somewhat different for gene cassettes because in this case PCR assays that target complete open reading frames (ORFs) (specifically those flanked by the integration elements that are part of integrons) can be used to select environmental DNA prior to cloning and sequencing (see Stokes et al. 2001). (In general and very roughly, open reading frames are parts of the genome that are potentially transcribable as units; in prokaryotes they can generally be assumed to code for a protein, though the situation is more complicated in eukaryotes [parenthesis added to this edition].)

discovered (because one gene was serendipitously close to a 16S ribosomal RNA gene being sequenced for phylogenetic purposes) via metagenomic analysis (Béjà, Suzuki et al. 2000; DeLong 2005). Further analysis involved the expression of these genes in a laboratory *E. coli*, and the correlation of proteorhodopsin with the light available at different ocean depths (Béjà, Aravind et al. 2000; Béjà et al. 2001). Venter's Sargasso Sea inventory also revealed the surprising abundance of proteorhodopsin genes in ocean waters.

This gene-based approach is most comprehensively captured by an innovative comparison of metagenomic data from microbial communities in farm soil and around the skeletons of decomposed whale carcasses in ocean waters (Tringe et al. 2005). Using bioinformatic techniques to predict the protein products of each metagenome, Tringe and colleagues compared their data to the acid mine drainage and Sargasso Sea metagenomes. They identified patterns of gene distribution that are specific to the environmental locations of the communities, thereby adding support to the argument for the importance of understanding microbial genome activity *in situ* (even though current techniques are biased towards the most common taxa). Rather than trying to reconstruct whole individual genomes from the metagenomic data, the study sought to elucidate the 'functional fingerprints' of the DNA of complex communities in a variety of nutrient-rich environments.[8]

Other functional studies of community genomes have uncovered versatile metabolic capacities,[9] such as the ammonia oxidizing activities of archaea in oceans and soils (a function previously known only in bacteria). The metagenomic identification of archaeal gene sequences involved in ammonia oxidation pathways provides a clear framework for deeper biochemical and physiological understanding of the role these mostly uncultured organisms play in the global nitrogen cycle (Treusch et al. 2005; Hallam et al. 2006; Francis et al. 2007). Amino acid analysis based on an Antarctic marine bacterial metagenome has led to further development of physiological hypotheses about cold adaptation (Grzymski et al. 2006). Another interesting investigation focused on 'reverse methanogenesis' or methane consumption (the reverse of the better-studied process of methane production or methanogenesis) in archaeal communities in deep-sea sediments (Hallam et al. 2004).[10] This study may have solved a

---

[8] Tringe and colleagues used partially assembled metagenome fragments, which they called 'environmental gene tags' or EGTs.

[9] Biotechnological hopes for applications arising from functional metagenomics are high, because the science opens up a previously concealed realm of microbial gene products (apparently inaccessible to standard cultivation techniques). Commercial applications of metagenomics focus on the discovery of 'novel natural products' such as enzymes, antibiotics, and other drugs (Courtois et al. 2003; Cowan et al. 2005; Lorenz and Eck 2005). Soils in particular are perceived to be rich sources of new biomolecules for industry and biomedicine (Voget et al. 2003; Daniel 2004). In cases such as acid mine drainage, metagenomic data is anticipated to lead to more effective bioremediation techniques (Pazos et al. 2003; Eyers et al. 2004). Even broader insight and applications are sought from metagenomics to address global warming (Committee on Metagenomics 2007).

[10] These communities were physically simplified before metagenomic analysis (to a greater extent than size filters during sampling would achieve), so are not truly 'natural' samples.

biogeochemical puzzle of why seabed methane does not escape into the water and is one of the many examples of the potential social and economic relevance of metagenomics to environmental issues such as global warming.

Many of these studies of function, however, are either of functions predicted from gene sequence on the basis of homology searches, or are based on low-throughput screening (Wellington et al. 2003; Johnston et al. 2005). Validating gene predictions was the bottleneck for monogenomic analysis, and there are fears this will be the case for metagenomics (Ward 2006), especially if expression studies can only investigate limited gene activity per assay (Sebat et al. 2003). Several metagenomic research laboratories are now extending their analyses to the transcriptome (comprehensive gene expression in a particular set of conditions as measured by the abundance of mRNA transcripts) and proteome levels (total protein expression). In parallel with the terminology of metagenome, these objects of investigation are referred to as the metatranscriptome and the metaproteome. Community microarray analyses of gene transcription are already extending bioinformatic inferences of function. Using microarray technology for studying metagenome expression and regulation is challenging but possible, according to early efforts (Wu et al., 2001; Dennis et al., 2003; Zhou, 2003; Poretsky et al., 2005).

Metaproteomic analysis is the next step towards achieving better understanding of functional gene expression in a range of environments and specific conditions (Kan et al. 2005; Powell et al. 2005; Ram et al. 2005; Schulze et al. 2005; Wilmes and Bond 2006). Using mass spectrometry, Ram and colleagues analysed the metaproteome of the same acid mine drainage biofilm as in Tyson et al.'s metagenomic study. Novel gene products as well as expected ones were quantified, and inferences made from metagenomic data were supported. Most crucially, a key iron-oxidizing enzyme involved in acid production was identified, which gave much greater depth to understandings of that particular ecosystem. Metaproteomics is most commonly used for low complexity environments and it is still very hard to apply to more complex ones or to extend the functional understanding gained from these early analyses (Wilmes and Bond 2006). Nevertheless, existing techniques plus combinations and extensions of them are being used to understand *in situ* function, and users are optimistic about further technical advances (Wilmes and Bond 2006).

Metametabolomics, the community-based version of individual organism metabolomics (the study of small molecules or metabolites in a particular physiological state of a cell) is only just beginning, and one of the first studies focuses on the metabolic networks of the microbial community in the human gut and the digestive capacities provided by microbes to humans (Gill et al. 2006). Such studies also investigate 'co-metabolomes' or the metabolites that can only be produced by host–microbial interactions (Nicholson et al. 2005). As well as intracellular metabolites, data on extracellular signalling molecules are needed, if communication and coordination processes within communities are to be understood (Buckley 2004b). Integrated analyses of metagenomes, metatranscriptomes, metaproteomes, and metametabolomes will be necessary for the ultimate metagenomic analysis, which will involve the investigation of all levels of community activity *in situ*.

Metaorganismal metagenomics, or microbial systems biology

The next ambitiously anticipated phase of metagenomics is microbial systems science, in which complex biological networks across multiple hierarchical levels are to be analysed by interdisciplinary teams (DeLong 2002a; Buckley 2004b; Rodríguez-Valera 2004). Systems biology more generally is now the most rapidly proliferating and heavily funded area of biology. It consists of large groups of molecular and cell biologists, mathematicians and bioinformaticians attempting to integrate huge bodies of 'omic' data with the aim of predicting and intervening in multiple levels of biological activities in wide-ranging environmental conditions. Although discussion of how community-based microbial systems biology will be accomplished has so far been limited, it is clear that a 'systems ecological' perspective would meld a molecular approach with a synthesizing perspective on ecosystems (including biogeochemical analysis) and would thereby extend the focus of current systems biology on intracellular interactions (Doney et al. 2004; Allen and Banfield 2005; DeLong 2005). More data is not the only requirement for a systems-biological approach to microbial communities. Advances in modelling and *in silico* simulation as well as serious interdisciplinary cooperation will be vital to the field's development (Lovley 2003). These requirements mirror challenges to systems biology generally (O'Malley and Dupré 2005).

Despite the limitations of currently available tools,[11] a great deal of metaorganismal analysis is already underway, both of purely microbial systems and of mixed microbial–macrobial systems. The dynamics between plant and nitrogen-fixing microbial communities or between squid and light-generating bacterioplankton are well-studied examples of associations in which there are mutual influences on gene expression and developmental processes (Xu and Gordon 2003). Particularly striking is the growing understanding that symbiotic bacteria are required for the proper development of many vertebrates. It was recently reported, for example, that environmentally acquired digestive tract bacteria in zebrafish regulate the expression of 212 genes (Rawls et al. 2004; Bates et al. 2006). In fact, for the majority of mammalian organism systems that interact with the external world—the integumentary (roughly speaking, the skin), respiratory, excretory, reproductive, immune, endocrine, and circulatory systems—there is strong evidence for the co-evolution of microbial consortia in varying levels of functional association (McFall-Ngai 2002). Germ-free (gnotobiotic) rodent studies indicate that removing all or part of a mammalian microbiome leads not only to abnormal physiological development, but also morphological abnormalities

---

[11] See Appendix.

and immunological depression (Berg 1996). The roots of plants lie in the midst of some of the most complex multispecies communities on the planet. Bacteria and filamentous fungi not only form dense cooperative communities in the soil surrounding plant roots (often described as the rhizosphere), but are also found within the roots themselves. The dynamic interactions—both intra- and extracellular—between roots, fungi, and bacteria provide nutrients, control pathogens and structure function, growth and community composition (Perotto and Bonfante 1997; Barea et al. 2005). These interactions, many of them obligate, break down the boundary between plant and non-plant, and also serve to illustrate that at the heart of every interface between multicellular eukaryotes and the external environment lies a complex multispecies microbial community.

Even if we insist on retaining a fundamentally monogenomic conception of the human organism, there are multiple ways in which studies of macrobe–microbe interactions can improve our understanding of human biology. A human body is a symbiotic system composed of 90 per cent prokaryotes and 10 per cent human cells (Savage 1977), a ratio that is made more dramatic when viruses (including the bacteriophages hosted by every prokaryote) are considered. Humans and their microbial partners (the microbiome) form a highly coordinated system that appears to be the product of co-evolution. Micro-organismal studies of human–microbial interactions conceive microbial communities to be at least as important for human health as they are for disease. Indeed, it may turn out that diseases caused by microbial pathogens are best seen not so much as an invasion by a hostile organism, but rather as a kind of holistic dysfunction of the microbiome.

Analyses of the diversity of the human gut metagenome (e.g., Gill et al. 2006; Zoetendel et al. 2006) allow richer understandings of how this many-celled 'organ' functions and maintains human life and health. Transcriptional studies further illuminate how this community modulates host gene expression and communicates with host cells (Hooper et al. 2001). Environmental perturbations of human microbiomes, such as may be caused by dietary changes or by antibiotic use, are linked to heart disease, cancers, obesity, asthma, and diabetes (Bäckhed et al. 2004, 2005; Ley et al. 2005, 2006; Ordovas and Mooser 2006). Human drug response is highly dependent on molecular interactions with microbially generated molecules (Nicholson et al. 2005). The human immune system recognizes a huge variety of prokaryotes organized in different communities as 'self' rather than non-self (the objects of immunological defence). Together, commensal prokaryote communities and the human host form an integrated immune system that provides benefits to all the organisms involved (Kitano and Oda 2006a, 2006b).

Such 'self-extending symbioses' argue Kitano and Oda (2006a), are the evolutionary norm because they are highly adaptive and robust against environmental perturbation. All of these insights, whether delivered by metagenomics or more general molecular microbiology, encourage us to see any animal as a composite of all three domains (bacteria, archaea,[12] and eukaryotes) and our own primary genome as a metagenome of microbial and human DNA (Gordon et al. 2005). It is even suggested that humans and other animals could be regarded as 'advanced fermenters', the main role of which is to house, nourish, and assist the reproduction of an enormous array of microbes (Nicholson et al. 2005). The original human genome sequencing projects were, from this perspective, about only a tiny and unrepresentative complement of our genes: a limitation that may ultimately be remedied by a human metagenome project (Relman and Falkow 2001).

If we think about symbioses (the fundamental condition of life) in more general terms than those symbioses limited to the outer surfaces of animal bodies, then metaorganismal metagenomics has very obvious contributions to make to understandings of ecosystem diversity, function, and dynamics. The applications of microbially grounded systems ecology are extensive. Bioremediation, one of the great hopes of genetic engineering, applied genetics, and microbiology, is unlikely to be successful without taking a systems-based approach (Cases and de Lorenzo 2005). The problem of previous approaches and the reason for their failures, argue some commentators, is that the genetic engineering of isolated strains took too limited a perspective on microbial interactions. This perspective can only be advanced upon by a multilevel 'eco-engineering' approach to the metabolic activities of indigenous microbial communities in their natural environments. There are already promising signs from systems-biologic approaches to microbial bioremediation of environmental pollutants (Pazos et al. 2003).

It is becoming increasingly clear that a range of fundamental questions about life on this planet will find their answers only with advances in system-based understandings of microbial communities in global environments (Hunter-Cevera et al. 2005). Will warming oceans disturb the world's primary oxygen producers, the marine cyanobacteria *Prochlorococcus* and endanger oxygen-dependent life-forms (everything apart from prokaryotes[13])? Will the thawing of the polar icecaps lead to intensified global warming as dormant methanogenic prokaryotes become active and release more methane (a contributor to global warming)? Will global changes in human habitat and diet modify the microbiome in human bodies and have significant health consequences? Will antibiotics still be effective in twenty years or will we see the return of high fatality rates from infections such as tuberculosis and pneumonia with the worldwide circulation of antibiotic-resistant genes in the microbial metacommunity? All of these are questions that metagenomics is beginning to address, and it offers real hope that they will be answered (Committee on Metagenomics 2007).

---

[12] Much less is known about archaeal symbionts than bacterial ones, but numerous recent studies are investigating their presence and role in all organisms, including humans. Methanogens (methane generating archaea) are the most frequently found symbionts, in host environments such as the human gut and mouth (Lange et al. 2005). There are still no known pathogenic archaea, but the limitations of detection tools may be the most appropriate interpretation of this non-finding (Eckberg et al. 2003).

[13] It appears unlikely that there are any truly obligate anaerobic eukaryotes (Bryant 1991; Lloyd 2004).

Dynamic system perspectives, which take networks of biological activity in ecological settings as their objects of study, have some of the most interesting implications for the philosophy of biology and its traditional considerations about the units of fundamental ontological importance. In system-level understandings of microbial communities, the metaorganism is conceived of as deriving causal powers from the interactions of the individual components from which it is constituted. At the same time, however, those components are themselves understood to be controlled and coordinated in various ways by the causal capacities of the metaorganism. Although metagenomics might be far away from the full implementation of such a causally dialectical research programme, every effort to think in systems-oriented ways and find techniques and approaches to answer the questions raised by this perspective inevitably points to the necessity of integrating multiple levels of separately realized biological insight. As mentioned earlier, this is one of the main challenges being addressed by systems biology generally.

Concepts of metagenome and metaorganism

For many of the field's practitioners, metagenomics is a technique that provides access to otherwise inaccessible microbial communities. We believe that a more complex line of thought informs metagenomics and has probably informed it since its inception (e.g., Rodríguez-Valera 2004). This perspective takes very seriously the proposal that metagenomes are communal resources and that the entity to which the resource is available is a coordinated, developing, multifunctional, multicellular organism composed of large numbers of cells of different varieties and capabilities, able to work in ways in which the collectivity regulates the functions of individuals.[14] Individual organisms, from this viewpoint, are an abstraction from a much more fundamental entity. A lot of evidence for this perspective has been generated by a variety of methods over the last two decades (e.g., Shapiro 1997; Kolenbrander 2000; Kaiser 2001). Metagenomic analysis is able to take these avenues of research into account and begin to treat the metagenome as the concept that most effectively captures the ways in which microbes survive and flourish in such a remarkable range of diverse environments.

Ultimately, metagenomics is about 'the community of all communities' on this planet, and, uncomfortable as the notion may be for many philosophers, for some practitioners metagenomics is perceived to be moving us inexorably in the direction of a Gaia-like concept of the world (Committee on Metagenomics 2007). Taking the interconnectedness of ecosystems into account does not mean taking on board all of Lovelock's philosophical notions or empirical predictions (in fact, going back to Baas

---

[14] There are, of course, some microbiologists who explicitly reject this perspective and argue for the retention of a single-organism focus, even if metagenomics has proved its use as a tool (e.g., Buckley 2004b).

Becking,[15] who applied the word in 1931, is probably a better idea), but it does mean that adequate understandings of climate change, global human health, and international economic prosperity will all depend on metagenomic knowledge.

Competition versus cooperation

It is plausible to conclude that the fundamental activity of cells, beyond self-organization and maintenance, is to form cooperative associations in a plurality of forms. This may suggest that the organization of life is determined not by competition but by the ability to cooperate (albeit competitively). To put it another way, although there is little doubt that competition and selection have been essential to the evolutionary process, it may be that the main respect in which organisms (or cells) have competed has been with regard to their ability to cooperate in complex single- or multispecies communities. If this is the case, then we might expect—contrary to the orthodox evolutionary view that altruism is exceptional and requires special explanation[16]—that the norm among organisms is a disposition to act for the benefit of other organisms or cells. Darwin was right about the general picture of the organic environment being fundamental to the determination of fitness, but his account of the relationships between organisms in those environments needs to be supplemented.

Process versus entities

A final crucial point about metaorganisms is that they are paradigmatically dynamic entities and therefore very clear illustrations of the ultimate necessity of a process-oriented approach to biological investigation. None of the entities that constitute organisms, or which organisms constitute, are static. Genomes, cells, and ecosystems are in constant interactive flux: subtly different in every iteration, but similar enough to constitute a distinctive process. The greatest significance of this point is perhaps that its appreciation will prevent us from taking too literally mechanistic models of biological processes. A good machine starts with all its parts precisely constructed to interact together in the way that will generate its intended functions. The technical manual for a car specifies exactly the ideal state of every single component. Though the parts of a machine are not unchanging, of course, their changes constitute a relentless, unidirectional trend towards failure. As friction, corrosion, and so on gradually transform these components from their ideal forms, the function of the car deteriorates. For a while these failing components can be replaced with replicas, close to the ideal types specified in the manual, but eventually too many parts will have deviated too far from the ideal, and the car will be abandoned, crushed, and recycled. No such unidirectional tendency towards failure characterizes biological processes: although perhaps all organisms die in the end, many exhibit high levels of stability for centuries and millennia. This is not,

---

[15] See Baas Becking (1931) as well as L. Margulis's work on this topic (e.g., Margulis 1998).

[16] Altruism and its interpretation have been much deliberated in the biological literature. See Lyon (2007) for a discussion of altruism and cooperation in relation to microbes.

however, a stability based on parts so durable that they take centuries to deteriorate, but rather the dynamic stability of processes that constantly recreate or maintain their essential constituents. Both the promise and the challenge of systems-biologic approaches is that they offer the hope of providing techniques for the modelling of such self-maintaining dynamic systems.

A further philosophical point about dynamically self-sustaining systems as opposed to mechanistic ones,[17] is that in the latter causation can be seen to run not merely upwards from part to whole, but also downwards from whole to part.[18] The behaviour of individual cells, for instance, whether in multicellular eukaryotes or complex microbial systems, is in fundamental respects determined by the features of the system of which they are parts. We suggest that if useful object-like abstractions can success-fully be produced from the flux of dynamic biological processes, then this reflects the fact that these 'objects' are temporarily stable nexuses in the flow of upward and downward causal interaction. So, for instance, a gene is a part of the genome that is a target for external (that is, cellular) manipulation of genome behaviour and, at the same time, carries resources through which the genome can influence processes in the cell more broadly. Since the analysis of biological systems into entities is not determi-nate, for some purposes of enquiry a monogenomic organism is the most appropriate, whereas for others the appropriate focus is on polygenomic systems. Answering the question, 'What is an organism?', requires seeing that there is a great variety of ways in which cells—sometimes genomically homogeneous, sometimes not—combine to form integrated biological wholes. The concept of a multicellular organism is, there-fore, a far more complex and diverse one than the simple, straightforward category we outlined in the introduction.

Conclusion

A specific and central philosophical aim of this chapter is to encourage scepticism about a concept that has often been treated as self-evident and unproblematic in theoretical biology: the monogenomic concept of an organism.[19] Its 'obviousness' is sustained by the distinction between unicellularity and multicellularity, where it is assumed that each organism has exactly one genome. If that genome appears in sharply differentiated but functionally interrelated cells, we think of the organism as multicellular; if cells

[17] Or at any rate machine-like ones. The concept of a mechanism developed recently by some philoso-phers (Machamer et al. 2000; Craver 2005) is in many ways well suited to the representation of dynamic biological systems.

[18] Craver and Bechtel (2007) provide an interesting but rather different perspective on top-down causation in complex systems from that assumed here.

[19] The concept is not wholly monolithic because of known exceptions, such as genomic mosaicism. [Added in this edition: See Chapter 7, above, for discussion of a number of further limitations of the idea of the monogenomic organism.]

with that genome seem more or less functionally homogeneous and more or less contingently functionally interrelated to other cells, we call it unicellular.

However, if we think of the organism as being simply whichever cooperative systems of cells are most usefully recognized for exploring biological function, then the assumption of 'one genome, one organism' starts to look like a poorly grounded dogma. As highly organized microbial communities such as biofilms illustrate, genomically diverse groups of cells can form organism-like communities (Costerton et al. 1995; O'Toole et al. 2000; Stoodley et al. 2002). Moreover, in understanding the functions of even such paradigmatically multicellular organisms as ourselves, many perspectives (including those of nutrition, development, and immune response) suggest that the relevant system is actually a much broader one, which includes many genomically different kinds of cells. Indeed, as we have noted, the majority of cells in the systems we think of as humans are actually microbial rather than what we have conventionally thought of as human cells. No doubt that which we are describing as a dogma is seen by many biologists to be firmly grounded in a conception of evolution associated with genetically isolated lineages. We have also argued, however (and in more detail in Chapter 10, above), that this view of evolution is itself based on a perspective that is dangerously macrobocentric. The prevalence of horizontal genetic transfer makes it seem increasingly implausible to see the evolution of microbes as fitting within this isolated lineages model, and indeed suggests that evolving units may turn out quite typically to be multispecies communities. Proper attention to microbes may force us to rethink some fundamental ideas about the evolutionary process.

Philosophers of biology have for many years raised concerns about central ontological categories: most notably the species and more recently the gene. We are urging that the organism should join this group of problematic biological categories. We also propose here a diagnosis of these ontological problems. Analysing biological processes into things is necessarily to make an abstraction. Life is not composed in a machine-like way out of unchanging individual constituents. Genomes, cells, organisms, lineages are all assemblages of constantly changing entities in constant flux. Unlike the case of a machine, the stability of life processes is not maintained by the constant interactions of unchanging parts, but by dynamic, self-sustaining, and self-repairing processes. There is no doubt that mechanistic investigations of life processes have provided profound insights, and this very fact is enough to show that quasi-mechanistic elements are fundamental constituents of life processes. But there are limits to how far conventional mechanistic investigations can take us in understanding the dynamic stability of processes at this hierarchy of different levels. Such understanding will require models that incorporate both the capacities provided by mechanistic or quasi-mechanistic constituents, and the constraints and causal influences provided by properties of the wider systems of which these constituents are parts. It remains to be seen how far we have the abilities to construct such models, but this is what enthusiasts for the rapidly growing project of dynamic metaorganismal metagenomics and systems biology should be aiming to offer.

Appendix: Problems and limitations of metagenomic analysis

At present, metagenomics is in an intensive discovery phase not unlike that of early single-organism genomics. It is more concerned with constructing inventories of environmental sequence and gene products than ready to give extensive insight into ecosystem function and physiology (Schloss and Handelsman 2003; Chen and Pachter 2005). Even if inventories were the only aim, the full metagenome sequence of the most complex and diverse communities (especially in soils) is still beyond the reach of current technologies because of the size and complexity of soil-based communal genomes, which require formidably high numbers of clones and sequence coverage to accurately represent their genetic composition (Riesenfeld et al. 2004). Shallower sequencing coverage means lower representation of less common taxa and functions (Foerstner et al. 2006). Various technical biases in regard to sampling, lysing cells for DNA extraction, cloning and expression systems are all being worked around and may eventually be overcome (Handelsman et al. 2002; Liles et al. 2003; Béjà 2004; Streit and Schmitz 2004).

There are additional sources of bias imposed by technological limitations. The extension of tools devised originally for individual genome and proteome analysis is not unproblematic. Reconstructing individual genomes from shotgunned metagen-omes (which would accord the latter more reliability) is very difficult, especially in complex communities with a lot of genomic microheterogeneity (DeLong 2005). New assembly algorithms and more extensive sequencing coverage are amongst the potential solutions (Allen and Banfield 2005; Chen and Pachter 2005; Edwards and Rohwer 2005), as well as combinations of small-insert and large-insert library construction (DeLong 2005). Comparisons of metagenomic libraries with monogenomic data are still crude, because of the low diversity of monogenomic data and (possibly) inadequate search tools. While comparative metagenomics, or the comparison of the genomic diversity and activity of different microbial communities has already begun (Tringe et al. 2005), it is currently shallow. Likewise, assigning functional roles to large numbers of environmental proteins is still impossible, which restricts understanding of metabolic activity in communities (Allen and Banfield 2005). Metagenomic expression methods developed so far analyse only bacterial gene expression because they are restricted to *E. coli*-based techniques (de Lorenzo 2005). Many metagenomic studies are more 'proof of principle' than wholly successful extensions of existing tools, but the potential rewards for overcoming such challenges are considered to provide more than enough motivation for investing time and effort in their development (Nicholson et al. 2004; Chen and Pachter 2005).

System-oriented analysis of the complex functions of microbial communities in their environments is still rudimentary (Torsvik and Øvreås 2002; Handelsman 2004; DeLong 2005; Ram et al. 2005). Advances in the integration of 'omic' data modelling techniques for single organism, lab-based, monogenomic systems biology (Palsson

2000; Joyce and Palsson 2006) are highly likely to benefit metaorganismal metagenomic analyses. The development of a systems-based molecular approach to microbial communities will also require tighter integration of microbial ecology, population genomics, phylogeny and biogeochemistry (Rodríguez-Valera 2002; DeLong 2005). More hypothesis-driven approaches (combining, for example, functional and sequencing aspects of metagenomics in the same investigation) and supplementary data from traditional microbiological techniques (e.g., culturing, microscopy, and other visualization techniques) are seen as desirable (Schloss and Handelsman 2005; Ward 2006). If metagenomic-based systems biology progresses even a little, it will revolutionize the way in which life is investigated.

12

Varieties of Living Things

Life at the Intersection of Lineage and Metabolism[1]

With Maureen O'Malley

It would seem that 60 years after Erwin Schrödinger wrote his book 'What is Life?' we should be able to answer the question. However, Nature never ceases to challenge the limits of our imagination.

—M. Y. Galperin (2005: 149)

This essay will not attempt to provide a definition that answers Schrödinger's question. We shall instead address it by describing a spectrum of biological entities that illustrates why no sharp dividing line between living and non-living things is likely to be useful. The more positive goal of these reflections will be to offer a flexible view of life that does in fact make good sense of why particular organizations of matter can be described as living. By identifying the different capacities exhibited by the various entities constituting our spectrum, especially problem cases such as viruses, we hope to address at least some of the issues that lie behind Schrödinger's question and its many earlier precursors and subsequent echoes. Such concerns have been raised in a striking way by recent attempts under the rubric of 'synthetic biology' to synthesize life from basic chemical building blocks.

In this chapter we shall highlight a tension in standard discussions of characteristics of life, which tend to prioritize one or other of two fundamental but very different features of living things: the capacity to form lineages by replication and the capacity to exist as metabolically self-sustaining wholes. We suggest that this tension can best be resolved by seeing life as something that arises only at the intersection of these two features: matter is living when lineages are involved—directly or indirectly—in metabolic processes. But also crucial to our argument and, we suggest, to many of the

[1] We wish to thank Mark Bedau, the Egenis Philosophy of Biology Journal Club, our two anonymous referees and our editor (Joan Roughgarden) for very constructive comments and advice.

difficulties that have confronted attempts to comprehend life, is the observation that the entities that form lineages are not always, or even usually, the same as those that form metabolic wholes. Metabolism, the transformative biochemical reactions that sustain life processes, we shall argue to be a collaborative affair. Life, we claim, is typically found at the collaborative intersections of many lineages, and we even suggest that collaboration should be seen as a central characteristic of living matter—a claim that also has implications for how we understand the origins of life. Further corollaries of this non-coincidence of parts of lineages with metabolic wholes are, first, that we cannot assume the identification of living things with organisms (at least as standardly conceived), and nor, second, can we assert traditional organisms to be 'the' biological individuals on which selection operates.

Collaboration and the diversity of life

The collaborative nature of living entities and processes is our essential starting point. Darwin's theory of natural selection has, quite appropriately, focused a great deal of theoretical interest on questions of competition. This focus, however, has had the less salutary consequence of diverting attention from the equally important topic of cooperation and has culminated in the assumption that altruism, understood as the conferral of a benefit by one biological entity on another, is a profound theoretical problem. Although this is generally seen as a problem pertaining to organisms, a similar argument has notoriously been applied to the topic of genes. Richard Dawkins (1976) made famous the idea that genes are fundamentally selfish entities in competition with one another. From this point of view, it is truly remarkable that the whole consortium of genes in an organism's genome can nevertheless manage to collaborate on a task as momentous as development.

In this chapter we place selfishness in a wider context and emphasize the broader perspective of life as a collaborative enterprise. We are not arguing that interpretations of selfishness are invalid but that, at best, they can only provide a limited perspective on life and evolution. Rather than reducing cooperation to selfishness, we suggest selfishness and cooperation might better be understood within a framework of collaboration. By collaboration, we mean interactions between components of a system that lead to different degrees of stability, maintenance, or transformation of that system. As in scientific collaborations, there may be some strongly selfish interests involved in such interactions (Hull 1988) but these selfish activities can only operate in a collaborative context. Defecting from collaboration is only possible if collaboration is the general default.

In every domain of organismal life, there are extensive sets of organisms that are problematic for standard evolutionary understandings of selfish individuals (Roughgarden 2009). Shared interests can lead to highly cooperative 'team' behaviour, described by Joan Roughgarden as 'cooperative teamwork' (2009: 13). Evolutionary payoffs for such team members may not be equal, but are distributed across the whole team. Collaboration, however, may also include the 'mere' coincidence of individual interests, and it is often in the interest of any individual to collaborate—at least to some extent. Collaboration from this point of view covers a range of interactive processes that may include both cooperative and competitive activities. At one end of this continuum the goals of participants may be completely aligned, while at the other end of the continuum, relationships may be largely or wholly hostile. We will try particularly to understand the evolutionary persistence of apparently 'parasitic' or selfish interactions between organisms, and the nature of the entities formed by what are usually conceived as separate biological individuals.

Part of cooperation is merely interactive combination. Thus atoms combine to produce molecules, and the latter have properties that are not found in any of the atoms of which they are composed. But certainly more than this is required to count as collaboration in the sense we are elaborating. In common with most who have considered the question of how living entities are constituted, we assume that there is one necessary condition for being a living thing that most combinations of atoms and molecules lack: the ability to reproduce. Though we take this to be a necessary condition, it is less obvious that it is sufficient. Living entities have also to be understood in relation to their capacity to sustain themselves through biochemical transformations. Metabolism in our account can be engaged in autonomously (this is the usual understanding) or collaboratively, through interactions with other biological entities. At any rate, as the microbial and microbe-like entities that we shall describe below illustrate, a very diverse group of things both reproduce and participate in metabolic systems.

Our empirically informed investigation of living matter will not be based on the animal, fungus, or plant life that has been the main interest of philosophers and scientists concerned with these issues; nobody questions the status of these as living things, and the problem is only one of deciding which of their characteristics confer the status of living on them. We shall focus instead on the realm loosely referred to as microbial, which includes some entities only contentiously afforded living status. Microbes are a group of organisms biologically and conceptually diverse to the point of incoherence, but then so are the macro-organisms or macrobes that loom large in most perspectives on life (Chapter 10, this volume). The category of microbes includes at least protists (unicellular eukaryotes, which have membrane-bound nuclei and other organelles), prokaryotes (which don't have such compartments but are highly organized in other ways), and viruses.

Viruses are the biological objects that are the pivot of our discussion because many biologists deny that they are living organisms. In fact, they are frequently considered to be prime examples of the boundary between life and non-life, organism and non-organism, and biology and chemistry (e.g., Stanley 1957; Wimmer 2006). They are most often assigned to the second of each of these pairs of categories. Viruses are often deemed not to be alive on the grounds that they cannot reproduce themselves autonomously, and nor can they metabolize. They can, however, carry out such biologically impressive activities as entering cells, co-opting the transcription and translation machinery of the cell, and picking up and moving about DNA from the organisms with which they interact. And by exploiting or collaborating with cellular organisms in these ways, they very effectively reproduce themselves and have no need of autonomous metabolism.

Thinking about viruses and their relegation to the realms of non-living and non-organismal entities necessitates a consideration of whether organism and living entity are identical categories, and whether a minimal account of life has to begin with cells. Such thoughts then invite further reflection on other biological entities that seem to have some autonomy but are almost never described as living organisms. Joshua Lederberg, a pioneer in molecular biology who first formulated the term 'plasmid' (Grote 2008), places these biological entities in the same category of 'symbiotic organisms' as he does mitochondria and chloroplasts. For him, they comprise part of 'the organic whole' (Lederberg 1952: 403). He argues more broadly that any scheme of life has to work out where to place prions, plasmids, integrons (gene capture and integration systems), and transposons—mobile genetic elements in a genome, sometimes called 'jumping genes' (Lederberg 1998).[2]

We will take our cue from Lederberg and start our examination of life with a discussion of some of the biological entities that inhabit this grey area between living and non-living, specifically prions, plasmids, organelles, endosymbionts, and reduced extracellular symbionts. As we move along this continuum of biological organization to entities whose living status is never questioned (micro- and macro-organisms), we will investigate whether these instances of entities possess some of the most frequently cited life-endowing characteristics, such as spatial boundedness, reproduction, metabolism, and evolvability, and how our criterion, collaborativity, relates to these characteristics. We will also argue that our account of cellular and subcellular entities fits very well with origin-of-life scenarios that stress chemical collaboration and community. Our bottom-up perspective, starting at the microscopic level of biology, rather than top-down from its most complex and undisputed exemplars, will suggest that much standard thinking is based on quite restricted and even covertly normative conceptions of what life is. This perspective will ultimately challenge the view that entities such as viruses are not alive and that the minimal definition of life must be cellular.

A spectrum of biological entities

Prions

Once thought of as 'slow viruses', prions are now commonly understood to be self-propagating proteins that are able to convert normal proteins of the same type into the

---

[2] Lederberg also includes in his 1998 list heterokaryon cells that have a diversity of nuclei in a common cytoplasm (see Rayner 1997 for details). We will leave these interesting entities out of our discussion.

pathogenic prion conformation (Prusiner 1998; Soto and Saborio 2001; Weissmann 2004).[3] They have a life cycle from induction to self-perpetuation (the conversion of another protein).[4] Prions are very robust and persisting entities, because their conformation makes them highly resistant to inactivation by chemical, heat, and irradiation treatments.

The central oddity of prions is that they propagate autocatalytically in a protein-only form, without DNA involvement.[5] For this reason, they are frequently referred to as protein-based genes (Uptain and Lindquist 2002; Wickner et al. 2004). Although best known as non-Mendelian hereditary elements in diseased sheep, cattle, and humans,[6] prions exist in unicellular organisms too. Yeast and other fungal prions share no amino acid sequence similarities with mammal prions, and they function and are transmitted very differently (Bousset and Melki 2002; Uptain and Lindquist 2002; Weissmann et al. 2002). Nevertheless, experimental work on yeast prions has provided deep insights into conformational change in proteins and their transmission (Wickner et al. 2007).

The Modern Synthesis does not cope well with prions, and this has led some commentators to propose that a more comprehensive theory of inheritance is needed for prions to be properly understood evolutionarily (Chernoff 2001; Jablonka and Lamb 2005). The prion-forming potential of the implicated yeast proteins is evolutionarily conserved, implying that it is adaptive (Chernoff et al. 2000). Diverse functions have been identified or proposed for prions in a range of taxa. There is some evidence that prions are associated with epigenetically enabling yeast cells to cope with fluctuating environments, and that they play a role in memory formation in sea slugs (Shorter and Lindquist 2005). The non-pathogenic isoform of human prion proteins (the functions of which are still largely mysterious) is linked to the prevention of Alzheimer's disease (Parkin et al. 2007).

These capabilities and characteristics do not give a ready answer to the question of whether the self-propagational status of prions gives them the status of being alive. Although genes are frequently given a special 'informational' role in accounts of heredity (e.g., Hood and Galas 2003), the conferral of a similar status on proteins—as information-bearing molecules—does not simultaneously make them into living entities. Genes and proteins are not classified as alive in their own rights,[7] despite the widespread 'selfish DNA' thesis that seems to confer autonomy on nucleotides

---

[3] PrP^C is the generic protein, and PrP^{Sc} is the pathogenic protein isoform. The designation of prion is made in relation to the pathogenic form's still hypothesized function (Weissmann 2004: 863). See Manuelidis (2004) for an argument against the protein-only understanding of prions and in favour of their viral status.

[4] Induction can occur spontaneously, through vertical inheritance and by lateral infection.

[5] The gene encoding the prion protein has to be expressed, of course, but the same nucleotide sequence can express either the pathogenic or non-pathogenic conformation of the protein.

[6] Prions are described as 'non-Mendelian hereditary elements' because they self-propagate by transmitting their conformational characteristics in a lineage-forming manner, but do not form Mendelian patterns of inheritance (Liebman and Derkatch 1999).

[7] See Muller (1966: 512) for an older view that the gene is a uniquely living material because of its capacity for reproduction, mutation, and enzyme production.

(Doolittle and Sapienza 1980; Orgel and Crick 1980), and despite the recognition of the absolute centrality of enzymes to life processes (Kornberg 1989; Lezon et al. 2006).

Prions exhibit collaborative behaviours that benefit themselves, as a class of protein isoforms, as well as their hosts. When low amounts of the non-pathogenic isoform are produced, the prion conversion process halts, and when high amounts of the former are produced, it may stimulate spontaneous prion formation in the previously prion-free cell (Chernoff et al. 2000; Derkatch et al. 2001). Prion propagation in yeast requires the involvement of chaperon proteins. Moreover, prions in yeast are associated with greater adaptability in yeast because they increase protein variation—a factor that may prove advantageous in variable environments and eventually be genetically assimilated (True and Lindquist 2000; Pál 2001; Masel and Bergman 2003). It is these abilities to interact with biological processes at different levels of organization that presumably explain evolutionarily the prion's powers of persistence.

Plasmids

Plasmids are small, stably inherited and self-replicating molecules of DNA (sometimes RNA) that are independent of the chromosomal DNA in bacterial, archaeal, and eukaryotic cells. Plasmids are prolific and diverse; they may be larger than some prokaryote genomes (del Solar et al. 1998). Many are mobile genetic elements that direct their own transmission to new host cells during conjugation (the unicellular equivalent of sex), thereby spreading themselves to closely related and evolutionarily distant prokaryotes (Thomas 2000, 2006; Sørensen et al. 2005). They are then trans-mitted vertically, from mother to daughter cells.

Plasmids have a two-stage life cycle of establishment and proliferation followed by a steady state that matches the cell cycle (del Solar and Espinosa 2000). Neighbouring plasmid-free cells are often killed by plasmids, and this leads to a very high rate of successful infection (Gerdes et al. 1986; Eberhard 1990; Bingle and Thomas 2001). The complexities of plasmid characteristics have led some biologists to describe them as 'subcellular organisms' or endosymbionts with distinct autonomy from their host (Perlin 2002: 508). Because of their many talents, plasmids have become a mainstay of laboratory genetic manipulation as vectors of gene transfer.

Plasmids are often described as selfish in the same way that other genetic elements are because they encode genes that are not essential for the host and may impose fitness costs (Kado 1998). Importantly, however, they also play cooperative roles in cells (Wegrzyn 2005). Plasmids often encode and express genes for a variety of functions apart from those for their own mobility and replication, such as antibiotic resistance, virulence, environmental protection (including biofilm formation), DNA repair, and supplementary metabolic pathways (Barton et al. 1995; Ghigo 2001). They can thus be seen as collaborative elements that enhance the functionality and adaptiveness of their host cells. The fact that these features favour plasmid survival has allowed these phenomena to be interpreted as instances of selfishness (Kado 1998), but in our framework they could equally well be interpreted as examples of (sometimes mutualistic) collaboration.

Organelles

Organelles are diverse membrane-bound compartments in eukaryote cells.[8] They carry out highly specialized biochemical functions and communicate between themselves to achieve this division of labour (Munro 2004; Lowe and Barr 2007). Major organelles include mitochondria and plastids (including chloroplasts, the organelles enabling photosynthesis in plants), as well as peroxisomes (compartments involved in metabolic activities that include the oxidative metabolism of fatty acids and the breakdown of hydrogen peroxide), Golgi complexes, and endoplasmic reticula. Apart from the nucleus, most organelles are primarily involved in energy generation, transport, and storage. They are often highly dynamic, mobile structures that react to relevant features of the environment to maintain cell function (Collings et al. 2002; Cutler and Ehrhardt 2000; Braun and Schleiff 2007).

Organelles are often considered to be 'autonomous structures' because of their semi-independent inheritance strategies (Nunnari and Walter 1996; Warren and Wickner 1996: 398). Organelles reproduce within cells and a complete set is passed on to the daughter cells during cell division. However, because most membranes have to be inherited from pre-existing membranes and are usually not constructed *de novo*,[9] organelles are templated from pre-existing organelles. They self-assemble on the basis of the information their membranes carry about membrane polarity, type, and location (Cavalier-Smith 2000; Lowe and Barr 2007).

Two of the most evolutionarily fascinating organelles were once free-living bacteria. Mitochondria and plastids functioned first as intracellular symbionts until most of their DNA migrated to the nucleus of the host over a billion years ago—a process that profoundly shaped the structure and content of the eukaryote genome and cell (Martin 2003; Timmis et al. 2004).[10] Now, to obtain the proteins they need for many functions, including their own metabolic activities, mitochondria and plastids rely on a protein import mechanism provided by the host's cellular machinery (Cavalier-Smith and Lee 1985; Thiessen and Martin 2006). This loss of genetic autonomy is not total, however, because plastids and mitochondria retain genes for translation and transcription machinery as well as metabolic function. They divide and grow independently of the cell cycle, although mitochondria gain some division assistance from the host cell

---

[8] There are increasing reports of a variety of compartments in prokaryote cells and the rising use of 'organelle' to describe these structures (e.g., Seufferheld et al. 2003; Niftrik et al. 2004; Kerfeld et al. 2005; Komeili et al. 2006).

[9] Peroxisomes and Golgi bodies can sometimes be reconstituted from other membrane types (Cavalier-Smith 2000; Lowe and Barr 2007).

[10] Mitochondria were incorporated into early eukaryote cells before plastids. As well as primary plastids, obtained in a single endosymbiotic event, there are also secondary and tertiary plastids gained from endosymbioses of plastid-carrying organisms (Archibald 2007).

(Osteryoung and Nunnari 2003). As well as inheriting their membranes directly, both organelles inherit their own organelle-specific DNA.

Mitochondria and plastids are not only essential to their cellular hosts, but are defining characteristics of them: there are no eukaryotes without mitochondria or plants without plastids.[11] Again, it is obvious that collaboration is happening here in ways that benefit—and make dependent—both organelles and the cells they inhabit. Indeed, the eukaryote cell could no more survive without its mitochondrial residents than the latter could survive in natural circumstances outside the cell.

Viruses

Viruses are typically very small packages of single- or double-stranded DNA or RNA (often just a few genes),[12] wrapped up in a coating of protein and sometimes an additional lipid envelope.[13] They are prolific, highly diverse, and ancient, although there is incomplete agreement about their evolutionary origins (as we shall see below). Viruses are generally excluded from organismal status because although they can synthesize some of their own proteins, they do not metabolize or reproduce independently (Van Regenmortel 2007). They either use their hosts, which probably include every organism past and present, or occasionally work in collaboration with other viruses to make necessary enzymes. Viruses do not reproduce by division but by self-assembly of the components that they manufacture with the help of the host cell. Some viruses influence host behaviour quite significantly by, for example, conferring either protection against other viruses or virulence properties (e.g., diphtheria or cholera toxins).

Viruses have well-defined life cycles that are often described as consisting of 'developmental' stages (e.g., Luria et al. 1978). The cycle begins with virions, the inert form of viruses, which are transformed into the next stage of adsorption, when viruses or phages (the viruses with affinities for prokaryotes rather than eukaryotes) 'dock' onto the outer cell membrane of their hosts and either enter the cell or have their DNA absorbed into it. Their protein coats dissolve or are discarded, after which the viruses co-opt the host's cellular machinery to express genes that lead to genome replication, maturation (in which the new genomes are wrapped in freshly synthesized protein), and, finally, release from the intact or lysed cell. A number of plant viruses move actively from cell to cell, using virus-encoded movement proteins (Boevink and Oparka 2005). Some viruses have an extra developmental stage in which they remain dormant in the host cell or genome as prophages or proviruses and are inherited (Casjens 2003; Bannert and Kurth 2004). Endogenous retroviruses, which are viruses that have integrated permanently into the host chromosomes and are inherited

---

[11] Subsequent loss or dysfunction notwithstanding. See Embley and Martin (2006) for a demolition of the 'amitochondriate eukaryotes' hypothesis.

[12] Single-stranded RNA viruses can be divided into positively and negatively stranded (sense and antisense) genomes, and retroviruses, which make DNA copies of themselves with their own reverse transcriptase before entering the host chromosome and being transcribed back to RNA (Ahlquist 2006).

[13] Viroids, which are tiny RNA viruses that infect plants, have no protein coat.

vertically, have left their mark on many organismal genomes, including our own (D. J. Griffiths 2001; Hamilton 2006). Included amongst these viruses are those that are crucial for the development of the placenta in mammals (Mallet et al. 2004).

The diversity and mutability of viruses makes them difficult to classify, although both genome sequence and protein structure analyses are constantly refining viral groupings, which were once based primarily on pathogenic effect (Bamford et al. 2005). The term 'species' is often applied, with many caveats, to subgroups of virus divisions (Hendrix et al. 1999; Lawrence 2002; Van Regenmortel 2007). The aim of such language is to 'bring the definition of virus species into line with the species definitions of cellular organisms' (Gibbs and Gibbs 2006: 1419). One earlier and another more recent division of life into superkingdoms give viruses a superkingdom (domain) of their own: the Acytota or Akamara, both of which are categories for acellular organisms possessing genomes (Jeffrey 1971; Hurst 2000; Weinbauer 2004). These domain-level classification schemas have the potential to identify viruses as genuine forms of life but have yet to gain many adherents.

There are three main hypotheses about the origins of viruses: primeval pre-cellular life (the virus-first or primordial hypothesis), degenerate intracellular parasites (the reduction or regression hypothesis), and as renegade prokaryote genes (the escape hypothesis). The most popular is currently the third one, which is that viruses are actually genetic elements that opted out of cellular organization and are thus true instantiations of 'selfish' genetic material (Hendrix et al. 2000; Campbell 2001). However, new versions of the primordial hypothesis are also being promoted. They shift the discussion back to the pre-cellular 'unselfish' gene pool and give viruses major roles as evolutionary innovators (e.g., Hendrix et al. 2000; Hendrix 2002; Claverie et al. 2006; Forterre 2006; Koonin et al. 2006). Whatever their origins, viruses have made extraordinary contributions to the evolution of non-viral life through their proclivity for mutation and recombination, and their ability to pick up and move genes from one organism to another (transduction) and integrate their own and other genetic material into host genomes (Lawrence et al. 2002; Villarreal 2004b; Weinbauer and Rassoulzadegan 2004; Hambly and Suttle 2005; Karam 2005). Moreover, their role as carbon regulators in the global oceans, for example (Suttle 2005), shows how a broader conception of collaboration is necessary to understand the evolutionary, biogeochemical, and ecosystemic contributions of viruses to all living systems.

The recently discovered Mimivirus (short for 'mimicking microbe') provides an additional challenge to some prevalent ideas about viruses and their capabilities. Mimiviruses are huge (larger in volume and genome size—over 900 protein-coding genes—than many of the smallest bacteria, some of which are described below) and, most surprisingly, they carry genes that are known to encode translation, DNA repair, and metabolic activities (Raoult et al. 2004).[14] They do not seem to have picked these

---

[14] While some other large viruses also carry translation and metabolic genes, Mimivirus greatly extends the known repertoire of these genes in viruses (Koonin 2005).

genes up from their hosts.[15] Although these viruses cannot synthesize their own ribosomes and do not metabolize (their metabolic pathways are incompletely coded), they can easily be conceived of as entities in transition from viruses to free-living organisms (Raoult 2005; Claverie et al. 2006; Forterre 2006). Mimiviruses certainly exhibit more independence than organelles and, moreover, seem to be in an 'evolutionary steady state' with no apparent signs of genome reduction (Claverie et al. 2006: 142).

Microbiologists and other biologists are highly ambivalent about the biological status of viruses. Although a strong line of thinking throughout much of the history of virus research and microbiology has advocated that viruses are alive and at least protoorganismal (Stanley 1941, 1957; Burnet 1945; Luria et al. 1978; van Helvoort 1992), the dominant view of viruses is still fixed by the assumption that only cellular entities are appropriately designated as living (Moreira and López-García 2009). According to virologist Marc Van Regenmortel,

> Only unicellular and multicellular organisms possess the property of being alive while the organelles, macromolecules and genes found in cells are not themselves considered to be alive. The difference between viruses [which are not alive] and various types of organisms is quite obvious when the functional roles of the proteins found in viruses and organisms are compared. (2007: 133)

Other microbiologists, however, believe that there are numerous reasons to give viruses the status of living matter. Because they 'have the intrinsic ability to mediate their own transfer from one host to another,' say Salvador Luria and co-authors,

> viruses are independent genetic systems. They are not accidentally separated fragments of a cell genome. They are endowed with genetic continuity and mutability, and contain sets of genes working in concert to make more virus. They have their own evolution, which is independent, to some extent at least, of the evolution of organisms in which they reproduce. (Luria et al. 1978: 481)

Some virologists go even further and argue that viruses exhibit the same primary features common to all life forms, such as internal homoeostatic controls that enable survival in changing environments, organization that is based on heritable nucleic acids, reproduction, exploitation of environmental resources, diversity of components and their functions, and the capacity to adapt and evolve (Stanley 1941, 1957; Mindell and Villarreal 2003; Mindell et al. 2003). Discovery of the debilitating effects of a minute 'virophage' on a huge virus has been argued as evidence for the aliveness of viruses: if they can be infected themselves, and respond in various ways to these infections, then the 'imaginary boundary' between viruses and true organisms seems to have been crossed (Claverie, and Koonin, in Pearson 2008).

A further stream of reflection sees no contradiction in regarding viruses as *alternating* between living and non-living phases.

---

[15] Although see Moreira and López-García (2005) for the opposite claim.

Outside the host cell, poliovirus is as dead as a ping-pong ball. It is a chemical that has been purified...and crystallized...with its physical and chemical properties largely determined...and its three-dimensional structure solved. Just like a common chemical, poliovirus has been synthesized in the test-tube. Once poliovirus, the chemical, has entered the cell, however, it has a plan for survival. Its proliferation is then subject to evolutionary laws: heredity, genetic variation, selection towards fitness, evolution into different species and so forth—that is, poliovirus obeys the same rules that apply to living entities. (Wimmer 2006: 56)

The inertness of virions outside the cell leads us to think that viruses are similar to prokaryotes with spore stages as well as to plant seeds and fungal spores. In our conclusion we shall (cautiously) endorse this perspective, and also suggest that it is helpful to distinguish the developmental cycle, which includes both active and inert stages, from the life cycle, which should be applied only to metabolically active phases of lineage-forming systems.

Historical echoes of the discussion of the status of viruses are amplified by recent practical achievements of creating synthetic viral genomes. Several of these have now been synthesized from scratch and used successfully to infect cells (e.g., Cello et al. 2002; Smith et al. 2003; Tumpey et al. 2005). Some of these researchers claim their achievements are the final nails in the coffin of vitalism, because their virus 'chemical' was resurrected in a cellular extract and not a living cell (e.g., Cello et al. 2002). However, those who do *not* see viruses as organisms perceive synthetic viral genomes as further proof that 'true' (cellular) life—still resistant to synthesis from the top down or bottom up—is something fundamentally different from the much more easily created biology of viruses or plasmids.

It is clear to us that leaving viruses out of evolutionary, ecological, physiological, or conceptual studies of living entities, would allow only an incomplete understanding of life at any level (Wilhelm and Suttle 1999; Weinbauer and Rassoulzadegan 2004; Suttle 2005). This deep and extensive interaction is too biologically important, from our perspective, to be considered as purely parasitic. Conceived of collaboratively, cellular life is constantly 'bathing in a virtual sea of viruses', within and without every cell, with evolutionarily significant consequences for the past, present, and future of all cellular life-forms (Bamford 2003: 232). In fact, says virologist Dennis Bamford (2003: 235), it is time to consider dividing life into two realms: the cellular realm and the viral one. He believes that only by dealing more thoroughly with a concept of life fully cognizant of the role of viruses will we be able to achieve an adequate view of life even as it applies to its cellular manifestations.

Endosymbionts

Endosymbionts are entities that live inside the cells of other organisms. Some are mutualists while others are more parasitic. Parasites are generally distinguished from other symbionts by their mode of collaboration with their hosts. While endosymbionts have a mutual give–take relationship with their hosts, obligate endoparasites are generally viewed primarily as receivers of benefits and not givers. Increasingly, however, these are being understood as more fluctuating and complex relationships (Valdivia and Heitman 2007). Numerous bacteria are obligate parasites that have reduced genomes and depleted cellular function. *Rickettsia, Chlamydia,*[16] and microsporidia are well-known examples. Microsporidia have lost so many genomic, biochemical, and morphological features that they were once thought to be the most primitive eukaryotes (Keeling and Fast 2002). Now, however, they are deemed to be fungi that are highly adapted to their parasitic lifestyles. Rather than relinquishing their genes to the host genome (as have organelles), obligate endoparasites have simply lost the genes that have become redundant due to reliance on host provisions (Tamas et al. 2001; Timmis et al. 2004). These are usually metabolic and mobility genes, although some of these symbionts retain capacities for intra- and intercellular mobility (Gouin et al. 2004).

Despite the ongoing reduction of their genomes, some of these parasites also acquire and exchange DNA via conjugation and transduction (Darby et al. 2007). Obligate bacterial parasites can be horizontally as well as vertically transmitted, and transmission between mammals and other animals often involves vector organisms such as ticks or fleas (Darby et al. 2007). Another form of symbiosis, 'reproductive parasitism' (Wernegreen 2004), is employed by *Wolbachia*. These are widespread hereditary endosymbionts of insects, crustaceans, spiders, and nematodes. The hosts do not depend on their endosymbionts for metabolism or defence,[17] but the bacteria significantly influence host lives and may induce speciation events by reproductively isolating insect lineages (Weeks et al. 2002; Charlat et al. 2003). *Wolbachia* control the reproduction and development of many of their hosts by biasing sex ratios and reproductive strategy (asexual rather than sexual), as well as feminizing genetic males (Werren 1997; Stouthamer et al. 1999). In addition to being inherited vertically via maternal transmission, *Wolbachia* spread themselves laterally, sometimes to evolutionarily distant insect hosts. Their genes are also transferred laterally (in one case ᵗʰᵉ ᵉⁿᵗᵢᵣe genome!) into insect host genomes (Dunning-Hotopp et al. 2007).

Many mutualist endosymbionts cannot live without their hosts and the hosts are frequently just as dependent on their endosymbionts. They are almost always transmitted vertically from host to host through the maternal line (Wernegreen 2002). Numerous insects are involved in obligate intracellular mutualisms with bacteria, to the extent that separate insect and bacterial lineages are fused into single, highly coordinated metabolic systems (Wu et al. 2006). These endosymbionts frequently live in specialized cells (bacteriocytes) created within the host organism and their primary

[16] Inert *Chlamydia* 'spores' (elementary bodies) exist outside cells but the 'live' form of the organism conducts all its activities intracellularly. Note the parallels with the developmental cycle of viruses, which *Chlamydia* were once thought to be.

[17] However, *Wolbachia* in nematodes do provide host-related metabolic and other physiological functions (Fenn and Blaxter 2006). There is also increasing evidence of insect host benefits from *Wolbachia* infections (Iturbe-Ormaetxe and O'Neill 2007).

endosymbioses are quite commonly associated with secondary endosymbioses (Baumann 2005; Douglas and Raven 2003).

One of the most intensively studied mutualist endosymbionts is *Buchnera aphidicola*, which lives in tight association with its aphid hosts (about a million *Buchnera* cells per aphid) and produces essential amino acids for them. It is vertically inherited from one generation of aphids to the next and its few regulatory genes appear to control its life cycle in relation to its aphid host (Moran and Degnan 2006). *Buchnera* have tiny genomes due to gene loss and no uptake of mobile genetic elements. They are about one-seventh the size of *E. coli* (although *Buchnera* cells are actually larger and contain many copies of the genome), with which they shared a common ancestor about 200 million years ago (Moran and Degnan 2006). Aphids and *Buchnera* co-evolve and co-diversify, meaning the phylogenies of associated lineages map onto each other (Moran 2006). *Buchnera* are commonly classed as endosymbionts but the depth of their dependence on their hosts means that some biologists see these bacteria as closer in status to organelles (e.g., Andersson 2000; Douglas and Raven 2003).

One key difference that is often said to distinguish endosymbionts from organelles is that endosymbiont genomes encode most of their essential proteins whereas in organelles, many of the genes for organelle function have shifted to the host genome and been replaced by a protein import apparatus (Cavalier-Smith and Lee 1985: 378; Thiessen and Martin, 2006). Not everyone accepts this distinction, however, and other commentators see variable degrees of biochemical and cellular integration between host and endosymbiont/organelle (e.g., Bhattacharya and Archibald 2006; Bodyt et al. 2007). There certainly appear to be numerous endosymbionts making the transition from organism to organelle status[18] and any definition of either will have to be based on a continuum of collaborative strategies rather than clear categories of distinct entities (Rodríguez-Ezpeleta and Phillipe 2006; Bodyt et al. 2007). Concomitant with observations about the occurrence of these evolutionary transitions from free-living organism to endosymbiont to organelle appears to be a shift in the language used by biologists: from autonomous 'invaders' to domesticated 'servants' to 'captives' or 'slaves' that have almost totally lost their bacterial identity (e.g., Dyall et al. 2004; Baumann 2005).

Reduced extracellular symbionts

A plethora of bacteria and other microbes live in intimate extracellular liaison with plants, animals, and fungi (sometimes these arrangements are called ectosymbioses or episymbioses). Cyanobacteria, as well as being ancestral to plastids, live in close symbioses with eukaryotes, providing nitrogen fixing and photosynthesizing capabilities through a variety of mechanisms (Douglas and Raven 2003). Some are vertically transmitted and a few free-living cyanobacteria exhibit trends towards genome reduc-

---

[18] Or from an endosymbiont with increasingly limited function to extinction. See Pérez-Brocal et al. (2006).

tion very similar to those in endosymbionts (Marais et al. 2007). Some ultimately obligate symbiotic arrangements have free-living stages, such as the *Rhizobium* bacteria that colonize plant roots and fix nitrogen for their partners.

Fascinating as many of these symbiotic arrangements are (e.g., bacteria that provide 'legs' for ciliates; others that oxidize sulphur for tube worms that lose their mouths and guts as juveniles when colonized by these ectosymbionts), we will focus here on 'transitional' organisms that seem to be on the very edge of 'independent' living. One example is *Nanoarchaeum equitans*, an exceedingly tiny archaeon, which is always described as an organism despite its extremely reduced genome and consequent inability to metabolize, grow, and reproduce independently of another archaeon, *Ignicoccus hospitalis* (Huber et al. 2002). A better-known example is the genus *Mycoplasma*, which consists of very small obligate parasites that are notable for having no cell walls (almost all bacteria do, as do plants and fungi but not animals or most protists[19]). They are usually regarded as the smallest free-living cell although they are heavily dependent on their hosts for amino acid and co-factor biosynthesis,[20] and fatty acid metabolism, especially sterols for membrane maintenance (Fraser et al. 1995; Rottem and Naot 1998). They have lost large numbers of their genes and are considered to have 'little adaptive capability' (Glass et al. 2006: 425).[21]

Because of this reduced genome and restricted function, *Mycoplasma* (*M. genitalium* in particular) have been popular candidates for minimal cell research, in which synthetic biologists attempt to recreate the simplest cellular form of life from synthetic or engineered components. One of the recent breakthroughs in synthetic biology involved 'rebooting' a *Mycoplasma* cell with a genome from a different *Mycoplasma* taxon (Lartigue et al. 2007). Although the experiment was successful, doubts were raised about the transferability of the technique to less closely related organisms and to those with cell walls (Pennisi 2007).

In none of this research, however, is it doubted that *Mycoplasma* is a living organism, so its dependent nature and restricted function are apparently insufficient reasons to consider it in the same light as a virus. One of the characteristics that tends to confer organismal status is genetic autonomy, or the capability of a biological entity to initiate and complete its own reproduction. This status does not obtain for plastids or organelles, however, which are usually perceived as mere parts of the cell in which they are found. The additional biosynthetic and metabolic capabilities of endosymbionts and exosymbionts, no matter how reduced, seem to be essential to the conferral of organismal status. However, given the complete dependence of these processes on

---

[19] All organisms have cell membranes, of course, but not the more rigid cell walls that plants and bacteria possess.

[20] There is increasing evidence, however, that *Mycoplasma* are also intracellular symbionts (e.g., Meseguer et al. 2003). And the genome of *N. equitans* mentioned above is smaller than that of any mycoplasma.

[21] Mycoplasmas do, however, have multifunctional enzymes that have taken on unusual roles.

contributions from the host cell, the grounds for this sharp distinction between viruses and (other) symbionts is far from clear.

Unicellular organisms and single cells

It might seem a strange turn in our discussion to interrogate unicellular organisms as to whether they are alive or not, when nobody has questioned that status. Our point here, however, is to continue to press the question of whether the boundaries of life are clear cut and, in particular, whether cellularity is enough in itself to confer 'aliveness'. Certainly, a single mammalian cell on a Petri dish, for example, is not normally considered a living entity in its own right,[22] in part because of the highly technical requirements for keeping this cell and its descendants alive (Bhardwaj et al. 2006). This ambiguous status is, we believe, the same ambiguity that bedevils our understanding of prions, plasmids, organelles, and viruses. Single animal or plant cells are only truly alive when they are collaborating with other cells. Whether prokaryote or eukaryote, micro-organismal or macro-organismal, cells work together in a great variety of ways, collectively structuring their activities through numerous mechanisms. In the same way that cellular life-forms are only fully functional when collaborating with other cells, so are viruses, plasmids, and prions. Is there a hard line worth drawing between different modes of cellular and subcellular collaboration—between collaboration and exploitation? We think not.

Moreover, even when single cells are considered in isolation, each cell is a complex of collaborating parts. In the case of eukaryote cells, those parts—as we saw in the discussion of organelles—may include once free-living cellular entities. A eukaryote cell, in the minds of some biologists, 'can be likened to a society composed of a nucleus and a crowd of subcellular organelles in which all members cooperate for the common good' (Eberhard 1980: 231). This is a complex collaboration, however, because competitive reproductive relationships may also exist between organelles or plasmids in a cell (for examples of such competition, see Eberhard 1980; Walsh 1992; Paulsson 2002). Such competition can also occur between cells in clones, as when somatic mutations occur in the meristems of vegetatively reproducing plants (Pineda-Krch and Fagerström 1999; Klekowski 2003). Although the philosophy of biology has directed considerable attention to the problem of conflict in the transition from single cells to multicellularity (e.g., Okasha 2004), it has not extended a similar level of scrutiny to intracellular cooperation and competition. We believe this is worth doing for a better understanding of these collaborative relationships between biological entities at multiple levels.

Multicellular organisms

Multicellular organisms, particularly plants and animals, and most notably ourselves, are considered to be 'paradigmatic' examples of living entities (Wilson 2000). Again, we

---

[22] See, however, Puck (1972) for an argument about the autonomy of mammalian cells.

think that this is far from clear, and that whatever aliveness consists of for an animal, for example, it is a much less autonomous state than is usually recognized in discussions of life (especially, but not only, philosophical discussions). The evolution of eukaryotes has largely been driven by micro-organismal interactions, and a variety of modes of dependence between eukaryotes and prokaryotes endures and diversifies in every existing eukaryotic organism. Vast numbers of eukaryotes cannot reproduce, develop, or metabolize without their prokaryote partners. We noted earlier that achieving organismal status is often understood to be the achievement of autonomy. This interpretation can easily mislead our understanding of life and what it is to be alive. Traditionally conceived biological entities are systems elaborated around unique genomes, but to consider them as autonomous individuals is a mistake, we argue: functional wholeness, the basis of any attribution of autonomy, is a characteristic of collaborative interactions, almost always involving diverse entities.

Not only are paradigmatic multicellular organisms more multicellular than is usually supposed (in that a multicellular organism should be understood as including all the entities that interact to achieve shared metabolic and reproductive goals), but even 'simple' prokaryotes could be thought to qualify for multicellular status on this basis. Take, for example, magnetotactic bacteria, which have organelles of magnetic crystals (magnetosomes) that line up inside the cell and are attached to the flagella of the bacteria. The magnetosomes function as compasses and guide the bacteria along local magnetic field lines (preferentially north or south, depending on which hemisphere the bacteria live in). As if this were not astonishing enough, some magnetotactic bacteria live in strictly multicellular arrangements. The individual cells form a spherical group of up to 40 bacteria, constructing an empty compartment in the middle of the group. As well as sensing magnetic lines together and moving in a fully coordinated manner, the groups reproduce together by coordinated cell division. They grow at the same rate (increasing in volume, not cell number) and then simultaneously divide into a new multicellular organism that swims off immediately after separation (Keim et al. 2004b, 2007; Abreu et al. 2007). Most multicellular organisms have a unicellular stage, whereas these magnetotactic bacteria have a strong claim to be exclusively multicellular throughout their life cycle.[23]

More variable in their organization than magnetotactic bacteria and other specialized multicellular structures of unicellular organisms (such as the well-known aggregating examples of *Dictyostelium* and myxobacteria) are other collaborative arrangements known as communities. Prokaryotes and other microbes seldom live as isolated single cells but cohabit in a variety of communal organizations such as biofilms. Micro-organisms that live as parts of biofilms express genes very differently from free-floating (planktonic) microbes, and in patterns that are structured at each stage of the biofilm's development (Costerton et al. 1995; Stoodley et al. 2002). Communities such as

---

[23] Abreu et al. (2007) assign the name *Candidatus* Magnetoglobus multicellularis to this organism ('*Candidatus*' indicates that it has not been cultured).

biofilms (which may be single or multi-taxa), as well as some populations of unicellular organisms, exhibit well-defined cell organization and a functional division of labour that includes specialized cell-to-cell interactions, the suppression of cellular autonomy and competition, metabolic collaboration, combined defence and attack strategies, and the coordination of movement, growth, and reproduction (Dworkin 1997; Shapiro 1998; Kolenbrander 2000; Crespi 2001; Kaiser 2001; Aguilar et al 2007; Cho et al. 2007). Many of these are activities that no individual microbe can accomplish on its own, and the collective behaviour is often achieved with a cost for individual 'altruistic' micro-organisms (if they are perceived through the lens of selfishness).

Some biologists and philosophers may prefer to define multicellularity in ways derived from reflection on animals and plants, and thereby exclude these microbial communities from that category. But certainly any general account of the varieties of biological organization will need to take account of them and explain how they conform to concepts such 'multicellularity', 'invididuality', and 'autonomy'. Do humans, for example, stop at their skin and have to be conceived of as tubular rather than solid in order to avoid incorporating large internal populations of gut microbes? Lederberg, with his concept of 'symbiome', raises the question of whether organisms are necessarily monogenomic or whether a multi- or metagenomic state is the usual state of organismal organization (Lederberg, in Hooper and Gordon 2001; this volume, chapters 7 and 11). Discussions of life and its organization have to take into account the fact that symbiotic relationships are ubiquitous and all organisms, when conceived as the functional wholes that interact with their surroundings, are multi-lineal and multi-genomic.

All multicellular organisms function with the inherited assistance of endosymbiotic partners in interplay with numerous other forms of partnership. All unicellular organisms are infected with phages and other unicellular organisms, and even viruses have their own phages, 'virophages' (La Scola et al. 2008). Although viruses are generally thought of as strictly parasitic, this view may owe more to preconception than to biological fact. The functions of micro-alliances with viruses are only beginning to be investigated, and one early investigative success has been delineating the contribution of cyanophages to cyanobacterial photosynthesis (Lindell et al. 2005). Similarly, the phages that infect the anthrax bacterium, *Bacillus anthracis*, play major roles in the bacterium's capacity to build communities and to produce the long-lived spores that ensure the perpetuation of the cycle of anthrax infections in animals (Schuch and Fischetti 2009). More broadly, the role of viruses as facilitators of genetic variation in multi-lineage communities and as fundamental agents in biogeochemical cycles (Suttle 2005) means they cannot be assumed to be exclusively parasitic and self-serving.

Overall, deep and extensive collaborations between biological entities blur—at the very least—any distinction between so-called individual organisms and these larger organismal groupings of which they are parts (Dyer 1989; Moran 2006). They also call attention to the non-discrete and highly dynamic nature of biological individuals (Rayner 1997). Although this is not our present focus, we should note that great evolutionary significance has been attributed to symbiosis (e.g., Sapp 1994). Symbioses have constituted innovations that have made possible some of the most significant transitions in evolutionary history, as our discussion of mitochondria and plastids made clear.

Characteristics of living biological entities

Common criteria of life

How is it usually decided which of these diverse entities is alive? All the definitions of life in current circulation emphasize particular life-bestowing properties. Some of these definitions take functional criteria (such as reproductive autonomy) to be the most important, whereas others emphasize evolutionary criteria, such as continuity or evolvability, or foreground metabolic or organizational characteristics (Koshland 2002; Pályi et al. 2002; Popa 2004; Szathmáry 2006; Zhuravlev and Avetisov 2006). The most inclusive range of criteria for deciding whether entities are alive or not is derived from exemplars already regarded as unquestionably alive. Animal characteristics often dominate these criteria, which are then modified to include plants, fungi, and unicellular organisms but to exclude entities such as fire and crystals (Chyba and Hand 2005).

Spatial boundedness is widely assumed to be a fundamental criterion of living entities, and is one reason larger biological systems, such as ecosystems, are seldom classified as living entities in their own rights. Boundaries usually consist of enclosing materials such as membranes, cell walls, and skin, which separate internal from external environments and enable internal activities such as metabolism (Popa 2004). Associated with spatial boundedness, and again almost inescapably connected to the project of distinguishing coherent subunits from encompassing systems, are stability and the ability to maintain a buffer against fluctuating environments. However, the interconnectedness of the diverse entities discussed in the preceding text points to obvious dangers in assuming that spatial boundaries can be straightforwardly and uniquely identified (Rayner 1997). The boundaries of a plant and animal are precisely the sites where complex interactions occur between entities generally considered distinct, but these interactions are so closely coupled that we are strongly tempted to see them as parts of the same system.[24]

Perhaps the most widely agreed criteria for being a living thing are *metabolism*, or energy transformation, and *reproduction*, the capacity of entities to make more of themselves. Biochemical transformation of energy from the environment, first to maintain their own structural and functional integrity, and second to reproduce themselves, is a plausible general account of what living things most fundamentally

---

[24] Of course we accept the importance of membranes in the origins and maintenance of life in general, as well as the epistemological necessity of imposing boundaries for both theoretical and experimental biologists. This epistemic function does not, however, require that such boundaries be uniquely and unequivocally identifiable.

do. Metabolism, then, is a basic means of survival for anything alive.[25] For many biologists, this is the most fundamental biological process and the true demarcator of living and non-living entities (Gánti 1997; Luisi 1998). An internal capacity for self-sustainability on the basis of the processing of external resources is a common understanding of organismal function (Luisi 1998).

Our reservations about this criterion are not about whether metabolism is a basic characteristic of living systems, but whether it can effectively be deployed to make the kinds of distinctions into discrete living entities that are generally expected by theorists of biology. The reason for this is that metabolism is typically a collaborative activity involving many of the things that are generally supposed to be discrete living entities. It is generally supposed, for example, that a human, qua discrete biological entity, consists of a lineage of cells deriving in a series of divisions from an original zygote. But a functional human consists also of very large numbers of symbiotic bacteria, in fact amounting to 90 per cent of the cells in the total human system. These microbial cells are deeply involved in the metabolic processes, most obviously digestion, that maintain the functioning of the system (Hooper and Gordon 2001; Gill et al. 2006). Hence, a human, conceived in the way just described, is not capable of performing *autonomously* the metabolic processes essential for its survival. If it is considered sufficient merely to carry out independently some metabolic processes, but not all those necessary for the survival of the entity, then organelles and endosymbionts will count as living entities.

While we noted earlier that reproduction is a necessary feature of life, we also mentioned its inadequacies for a full understanding of life. As we have shown, viruses, organelles, and even prions reproduce themselves. The reproduction criterion is sometimes tied to autonomous reproduction, so that viruses and the like, though they are very effective replicators, are often taken to fail this criterion because they do not reproduce independently and must use 'true' organisms from different lineages to achieve their reproduction. However, it is doubtful whether even paradigmatic multicellular organisms can meet the criterion of lineage-exclusive autonomous reproduction. Those insects in which reproduction is substantially under the control of endosymbiotic *Wolbachia* are one obvious counterexample. But more generally, in so far as reproduction requires the deployment of metabolic processes, as it surely must, it depends also on endo- and exosymbiotic microbes.

Another criterion sometimes proposed as definitive of life is *evolvability* (Ruiz-Mirazo et al. 2004). One highly cited definition of evolvability is that provided by Marc Kirschner and John Gerhart, in which 'evolvability is an organism's capacity to generate heritable phenotypic variation' (1998: 8420). A consequence of taking this as criterial for living entities is that it would include all the entities we have described down to viruses and prions. Our interest in the concept, however, is rather to make a much more general point which, we think, cuts to the heart of the difficulty in

---

[25] The necessity of biochemical transformation rules out phenomena such as computer viruses as candidates for life, because they do not sustain themselves through biochemical means.

defining a living entity. Evolvability, in the sense of Kirschner and Gerhart at any rate, is a characteristic of lineages. Viruses, prions, organelles, unicellular organisms, and multicellular organisms conceived as monogenomic wholes, all form the appropriate kinds of lineages. So, although we agree that these criteria of spatial boundedness, metabolism, reproduction, and evolvability are truly important to understand life, we believe that they are being understood within a framework that misconceives living entities in a fundamentally important way.

Reframing the criteria of life

Our reservations about the above criteria arise from the fact that none of the entities we discuss are the functional entities that interact with their environment and whose success or failure in such interaction determines the success or failure of these lineages. These functional entities are, rather, associations of a variety of such lineage-forming entities. A typical large eukaryote, for instance, is constituted by entities of all the kinds we have distinguished above. We might invoke here David Hull's (1980) well-known distinction between replicators and interactors, but in a very different way from that originally supposed by Hull. Interactors, in our view, are complex systems involving the collaboration of many highly diverse lineage-forming entities. This sort of inter-actor, we also suggest, is the most fundamental unit of selection. This perspective has radical implications for the way we think about evolution. It would entail but obviously go beyond contemporary concepts of group selection in multi-level selec-tionism (Sober and Wilson 1998).

Amongst those implications is the importance of the notion of collaboration, which is seldom proposed as a criterion of life (although see, e.g., Lezon et al. 2006, for an emphasis on cooperation). It is hard to imagine life that is not collaborative, in the sense described above, both at the intracellular and the intercellular levels, and we suggest that collaboration is, therefore, one of the central characteristics of life. To treat it as such it will be necessary to specify more carefully what the relevant sense of collabora-tion entails. We do not want to rule out automatically even simple chemical systems. Some chemical aggregations exhibit growth, reproduction (leading to lineage forma-tion of varying persistence), error correction, and environmental sensitivity (Schulman and Winfree 2007; Weber 2007). It would be surprising if these features were not to be found in the chemical world, because otherwise it would be hard to imagine how life could have originated. Our continuum view of life is open to chemical systems being sometimes describable as living systems, though perhaps it is likely that they will meet the relevant criteria only transiently. Because biological entities in our conception are series of dynamic and diverse collaborations, boundaries are flexible and unfixed. Any claim that something is a living thing needs to be assessed in relation to the general characteristics we describe.[26]

---

[26] Interactors thus conceived also rule out the whole planet as a candidate for evolving life, because although the planet could be conceived as metabolizing (in a highly collaborative way), it does not interact

But more important than any attempt to specify limits on what is or is not alive will be to emphasize the contrast between this perspective on life as collaborative and the much more familiar assumption that life is fundamentally selfish and entails competition between reproductively and metabolically autonomous organisms. The outlines of our response to the view that only selfish entities will win out in the battle of all against all should by now be clear: the unit of selection, the entity in which selfishness may perhaps be expected as the norm, is a collaboration of many different lineage-forming entities.

The context in which the latter evolve, then, is quite typically one of collaboration. We said that the collaborative whole may perhaps be expected typically to display selfishness. But, of course, this assumes that there is some natural terminus to the process of collaboration. We hypothesize that competitive activity is a transitional rather than a terminal state and that such temporarily competitive wholes will exhibit a strong tendency ultimately to compete most successfully by engaging in new levels of collaboration with similar or different entities. We see the emergence of sociality as an instantiation of such a process as are, more generally, the evolutionary transitions that have been highlighted by the work of John Maynard Smith and Eörs Szathmáry (1995). Our spectrum of biological entities exemplifies a number of different forms of collaboration that are central to such an evolutionary schema. As we have emphasized, our concept of collaboration assumes no sharp boundary between selfish and cooperative interactions, something surely to be expected if the former is inclined to evolve into the latter.

We have certainly not exhausted the criteria that have been proposed as characteristic of living entities. We have not explicitly considered, for example, environmental responsiveness, the ability to detect and respond appropriately to salient features of environments, or development, the recurrent production of the characteristic stages of a life cycle (although much that we have said has addressed these criteria implicitly). We do not mean to minimize the significance of these, and perhaps other, distinctive characteristics of living systems. What we do argue is that a focus on metabolism and reproduction, widely agreed to be fundamental features of life, has the additional virtue of drawing attention to a characteristic that has been greatly underemphasized: that of collaboration. That this has been downplayed is a readily intelligible consequence of the importance that has been attached by biological theorists to competition (Roughgarden 2009). But for this very reason giving collaboration proper emphasis could provide important fresh insight into the nature of evolutionary processes because it affects how we conceptualize the entities and activities central to evolution.

---

with other such wholes. It also lacks any means of reproducing itself. We are more ambivalent about ecosystems, which may frequently interact, but a case would have to be made for ecosystems forming a lineage that was more than mere continuity (as a substitute for replication).

Autonomy and the origins of life

Our collaborative interpretation of life suggests that it is possible to sidestep the usual problems associated with defining life. Although we do not claim to have provided a definition of life, we do believe we have offered a view of living matter that offers a flexible resource for understanding the many ways in which life can be organized. The tension between replicating lineages as one criterion of life, and metabolic self-sustainability as the other, can be reconciled by taking a much more interactive view of metabolic processes and by reconceiving cooperation and competition within a broader framework of collaboration. Life, according to our analysis, occurs at the intersection of lineage formation and (typically collaborative) involvement in metabolism. Entities that are problem cases, such as viruses, can be understood as alive when actively collaborating. When not collaborating, they have at most a potential for life. We invite our readers to apply our framework further along the spectrum than we have gone, to various chemical and physical systems and to ecosystems.

What of the autonomous individual organism, often the conceptual target of attempts to define life, and the thing that is assumed by models of evolution through competition and selection? To the extent that such individual autonomy requires just an individual life or life history, then it surely applies much more broadly than is generally intended by biological theorists. Countless non-cellular entities have individual life-histories, which they achieve through contributing to the lives and life-histories of the larger entities in which they collaborate, and this collaboration constitutes their claim to life. But—and this is our central point—no more and no less could be said of the claims to individual life-histories of paradigmatic organisms such as animals or plants; unless, that is, we think of these as the collaborative focus of communities of entities from many different reproductive lineages. In much the same way, whatever sense we might try to make of the Dawkinsian idea of selfish genes, molecular replication is *always*, and has always been from the pre-cellular molecular community to the present, the achievement of ensembles of molecules, not of individual molecules (Segré and Lancet 2000).

It is entirely reasonable to think of autonomy as centrally exhibited in collaboration rather than just rugged independence.[27] Assuming that this kind of autonomy is what is needed to be a living thing, our account therefore includes viruses as not only living matter, but as full-blown living entities when they enter cells and interact with the cell's metabolic capacities. As virions, they are still lineage elements but are temporarily disengaged from metabolic collaboration (likewise bacteria such as *Chlamydia* in their inert spore-like state and perhaps even many plant seeds and fungal spores). This is why we suggested above that viruses should strictly be described as having developmental cycles rather than life cycles.

---

[27] It is tempting here to invoke Kant's analysis of autonomy as the possibility of conformity to duty, an essentially socially defined concept, and his rejection of the notion of autonomy imagined as following no more than the pursuit of contingent individual interests or desires.

Taking this perspective not only renders unproblematic the idea of an entity being sometimes living and at other times non-living but also reinforces the idea of life and the evolution of life as a continuum of collaborativity. Given the acceptance that life has evolved from a chemical context, ruling out self-replicating complexes of chemicals and molecules on the grounds that they are not cells seems misguided. A commitment to life as exclusively cellular and monogenomically organismal would mean that the origins of life must involve a single leap from fully non-living to fully living, something that is conceptually difficult to accept and, for that matter, provides a natural target for creationists to insist on the need for supernatural intervention. The spectrum of biological entities we have described shows that an inflexible dichotomy of life and non-life is, in any case, highly problematic, even for making sense of the entities that now exist. Our more generous framework can encompass a range of theories about the organization and evolution of pre-cellular life which give prions, plasmids, and viruses important roles, as well as other macromolecular complexes (e.g., Eberhard 1990; Koch 1995; Kado 1998; Rode et al. 1999; Lupi et al. 2006, 2007).

We also think that our thesis of multi-modal, interconnected and overlapping life processes suggests a more continuous vision of evolutionary history. Many discussions of early life posit a radical transition from a community of genetic exchange to one of restricted vertical inheritance, cellular autonomy, and stable genealogy (Woese 2005; Dawkins 2008). Although some biologists believe that pre-cellular life is best conceived as 'unselfish' communality in which genetic resources are shared (Woese et al. 2000), others such as Dawkins presume that pre-cellular life was driven by selfish replication, and that promiscuous horizontal exchange simply *extends* the opportunities for selfishness (Dawkins 2008). Rather than restricting some evolutionary processes to a discontinued past, we prefer to incorporate them into a schema that allows for the continuity of lateral gene transfer as an important characteristic of today's collaborative evolution.

We find here the reflections of Norris et al. (2007) and Hunding et al. (2006) very helpful. They argue that life evolved as a 'diverse interacting community of molecules'—a 'pre-biotic ecology' that implies a more ecological and community-based view of any biological entity, pre- or post the evolution of cells. Their model describes

the emergence of life as a functional ecological system through a process of integration from diverse components, not as a single entity . . . there is no identifiable point at which life emerged. Rather, [it is] a continuous *process* by which increasingly complex, integrated, self-replicating, autocatalytic, module systems evolve new properties in tandem with their environments. (Hunding et al. 2006: 409–10)

We believe this sort of dynamic system-based scenario fits more appropriately what we know of the rest of the evolution of life.[28] Evolutionary history suggests that life

---

[28] Dynamic system-based scenarios of life are a general idea more precisely formulated in work by theoretical biologists such as Tibor Gánti, Robert Rosen, Stuart Kauffman, and Humberto Maturana and Francisco Varela. Our emphasis on collaborative interaction suggests congruences with this sort of work, but we have yet to explore these in any specific way.

involves a range of co-evolving hierarchies, and that non-life and life share a huge and biologically significant territory that buffers and makes more complex any account of either. Ecology presents us with scenarios of collaboration at least as compelling as those that highlight competition, and the former are rapidly increasing our understanding of the macrobial and microbial world (see Chapter 11). Thinking of life as the result of the intersection of lineage-forming, metabolically collaborative matter, organized within different interacting levels, allows a smooth transition from the earliest living matter to standard examples of life and beyond them all the way up to contemporary ecosystems. A general account such as ours is not, and need not be, definitional. It is, however, sufficient to encompass what is known about an ever more striking variety of biological entities and their evolutionary histories, and to reorient approaches to life around a biologically realistic interpretation of collaborativity.

13

Emerging Sciences and New Conceptions of Disease: Or, Beyond the Monogenomic Differentiated Cell Lineage[1]

Summary

This chapter will begin with some very broad and general considerations about the kind of biological entities we are. This exercise is motivated by the belief that the view of what we—multicellular eukaryotic organisms—are that is widely assumed by biologists, medical scientists, and the general public, is an extremely limited one. It cannot be assumed a priori that such a general view will make a major difference to the science or practice of medicine, and there are areas of medicine to which it is probably largely irrelevant. However, in this case there are important implications for medicine, or so I shall argue. In particular, it enables us to appreciate fully the potential medical significance of some of the most exciting contemporary advances in general biology, in such fields as epigenetics, metagenomics, and systems biology; and part of this significance is that these advances have raised serious doubts about how we should understand the biological individuals that medicine is generally assumed to aim to treat.

The monogenomic differentiated cell lineage

My subtitle is not pretty. The phrase 'monogenomic differentiated cell lineage' does refer, however, to a very important idea, and indeed a standard scientific view of ourselves. We are, of course, multicellular organisms. One of the things, we assume, that make the many cells of which we are composed parts of us is that they form a lineage: our bodies are composed of multiple cells all originating from one single cell formed in the act of conception through the fusion of two gametes. This lineage structure provides more than merely historical connections between our constitutive

---

[1] This chapter has benefited from comments from Sabina Leonelli, Pierre-Olivier Méthot, Staffan Müller-Wille, and Maureen O'Malley.

cells. It also provides them with unique characteristics, notable among which is the unique (except, briefly, for monozygotic twins) sequence of nucleotides in their genomes. But, finally, it is the highly consistent and structured differentiation of these cell lineages, human as well as countless others, that provides the remarkable array of wonderfully structured organisms that have understandably fascinated and obsessed generations of biologists.

The monogenomic differentiated cell lineage (MDCL) is a powerful image of a vast number of forms of life because it provides a rich explanatory framework. It's true, and something I have been much concerned to emphasize in recent years, that there is a vastly greater number of monogenomic cell lineages that are not differentiated, namely the countless single-celled bacteria, archaea, and protists. But given that the present topic is medicine, and given also the centrality of ourselves and creatures somewhat like ourselves in so much of biological theory, the MDCL is an appropriate focus for this chapter. It is at the core of a highly developed and sophisticated approach to the question why all these life-forms exist, in evolutionary theory; understanding the replication and differentiation of cells within MDCLs is the central project of contemporary developmental biology. MDCLs, then, are an important and influential idea.

Why do I use this clumsy term? What's wrong with 'multicellular organism' or, since these are evidently what we are most interested in, just 'organism'. In recent work Maureen O'Malley and I have questioned the assumption that organisms just are MDCLs or, in the case of simpler life-forms, just cell lineages (Chapters 10–12, this volume). In particular we have stressed the fact that MDCLs typically depend for their successful differentiation and survival on reciprocal interactions with a highly diverse set of symbiotic microbes. In our own case, complex microbial communities cover the entire area of our boundaries with the world outside. They swarm on our skin and in all our body cavities (oral, nasal, genital, etc.) and trillions occupy the length of our digestive tract.

This still does not show that there is anything wrong with the MDCL concept. What it does indicate, however, is that the MDCL is an abstraction from a more complex reality. Again there is nothing wrong with this: in fact this is plausibly the inevitable character of biological concepts. Living things are complexly intertwined and interconnected in multiple ways, and studying them requires identifying parts of this interconnected whole that have some kind of independent coherence. Consider, for instance, a cellular process such as the Krebs cycle. This is an essential component of the processes by which cells derive energy from carbohydrates and other energy sources and its elucidation was rightly seen as a major achievement in the understanding of how cells work. Not only did it illuminate a particular metabolic pathway, but it also served as a paradigm of the biochemical analysis of living processes. But no one imagined that the Krebs cycle was the sort of thing that could in theory exist independently of its cellular context. One reason for this is that it is in constant interaction with other metabolic pathways that provide its essential chemical resources and to which it contributes its own products. The particular series of reactions distinguished as the

Krebs cycle are sustained by this dynamic interaction with a variety of simultaneous and interlocking processes.

This finally brings us to the real problem with the MDCL, that it *is* liable to be understood as just such an independent entity, and indeed it is often treated as what the organism essentially is. Philosophers will be familiar with the idea promoted—or perhaps just assumed—by Saul Kripke that the essence of an individual human is the origination from a particular zygote, or fertilized egg (Kripke 1980). This is a paradigm case in the revival of essentialism over the last 40 years. But similar assumptions can be found in more biologically sophisticated thinkers than Kripke. Particularly salient here is the gene-centred view of evolution famously promoted by Richard Dawkins (1976).[2] In this picture evolution is fundamentally a succession of genomes passing down through evolutionary time. The cell provides the context in which the genomes can function to guide the development of organisms, and the environment determines which organisms, and hence which guiding genomes, will thrive and multiply. The function of MDCLs in this picture is to determine the trajectories of a different kind of cell lineage, the sequence of successful zygotes that carry evolving lineages into the future. The monogenomic character of the MDCL is essential for it to play this evolutionary role.

Even this abstraction has its uses. To its credit are a widely admired tradition of modelling in population genetics and a rigorous analysis of such evolutionary processes as kin selection. It does have some very serious limitations, however, not least the tendency to reinforce the idea of the MDCL as the fundamental—real, true—account of the multicellular organism. Since, as I shall argue later, the MDCL really is an inferior or even inadequate conception of the medical individual, the tendency to essentialize it, thus making it unavoidable even for purposes to which it is ill-suited, must be robustly resisted. In the attempt to weaken the hold of this picture let me offer a quite different perspective on evolution.

For over three billion years of evolutionary history, 80 per cent of the history of life, there was nothing but microbes. This is a self-evidently important point to bear in mind for assessing general views of evolution; we would do well to avoid general theories that depend on properties that microbes do not possess. Microbes do not evolve as MDCLs. They have often been thought of, however, as MCLs, or clones, monogenomic but undifferentiated cell lineages. As a matter of fact they do not solely evolve this way, but engage in frequent transfers of genetic material across lineages, a realization that is transforming a great deal of evolutionary thinking. But this will not be my main concern here. I want rather to emphasize the importance of the background of these aeons of microbial evolution for understanding the emergence of multicellular organisms. Contrary to what seems sometimes to be assumed, multicellular organisms did not just separate themselves from the microbial background and

---

[2] There is a further problem here in the tendency to reduce the cell to the genome. For present purposes I shall treat Dawkins's view (charitably) as acknowledging the whole cell as the minimum unit of reproduction.

begin to evolve their own transcendent multicellular lifestyles. Rather, they have evolved as deeply embedded components of the complex microbial consortia that long precede them.

The earliest multicellular organisms may perhaps be seen as little more than homes for microbial consortia. Sponges, some of the earliest multicellular organisms to evolve, are hosts to populations of microbes that may make up over half their weight. Some species host photosynthetic bacteria, which provide them with energy, but in most cases the nature of the interaction is poorly understood (Webster and Blackall 2009). However, it is known that these microbial populations are often very consistent in their constitutions and this is despite the daily passage through them of thousands of litres of sea water, itself teaming with bacteria. The stable bacterial populations are presumably actively maintained by the system as a whole. Or consider the tube worms that live around hot vents in the deep ocean, and live on products generated by resident microbes that subsist entirely on the metabolism of sulphides. Even these, I should stress, are fully developed MDCLs. The structure of the MDCL is, however, somewhat less likely to obscure awareness of the reciprocal relation between MDCL and microbial symbionts than is the case for more immediately impressive organisms.

MDCLs and their symbiotic partners

My point in this chapter is not primarily to consider issues in deep evolutionary history. However, the recollection that MDCLs evolved against an ancient and entrenched background of microbial life should help us to appreciate that even today they exist typically in intricate and obligate relationships with vast numbers of microbial symbionts. It is easy to underestimate the importance of the fact that microbes are almost all invisible. Imagine for a moment that the world was suffused with a lurid green light the intensity of which was proportional to the prevalence of micro-organisms. Bare soil would glow brightly, registering the billions of microbes in every cubic centimetre, as would damp surfaces—streambeds and suchlike—with their inevitable coverings of microbial biofilms. Other less hospitable inert surfaces would probably display only a much lighter dusting of transient cells. Water would tend to show a diffuse and variable glow from the many planktonic bacteria dispersed throughout. But of particular importance would be the strong light suffusing all or almost all macro-organisms (or macrobes), whether plants or animals. Leaves, for instance, are homes to diverse communities of microbes, changing with diurnal, seasonal, and occasional fluctuations in the physical environment (Hirano and Upper 2000). Similarly the skin of animals, including our own, swarms with diverse and rapidly changing communities of microbes—about one trillion on a typical human skin. Still invisible, for obvious reasons, would be the much larger communities occupying the digestive tract. This vision might reduce the temptation to see macrobes as having diverged from, and somehow left behind, their microbial ancestors: everywhere macrobes are deeply and inextricably associated with microbial life. It might also problematize the even more widespread assumption that microbes are primarily dangerous things that constantly threaten our well-being. I shall say more about that in a moment.

To emphasize the fundamental importance of the relations between MDCLs and microbial symbionts, Maureen O'Malley and I have argued that the organism, understood as the functional entity that interacts with the rest of the biotic and abiotic environment, is not the MDCL, but a symbiotic whole (Chapters 11 and 12). A particularly significant implication of this is that it draws a wedge between the elements that make up genetically cohesive evolutionary lineages (micro-organisms and MDCLs) and the organisms that interact with the environment and which are, therefore, also the primary units of selection.[3] Again, cell lineage looks less fundamental to the distinguishing of biological things than is generally supposed. That, however, is a matter for another time.

For my present purposes a slightly different point about the picture I have been sketching is crucial. Microbes do not just live on human skin, for instance, because it is a convenient living space. Or anyhow, if they do so, it is made a convenient living space because the human MDCL requires that it be occupied by appropriate microbes. Although the benefits that humans derive from their relationship to microbes is better understood for the case of gut bacteria than for those on the skin, information on the latter is beginning to emerge. For example it appears that bacteria produce a substance that plays a vital role in suppressing inflammation following injury (Lai et al. 2009). More generally, it seems that many skin diseases involve some kind of disorder of the microbial skin communities. This ties in with the broader hypothesis that the microbial communities that occupy pretty much the entire interface between the MDCL and the outside world should be seen, among other things, as an essential part of the immune system, as deciding, in effect, which microbes are allowed to settle in the broader system and which are a threat to it.[4] It is worth recalling that it has been traditional to consider the human organism as a tube, and therefore to maintain that the digestive tract is on the outside of the organism. On this view gut bacteria are, like skin bacteria, populating the outside of the organism. As I have indicated that for most purposes I take the symbiotic microbes to be *part* of the organism they occupy, I must clearly take a somewhat different view: either the surface of the tube extends a little further than previously thought or, perhaps better, the inside of the digestive tract is in fact part of the organism. It doesn't really matter which; the traditional view is at least useful in

---

[3] In terms of Hull's well-known distinction between replicators and interactors, we are saying that an interactor typically contains many different kinds of replicators. However, I am sceptical of the ultimate viability of this distinction. From the developmental systems perspective that I advocate the whole interactor replicates itself, but it does so, in part, by virtue of the replication of many other replicators (which are also at their own scales interactors) of diverse kinds.

[4] Researchers concerned with the incidence of serious fungal infections of the skin that affect amphibians have found that application of a bacterium with fungicidal capacities to the skin can be an effective treatment (Harris et al. 2009). Comparable therapeutic uses of microbes in human medicine are at least an interesting possibility.

pointing to continuities between gut and other symbiotic microbes in the human system.

For the case of gut bacteria, at any rate, the interdependence of microbes and MDCL is increasingly widely recognized on the basis of a rapidly growing body of evidence. The role in digestion has long been appreciated, but increasing interest has focused on contributions to the immune system and to development. Studies of animal models confirm that symbiotic microbes regulate the expression of many genes during development (Rawls et al. 2004). The point I particularly want to stress, and which is illustrated in both these cases, is that the boundary between the MDCL and the outside world is a complex symbiotic space and that adaptive responses to the outside environment by either the MDCL or the entire system, is routinely mediated by the behaviour of this symbiotic space.

Epigenetics

This last point makes a connection with the other major development in biology generally and human biology in particular that I want to emphasize, epigenetics. Epigenetics is an elusive concept, defined in more or less subtly different ways by different biologists. Etymologically it appears to originate in the contrast between preformationist theories of inheritance, in which development was the pure unfolding of a pre-existent pattern, and epigenesist, in which the elements of an organism were accreted one by one throughout the developmental process. Contemporary usages are generally traced to C. H. Waddington, and especially his metaphor of the epigenetic landscape. This metaphor compares the increasingly determinate fate of cells in development to the topography of a ball rolling down a complex landscape. Ridges separating valleys represent degrees to which the fate of a cell lineage has been determined one way or another; the existence of such separate valleys emphasizes the existence of distinct possible developmental pathways. This, finally, leads to the narrower contemporary usage, in which epigenetics studies modifications to the genetic material but, crucially, not to the DNA sequence, which influence the expression of genes. That is to say, epigenetics is now understood as embodying a theoretical understanding of how the fate of cell lineages within an MDCL is determined.

The acceptance of this contemporary version of epigenetics still leaves a central question open. Epigenetics could be a theory of the way that initial gene sequence determines the differentiation within a monogenomic cell lineage. Initial gene products, for example, could produce epigenetic changes in chromosomes that caused changes in the subsequent history of that lineage. No doubt this is part of the story. If it were the whole story then, paradoxically perhaps, epigenetics would have turned out to be an explanation of preformationism. But it is not the whole story. In fact one of the fascinating aspects of epigenetics is that it appears that we need to reconcile its roles in two almost opposite processes. On the one hand epigenetics clearly has a fundamental role in the explanation of the consistency of biological form, the ability of

MDCLs to produce basically similar MDCLs exhibiting, therefore, essentially the same pattern of cell lineage differentiation. But on the other hand it seems increasingly clear that epigenetic processes also serve to mediate the response of the developing organism to the environment, i.e. to provide different patterns of development in response to different environments.

Perhaps there is nothing deeply paradoxical about these contrasting roles for epigenetics. Complex biological systems arise only from other similar systems. They could not exist if they did not have the ability to generate systems extremely similar to themselves. But the replication of absolutely identical systems starting with a system reasonably well adapted to its environment would, assuming a changing environment, inevitably produce systems that gradually became less and less well adapted. So the conflict between the need to reproduce accurately and the need to track environmental change is an inescapable one. A familiar answer to this problem is, of course, the Darwinian appeal to evolution by natural selection. In its currently orthodox neo-Darwinist version organisms are seen as adaptively tracking their environments by endogenously generating random changes some of which will, with luck, produce adaptive advantages.

But of course this is not the only way that organisms adapt flexibly to their environments, and much adaptation takes place on a far shorter timescale than the Darwinian process just mentioned. Many readers will be familiar with Mary Jane West Eberhard's (2003) classic work on developmental plasticity as not only a way for individual organisms to adapt to the environment, but even as a crucial driver of evolution. That organisms do exhibit some adaptive plasticity in development is hardly news. The phenomenon is perhaps most familiar in plants, which can adopt quite different forms according to the availability of light, nutrients, and so on. Or, for that matter, the fact that humans growing up in England learn to speak English while those growing up in France learn to speak French is an example of the way that nervous systems not only enable immediate adaptive responses to salient features of the environment (food, predators, mates, etc.) but also make possible the acquisition of different behavioural dispositions or capacities honed by the particular environments they have experienced.

But although the phenomenon of developmental plasticity is not news, epigenetics indicates hitherto unperceived subtleties. A much-discussed example is of the impact of food scarcity on early stages of foetal development. A classic case has been the epigenetic consequences of the Dutch hunger winter of 1944–5, which appear to have had epigenetic effects on foetuses that remain detectable in individuals 60 years later (Heijmans et al. 2008). Although it has been reported that these effects are heritable to subsequent generations, this claim remains controversial (Morgan and Whitelaw 2008). Environmental epigenetics is beginning to explore the ways in which exposure to a variety of toxins can cause developmental changes that may be transgenerationally inherited (Bollati and Baccarelli 2010). And neurogenetics is beginning to explore the ways in which parts of the brain, especially the hippocampus, react with highly dynamic epigenetic changes in response to various environmental conditions (Covic et al. 2010). As I shall discuss in more detail below, a very important set of features of the environment to which epigenetic changes are reactions are states of the complex cellular communities that constitute the boundary between an organism and what is external to it.

What is the central point so far? Although nobody much subscribes officially nowadays to preformationism, it is very easy to maintain central aspects of that philosophy. It is easy to think of the organism as something that has an intrinsic nature that is realized in development and then finds itself in some kind of interaction with the natural environment that surrounds it. No doubt the revival of essentialism in recent years will lend support to such ways of thinking. And even without essentialist prejudices, the phenomena of reproduction, the constant reappearance of strikingly similar forms, will naturally encourage the thought that the form must somehow have been there all along. The neo-Darwinist account of evolution, finally, as composed only of random, internally generated changes, also assumes a fundamentally endogenous view of development.

There are many things wrong with neo-Darwinism, but the point I need to stress here is the error of this treatment of development as endogenous. This is something that has been emphasized for a number of years by developmental systems theorists, who have insisted on the importance of non-genetic aspects of inheritance including the construction of an external niche in which the organism develops (Oyama et al. 2001). What I have said so far is entirely congenial with that framework. What I have particularly wanted to stress is the need for a very close scrutiny of the complex and dynamic interface between an MDCL and its surroundings. And this, finally, brings us to the implications for medicine.

Medical implications of going beyond the MDCL

I could begin by noting that of course medicine should care about how its primary object—the human body—is to be understood. However, this is more than just a philosophical matter. Many of the most interesting developments in our understanding of disease and disease aetiology are emerging in the context of a growing awareness of the complexity of the boundary between the organism and its external environment.

Consider the human microbiome project, what might be considered the remaining 99 per cent of the human genome project after the completion of the sequence of the human MDCL (see http://nihroadmap.nih.gov/hmp/). One explicit aim of this project is the discovery of correlations between human health and characteristics of, or changes in, the human microbiota. And some such correlations are already being explored in detail.

The intestinal bacterium *Enterococcus faecalis* has a dichotomous metabolism: under normal circumstances it has a respiratory metabolism, but when not provided with exogenous haematin it reverts to a fermentative metabolism that generates potentially harmful superoxide (Allen et al. 2008). Further exploration of the consequences of this change revealed substantial changes in gene expression in the colonic mucosa, the internal lining of the intestine. These changes affect inflammation, apoptosis (or cell death) and cell-cycle regulation (Allen et al. 2008) and there is a natural suggestion that these phenomena may be implicated in colon cancers. But crucially, this is not a case of invasion by hostile alien organisms, but a potentially pathological behaviour of organisms that are a normal, and generally desirable part of the overall system.

That colon cancer or inflammatory bowel disease might be linked to conditions of the gut bacteria may not seem so surprising, though the fact that this appears to work through modulation of gene expression in the 'human' gut cells might be more unexpected. But this is by no means the limit of the implications of gut microbiota for health. Currently research is ongoing on a suspected link between microbiota and breast cancer,[5] a hypothesis that was suggested, interestingly enough, almost 40 years ago (Hill et al. 1971). If this seems surprising, one should reflect on the very familiar suggestion that many cancers and indeed most of the most serious non-infectious diseases are strongly linked to aspects of the diet. If the excessive consumption of the wrong kinds of fat or the inadequate ingestion of fruit and vegetables are indeed, as we are increasingly told, the major determinants of disease, then the suggestion that that influence might be transmitted by the population of microbes in our digestive tract would hardly be shocking. Put together with the observation that the gut microbiota can modulate the expression of genes within the human MDCL it becomes at least plausible that this may be a fundamental site at which disease or health is determined throughout the body. But again I must emphasize disease *or health*. The suggestion is that a functional, indeed essential, part of the system can malfunction and cause illness. The fact that pathological effects are prima facie distant from the parts that are causing the malfunction is indicative of the depth of involvement of the latter, the microbiota, in the overall system.

Consider the question how we should best allocate cancer to one of the various branches of contemporary biomedical science. Prima facie, one might think, it was a cytological disorder. Observations of the close similarity between stem cells and cancer cells (both involve cell lineages the fate of which is not yet fully aligned with the orderly development of the MDCL) fit within this disciplinary domain. For a long time, on the other hand, we have been told that cancer is turning out to be a genetic disease. Many genes (that is, particular genetic variants, or alleles) have been identified as being implicated in the aetiology of cancers and there is certainly no question but that there is a genetic aspect to all or most cancers.

Cancer-causing genes are generally genes supposed to be involved in the suppression of tumours that have been damaged by mutations, deletions, insertions, and so on. However, the failure of a gene to be transcribed may also have epigenetic causes as may

---

[5] See http://www.rush.edu/rumc/page-1262026413817.html.

also up-regulation of gene expression. So epigenetics is very likely to be a parallel possible cause of cancer, and may well be a more important one than gross genetic damage; certainly a great deal of current research on cancer aetiology is focused on epigenetic issues.

It is also a very familiar fact that much cancer is environmentally induced, by toxins such as tobacco smoke or asbestos. This could be mediated through genetic changes (mutation, etc.) or through epigenetic effects. As mentioned above, one active area of research in epigenetics concerns the epigenetic response to toxins, and it is very plausible that this will start to provide some mechanistic substance to epidemiological work on the carcinogenic properties of toxins. Saturated fats and other dietary elements currently seen as unhealthy may not, perhaps, be properly classified as toxins, but their impact directly on human microbiota and indirectly thereby on epigenetic features of cells in the human MDCL may turn out to be an important pathway from environment to cancer. I don't, of course, propose to advocate one approach to the study of cancer as of pre-eminent importance. The point is rather to note the complex interactions between environment, symbionts, genetics, and epigenetics that are likely to be implicated in the misregulation of cell replication.

Pathological epigenetic effects are surely not the only medically significant feature of human symbiotic microbial communities. One fascinating thought is that the microbiota are active at various timescales in maintaining a stable and sustainable relationship between the organism as a whole and the biotic environment, in part by recruiting organisms or genetic material from the environment. One striking example of this has been the recruitment by Japanese human microbiota of genes from generally marine microbes, which facilitate more efficient metabolism of raw fish (Hehemann et al. 2010). This is an example of microbes mediating a rapid evolutionary response to an environmental pressure. Rapid generation times, high mutation rates, and lateral gene transfer are several convincing reasons why we should expect the microbiota to be the part of the whole human system best able to respond rapidly (and evolutionarily, i.e. heritably) to environmental changes. The possibility of recruitment of new types of cell to the system is another potentially powerful such mechanism. This, as opposed to the rather dubious programme pursued as Darwinian medicine, is arguably the really interesting area for evolutionary medicine. This idea has been developed in recent work by Pierre-Olivier Méthot (2010).

A fascinating and related topic for further study, but one which seems hardly yet to have been broached in the scientific literature, is the potential role of viruses in mediating even more rapid such responses. Here again cancer research provides remarkable if tantalizing insights. A lot of recent research has suggested an important role for viruses in the aetiology of cancer, for example high incidences of human papilloma virus have been found in lung tumours (Klein et al. 2009). However, there is also quite promising research on a virus known as the Seneca Valley Virus-001 that appears to have considerable potential for selective destruction of tumour cells, and is currently being investigated as a possible therapy for cancers of the lung (Venkataraman et al. 2008). It is at any rate an intriguing possibility that viruses may have far more diverse roles in biological systems than merely the pathological, parasitic ones with which we are most familiar.

Returning to the parallel consideration for cellular microbes, consideration of the inclusive organism that comprises all of the cells that together are required for successful and sustainable functioning should make one worry about the extent to which it is still assumed that microbes are invariably the enemies of health; or as the point appears in typical advertisements for cleaning products, the only good bacterium is a dead one. This idea has been aptly referred to as the 'Pasteurian Paradigm'. It is in fact difficult to find serious questioning of this perspective outside the realm of alternative and not always evidence-based medical practices; and of course no one should question that some very dangerous micro-organisms are responsible for some very nasty diseases. One serious discussion explicitly questioning the Pasteurian Paradigm focuses, interestingly enough, not on medicine, but on craft cheese-making with unpasteurized milk (Paxson 2008). This paper is mostly concerned with the ethnography of small-scale cheese-making, but it clearly raises the thought that food products properly made from local microbes might have beneficial effects in facilitating adjustment between organism and environment. (It should be stressed that this is not a naïve microbophilia: makers of unpasteurized cheese take great pains to ensure the access by only welcome microbes to their products.)

Consideration of this relation between organism and microbial environment raises the whole question of immunology. The major tradition in immunology has unsurprisingly begun with the concept of the MDCL, and has seen its major problem as explaining how the organism identifies alien cells—pretty much any genomically different cell—and eliminates them. This perspective clearly has a problem in understanding the relation of the MDCL to symbiotic microbes but, recalling the idea that the human (or other vertebrate) body is a tube, and the inside of the digestive tract is therefore on the outside, the solution is not too difficult: it is simply a matter of maintaining the boundary and making sure alien (i.e. microbial) material remains on the outside. Important recent work by Thomas Pradeu (2012), however, shows how much work in immunology supports a rather different picture, in which the immune system consists of a continuous and dynamic process of monitoring all the cells in the system, and neutralizing and disposing of whatever biological objects fail to pass rigorous quality control tests. This perspective reminds us that the immune system doesn't merely target alien material, but is also busy disposing of dead and diseased material from the human MDCL. Much more interestingly, it presents the relation between the MDCL and what I would describe as other parts of the human system (symbiotic microbes) as an active process of maintaining an appropriate equilibrium at the organism's boundaries, while constantly monitoring and responding to the presence of potentially harmful cell-types.

This alternative perspective on the immune system is an important ingredient in the view of the organism that goes beyond the MDCL. On the MDCL conception, the organism is a set of relevantly homogeneous cells warding off attacks by other, hostile, cell lineages. The extended conception looks at the functioning of the system in relation to what is outside it, and observes that a proper account of functioning requires inclusion of much more than the MDCL. The immunological perspective adds another criterion for distinguishing what is and what is not a part of the system, namely by analysing a process through which the system itself determines what biological entities are and are not welcome constituents of the whole.

Metaphysical aside

I have advocated a view of the human organism that includes an extended set of elements that are together required for proper functioning and sustainability of the whole. Though I have suggested that immunological considerations may provide some convergent criteria for deciding what is and what is not a part of the organism, I do not want to suggest that such decisions reflect unequivocal matters of fact that science will eventually resolve one way or another. On the contrary, my view is that drawing boundaries round biological objects is to an important extent a matter of human decision driven by particular human goals, practical or theoretical. I propose a 'promiscuous individualism' parallel to the promiscuous realism I have advocated for many years with reference to natural kinds. That is to say, there are various ways of drawing such boundaries, reflecting real biologically salient aspects of the multiply interconnected systems that make up the biological world.

So I have not wanted to say that the MDCL is an erroneous conception. It is arguably an inevitable conception for pursuing certain kinds of evolutionary question. The mistake is to think that it involves a discovery of what the organism really is, and must therefore be the right conception for all purposes. Against this, I have suggested that a more inclusive, functional and polygenomic, understanding of the organism is better suited to many or most medical purposes; in this context commitment to the fundamental importance of the MDCL does more harm than good. It is an interesting thought that within this broader perspective various important processes such as metabolism, development, and immunology, will not necessarily provide the same answers to questions about the boundaries of the organism. I don't see that this should be a cause for concern. Indeed, appreciation of a degree of arbitrariness in any boundary between ourselves and our biological environment should help us to move away from the isolated, antagonistic view of our relation to our biological environment, towards a more nuanced one which, while recognizing that nature is full of threats, also appreciates the deep interconnectedness of ourselves and our environment.

This interconnectedness is historical as well as functional. One common idea is that after billions of years in which only simple primitive organisms were evolving, something new and better appeared, and the primitive precursors remained either as lowly chemical operatives disposing of waste, or as hostile predators, bent on turning us into further waste for their chemical maw. But we have not transcended our ancestors, but rather have evolved as fully interdependent enhancements of the pre-existing biological systems. Eventually, it may be expected, this insight will have a profound effect on our view of health and disease.

Conclusions

Research in many areas of biology is changing our basic understanding of living processes. Epigenetics has undermined earlier deterministic ideas about the action of genes; microbial ecology and metagenomics are demonstrating the massive interconnectedness of the diverse elements of functional biological systems; systems biology is offering a few glimpses of ways that we might eventually learn to analyse these terrifyingly complex systems. And there are many other advances that I have not considered in this brief discussion. It is self-evident that medicine must attempt to come to grips with these new insights into the nature of the biological.

But more specifically, the diseases of the wealthy nations—cancer, heart disease, diabetes, etc.—seem increasingly to require this broader perspective as we try to understand the relations between the environmental, cytological, genetic, and epigenetic perspectives on these maladies. And the infectious diseases that continue to plague poor countries may well return to afflict the wealthy as the policy of massive and indiscriminate microbicide is gradually being defeated by the resilience of the microbes themselves. It may be hoped, though perhaps it is still little more than a hope, that a better understanding of the relation between the human MDCL and the microbes together with which it constitutes the functional human system, may provide us with new ways of responding to the catastrophic imbalances of the system that can result from the wrong microbes in the wrong place. There are, at any rate, massive challenges and opportunities for medicine in emerging biological science.

PART IV

Humans

14

Against Maladaptationism: Or, What's Wrong with Evolutionary Psychology[1]

This chapter is not directly concerned with topics for which Barry Barnes is mainly known, but it is indebted to him nonetheless.[2] The chapter updates criticisms of Evolutionary Psychology that I have been intermittently engaged in for the last 20 years or so with reference to much more recent work I have been doing on genomics. The latter has been carried out in collaboration with Barry from the start, and ultimately resulted in a book co-authored by Barry and myself on the sociology and philosophy of genomics (Barnes and Dupré 2008). So what is in fact the central topic of the chapter, recent developments in genomics, is the one around which the work of Barry and myself has been most closely interconnected.

This chapter also reflects my long-term interest in critical science studies. Here the connection is more diffuse. However, the sociology of scientific knowledge provides a fundamental perspective from which to ask awkward questions about what one is really doing in proposing critical analyses of allegedly scientific projects. Like many philosophers I have sometimes been tempted to seek a God's-eye view from which to pontificate about what is good and bad about particular scientific projects. However, on more mature reflection I'm quite content to frame my arguments in more naturalistic terms: Evolutionary Psychology, specifically, fails to meet the standards generally agreed among enthusiasts for Western science, and fails to cohere with scientific views that are much more deeply grounded in our epistemic standards than it is. Moreover its pronouncements may very well be bad for us.[3] That's enough to be getting on with, I think.

My title may surprise some. One of the common accusations against Evolutionary Psychology is that it is panadaptationist, or Panglossian; it supposes that everything

---

[1] A version of this chapter was presented at Princeton University as a Decamp lecture, in February 2005. I am grateful for comments on that occasion by Robert Wright, though I fear he will not think I have properly appreciated them.

[2] The original version of this chapter was published in a Festschrift for Barry Barnes (Mazzotti 2008).

[3] This possibility has been confirmed by work carried on in collaboration between Egenis and the Exeter Department of Psychology, showing that sexist assumptions, in particular, are liable to be strengthened by exposure to claims that sexual differences are biologically grounded (Morton et al. 2006).

about us is an adaptation, perfectly shaped for its conditions of life by the all-powerful hand of natural selection. But there is no mystery here, and both criticisms may be correct. The explanation is just that Evolutionary Psychology, or the variant of it with which I shall be concerned, supposes that we are adapted to the environment of the Stone Age. Thus we are adapted, perhaps Panglossianly adapted, by natural selection, but not to the environment in which we have the misfortune to find ourselves but to one long past. Hence the maladaptation. To take one familiar example, in the Stone Age fat and sugar were rare and excellent sources of energy, so we became adapted to consume them voraciously whenever the chance presented itself. But in a world full of Krispy Kreme doughnuts and deep-fried Mars bars this trait is highly maladaptive and leads to heart disease, diabetes, and all the other woes of the age of obesity.

The brand of Evolutionary Psychology I'm considering here is the programme developed by Leda Cosmides, John Tooby, David Buss, and a few others, and popularized with great effect by Stephen Pinker and, with rather more sophistication, by Robert Wright.[4] It is not the only version, but it is the most prominent. It can be quickly summarized. It holds that our minds consist of a large set of modules, shaped by natural selection to solve particular problems set in our evolutionary history. These are problems from the period known as the Pleistocene, roughly the one to two million years preceding the emergence of human civilization, which I shall refer to loosely as the Stone Age. The view that we are adapted to the Stone Age rather than to modern life is the maladaptationism of my title.

It is quite surprising that we should be, in this way, systematically maladapted. We are, after all, probably the most successful large organisms in the history of life, and this success has accelerated as the conditions of our existence have diverged ever further from those of the Stone Age. It is surprising, at least, that this should have happened if we are systematically adapted to a quite different environment from the one in which we appear to have thrived so spectacularly. Fortunately, there is no good reason to accept this maladaptationist thesis. It is based on bad biology: an obsolete view of genetics and a dubious and probably unsupportable view of evolution. There is much else wrong with Evolutionary Psychology, and its errors have been thoroughly documented by myself (Dupré 2001) and others (e.g., Buller 2005). I shan't discuss here, for instance, the Panglossianism mentioned above, the assumption that any feature of an organism, including the cognitive structure of the human mind, is likely to be an optimal response to some conditions at some point in evolutionary history, or the even less defensible obverse assumption, that if something would have been a good idea, it almost certainly evolved. I shan't consider the controversial issue of the modularity of the mind, often described by Cosmides and Tooby as analogous to a Swiss Army knife. And I won't say anything about the endemic evidential weaknesses of the project, the ways in which evolutionary speculations or conveniently hand-picked animal

---

[4] The classic source is Barkow et al. (1992). See also Buss (1999). More popular accounts are Pinker (1997) and Wright (1994).

analogies so often make up for thin and controversial empirical grounding of the claims about what has actually evolved in the human case.

Given the widespread criticisms of Evolutionary Psychology and, perhaps more importantly, the fact that there are other perhaps more credible evolutionary approaches to understanding human behaviour, it may well be wondered why I should spend any time on it at all. So let me offer some broader context. I think quite generally that the ways in which evolution can illuminate biological questions is often misunderstood. Evolution provides one essential kind of explanation of biological phenomena, but its ability to predict or discover phenomena is limited. Attempts to do so generally involve extremely simplistic evolutionary models, and their apparent outputs can be almost entirely traced to these simplifications. The one important exception to this sceptical suggestion is the extent to which evolution legitimates comparative biology. Detection of homology, the common evolutionary origin of a feature, can provide defeasible but valuable clues about function. Despite the hype about the Human Genome Project, it has been well understood from the start that the most interesting information that might come from genome sequencing technology was comparative, the ability to detect similar and evolutionarily related genomic elements in different biological contexts. An extraordinary difficulty for any form of evolutionary psychology is that there are no relevant species for evolutionary comparison. To the extent that cognitive mechanisms evolved, as Evolutionary Psychologists propose, several million years after the division of the human lineage split from that of the chimpanzees, and given that everyone agrees that all contemporary humans belong to one species, this lack is indisputable.

Evolutionary Psychology proposes to fill this gap by claiming to know when we evolved our distinctive psychology and what were the environmental conditions that our ancestors faced at that time, and by offering a priori arguments about what would be the best psychological mechanisms to deal with those conditions. Such arguments then provide epistemological depth to thin and controversial evidence about what humans are actually like. I claim that all of these steps are invalid. We don't know when our distinctive psychology evolved and much of it is likely to be well adapted to contemporary conditions. We don't really know a great deal about the conditions of the Pleistocene and even if we did this would provide the most doubtful grounds for inferring anything about our adaptive responses to them. Psychology should be empirical not a priori.

As I have indicated, my main focus will be on the first part of the argument. Much of this, however, is an excuse to discuss some remarkable developments in recent biology which should quite generally invite reconsideration of some common broad assumptions about the mechanism of evolution. Complexities emerging from recent molecular biology point to a much wider range of possible evolutionary mechanisms than have been generally recognized. Though in one way this makes evolutionary theory an increasingly rich and exciting field, it also makes attempts to infer biological fact from evolutionary theory increasingly risky. And while the plurality and complexity of evolutionary mechanisms greatly increases the resources for evolutionary explanation, it correspondingly decreases the possibilities for evolutionary prediction. In fact, and perhaps for this I should apologize, I won't have a lot to say at all directly about Evolutionary Psychology. I do hope, however, to show what an increasingly implausible project it is becoming in the light of recent biology. As evolutionary thinking begins to catch up with the revolution in molecular biology, the decades-old evolutionary theory on which Evolutionary Psychology has been built can now be seen to be of largely antiquarian interest.

Why are we thought to be adapted to the conditions of the Stone Age? Let me quote Leda Cosmides and John Tooby:

Our species spent over 99 per cent of its evolutionary history as hunter-gatherers in Pleistocene [Stone Age] environments. Human psychological mechanisms should be adapted to those environments, not necessarily to the twentieth-century industrialised world. The rapid technological advances of the last several thousand years have created many situations, both important and unimportant, that would have been uncommon (or nonexistent) in Pleistocene conditions. Evolutionary theorists ought not to be surprised when evolutionarily unprecedented environmental inputs yield maladaptive behaviour. (Cosmides and Tooby 1987: 280–1)

Here we see not only the explicit maladaptationism but also the implicit panadaptationism: 'human psychological mechanisms should be adapted to those environments'.

Why should we not expect that psychological mechanisms have adapted to more contemporary conditions? It is not enough to say merely that our ancestors have spent more time in the Stone Age. Our ancestors have also spent a great deal more time as single-celled organisms. But this does not show that we are adapted to life in the primordial slime. The answer widely assumed by Evolutionary Psychologists is that there has not been enough time since the Stone Age for us to have adapted significantly to more recent conditions (and, of course, that there was enough time for our early human ancestors to adapt to the conditions they encountered, whatever those were).

So how much time is enough? How fast is evolution? It is still commonly believed, and underlies this part of the Evolutionary Psychologists' argument, that evolution consists in change in gene frequency. The whole story goes something like this. Psychological adaptation amounts to the existence of neurological structures in the brain. These structures are built by genes. The necessary genes are acquired by random mutation of existing genetic material and selection of advantageous mutations. Since a random mutation is almost certain to be disastrous unless its consequence is fairly similar to that of the unmutated state, each mutation is assumed to provide only a small change. A series of these small changes, each of which will take a substantial number of generations to reach fixation in the population, can eventually produce complex adapted structures.

Richard Dawkins (1996) gives a celebrated illustration of this way of thinking in his discussion of the evolution of the eye in *The Blind Watchmaker*. Provided we can think of 1,000 or 10,000 steps between no eye and fully functional eye, geological time is long enough for each of these steps to have appeared by chance mutation and spread to fixation through the population. Dawkins in fact seems to think that this development is almost inevitable, but we need only assume that it is possible.

I'm not at all sure whether, if this picture is right, a million or two years is long enough for the evolution of the human mind. Our ancestors two million years ago had brains about one third of the size of our present brains, so it is reasonable to assume, as Evolutionary Psychologists generally do, that important contemporary human neuro-logical structures evolved in those two million years. For Evolutionary Psychology this amounts to the generation of genetically determined neurological structures, mutable only by thousands of generations of genetic trial and error.

One crucial idea behind this argument, then, is that adaptive traits are carried over from the periods in which they evolved by genes. And the random mutation and selective retention of genes is a process that requires thousands of generations. So let me begin by saying something about genes. For the reason just noted, genes figure prominently in Evolutionary Psychological writing. Although they reasonably enough protest when accused of holding that genes determine behaviour, they do generally hold that genes determine psychological mechanisms. To quote Robert Wright: 'They boil down to genes, of course (where else could rules for mental development ultimately reside?)' (1994: 9).

So what are genes? It is not sufficiently widely known how difficult this question has become to answer.[5] One possible answer goes back to the history of genetics and the Mendelian research programmes, particularly on fruit flies, of Morgan, Mueller, and others. This programme investigated hypothetical factors that were the heritable causes of differences between organisms. It became clear that these causes had something to do with chromosomes, and experiments on linkage, correlations between inherited traits, enabled the mapping of these factors as quasi-spatially related. When Crick and Watson famously published the chemical structure of DNA it was natural to suppose that these hypothetical factors could finally be identified with concrete material objects, parts of chromosomes or, that is, sequences of DNA molecules.

This conclusion has turned out to be highly problematic, however. Certainly the phenotypic differences studied by classical geneticists could generally be identified with differences in the DNA sequence somewhere in the genome. Alternative bits of sequence with identifiable phenotypic effects are referred to as alleles, so that for example, we can talk about alleles for blue eyes or brown eyes which, more or less, follow the familiar Mendelian laws. However, we should not assume that these alleles are readily identifiable objects. We can see this by looking at what may be the most important upshot of allelic selection, elimination of genomic errors. Medical genetics, because it is concerned precisely with harmful genomic errors, retains a strong connec-tion with the tradition of classical genetics. But, to take one of the best known genetic

---

[5] For discussion of the problems see Moss (2003); various papers in Beurton et al. (2000); Dupré 2005, and this volume, Chapter 6.

diseases, cystic fibrosis, there is no well-defined object referred to by the expression 'gene for cystic fibrosis'. Cystic fibrosis is caused by a dysfunction in a protein that controls ion transfer across cell membranes. Hundreds of mutations have been identified in the genomic region that codes for this protein, with different mutations determining varying severity of symptoms in cystic fibrosis patients. The gene for cystic fibrosis, then, is a set of errors. Though in this case it would be correct to say that any of these mutations causes cystic fibrosis, it would be highly misleading to describe the unmutated sequence as a gene for not having cystic fibrosis.

As a matter of fact a very similar story can be told about the gene for blue eyes. Again, this is not a piece of DNA that somehow produces the blueness of eyes, but any of a range of errors in the DNA sequence that subverts the production of brown pigment in the eyes. And again, though there are therefore genes that cause blue eyes, it is at best misleading to think of the functional alleles at blue eye loci as causes of brown eyes. The complexity of causal paths from bits of DNA to features of organisms makes the project of correlating things of these two kinds largely futile. Many different bits of DNA sequence and much else besides are involved in the normal production of a phenotypic trait. We can confidently assert that a bullet in the head was the cause of death, but it is problematic to suggest, except under very unusual circumstances, that the absence of a hole in the head is the cause of someone staying alive.

One might say that genes for brown eyes are parasitic on genes for blue eyes: if there were no identifiable effects of mutations in the relevant bit of DNA there would be no classical genes for either blue or brown eyes. And this follows merely from the quite uncontroversial point that the Mendelian concept applies only to differences. There is an irony here with some of the more acrimonious debate around Evolutionary Psychology. Critics of EP have suggested that Evolutionary Psychologists are involved in providing genetic explanations for human differences, between males and females or between homosexuals and heterosexuals, for example, and thereby reifying what are in fact superficial and malleable distinctions. Evolutionary Psychologists retort indignantly that their central concern is with the genetic basis for human universals. But in the only really clear sense of the word 'gene' there are no genes for universals. And that by definition, since genes are defined only by the differences they cause.

Am I suggesting that there is no genetic basis for normal development? Only, admittedly, in the pedantic sense that there are no well-defined entities, answering to the concept of genes, involved in normal development. But we can then say that there is a genomic basis for development: parts of the genome are crucially important. So why not just call those parts of the genome genes, and stop quibbling? To answer this we need to look a bit more closely at the quite different concept of the gene employed in genomics.

When analysts of data from the Human Genome Project report that there are about 23,000 genes therein, this estimate has nothing to do with relations to phenotypic traits. Very roughly speaking, what they mean is a sequence of coding DNA between a signal to start transcribing (that is, generating RNA sequence that may later be translated into amino acid sequence that may become part of a functional protein or enzyme) and a signal to stop transcribing. It is instructive that for several years there was serious controversy, with wildly different estimates circulating, about this number, though now something like consensus seems to be developing. A closer look at genomic activity makes it easy to see why this was not a simple matter of counting.

The number of proteins produced in human cells is a more controversial issue than the number of genes, but typical estimates start at around 100,000 and range up to around a million. Obviously this indicates that a gene, in the sense used by molecular biologists, can be involved in the production of many proteins. In fact these molecular genes are known usually to consist of alternating segments, known as exons and introns. In the simplest case, after the gene has been transcribed into RNA the introns are edited out and only the exons are translated into a protein. But in many or most cases different sequences of exons are composed by different editing processes, genes are 'altenatively spliced', and the same gene may give rise to many different proteins. In some cases products from parts of other genes, even the introns from other genes, are included in the splicing process. And further modifications to proteins occur after the edited RNA product has been translated into a protein. Cases are known in which several hundred different protein products are derived from the same gene.

Why does all this complexity matter? In the first place it contributes to dislodging a picture of the genome that still informs a good deal of thinking about evolution, not least human evolution, of the genome as some kind of blueprint or programme for the production of an organism. It begins to suggest, instead, something quite different, a repository of informational resources upon which the cell can draw in making a huge range of functional products. I think this is still misleading, because like the blueprint or programme metaphors it aims to replace, it sounds too static. The genome is located in a complex structure and the various forms that this structure adopts in the life of a cell are important to its functioning and its interactions with other components of the cell. My genomicist colleague Steve Hughes likes to define the genome as 'a space in which genetic events happen'. But for now the important point is to dislodge the metaphors that somehow suggest that the whole organism is somehow encoded in the genome, the idea that Lenny Moss (2003) has appropriately related to the preformationist tradition in the history of biology.

To get beyond this picture we need to look at another bit of biological dogma, the demise of which is perhaps less universally acknowledged, the so-called Central Dogma of molecular biology. This dogma holds that information flows only from DNA to RNA and finally to amino acid sequence, never in the other direction. This may sound like nothing more than a characterization of the basic steps in the production of functional proteins, but in fact it is widely used to lend support to the preformationist picture of the genome: since information only flows outwards from the DNA, it must all be contained in the DNA. At any rate, interpreted in anything more than the narrowest sense just indicated, it is completely false. In a way it seems obviously false. For what matters to the functioning of a cell is that the right functional products get produced at the right time and in the right place and this is certainly affected in part by changes in the chemical environment of the cell. Still, if one held to the view of the genome as a programme, one could think of the cellular environment as something fully controlled by the genome and thereby effecting the appropriate expression of the necessary bits of DNA sequence at that point in the programme. To see that this picture cannot be sustained we need to look at more direct ways in which the Central Dogma is mistaken.

The most important, or at least the best understood, of these is methylation. This process, which has led some to refer to methylation (or more precisely 5-methylcyto-sine, the chemical result of methylation) as the fifth base in the DNA code, is a modification of the DNA structure that suppresses the expression of modified bits of sequence. This is a process that occurs throughout the life of a cell, and is certainly one of the crucial determinants of gene expression. It is clear that methylation is not just part of a system by which the genome regulates itself. For example, it has been established that maternal care or its absence affects methylation patterns in the brains of juvenile rats and, in turn, this affects the disposition of the affected rats to provide such care when adult (Weaver et al. 2004). While it was once thought that methylation was removed during the production of gametes, it is becomingly increasingly clear that this is by no means always the case. This leads us to what might well be called the Central Dogma of evolutionary biology, a dogma closely related to the previous Central Dogma, and one that has also become wholly untenable, the assumption that the only thing that is inherited is DNA. And it is essential to note that it is not crucial whether this inheritance is transmitted through the germ plasm: the sequence of events sketched for rats is an entirely feasible mechanism of intergerational inheritance.

One of the points of problematizing the gene concept is to raise the question what kinds of genomic difference are in fact the important targets of selection. It seems increasingly likely that the importance of selection between alleles has been greatly exaggerated in recent evolutionary theory and it may indeed turn out, as most geneticists believed in the heyday of classical genetics, that this is largely a process of error elimination and not one capable of creating major evolutionary novelty. Getting rid of the idea of genes for traits, too easily interpreted as objects with the specific function of causing those traits, should at least raise doubts about this idea.

Perhaps more importantly there is increasing awareness of a much greater range of possibilities for genomic changes that may provide far more promising bases for major evolutionary change. We have seen that a DNA sequence comprising a set of exons and introns may provide the basis for production of a large number of distinct products. Which, if any, of these it produces at a particular point of development and in a particular tissue will depend on a wide variety of factors: chemical modification of the genome, as for example in methylation, structural changes such as greater or lesser condensation of the chromatin, and the chemical species present in that cell at that time. There are parts of the genome capable of initiating cascades of developmental changes, and interestingly, these genetic triggers are generally extremely ancient, found in very distantly related organisms. As I shall explain further in a moment, some of these factors, and not merely those consisting of DNA sequence, can also be passed on to offspring.

Genomes themselves evolve in a great diversity of ways. Recombination, the result of random sampling from the genomes of two parents in producing an offspring, is standardly recognized as an important source of variation. But there are many other processes. Whole chromosomes and even genomes can be duplicated. These duplications are thought to be important in providing redundant genetic material in which large changes of organization or sequence can occur without loss to the organism of essential functions. Smaller parts of genomes can be duplicated within or across chromosomes by inserting copies of themselves into the genome. Retrotransposons, a very important class of such genomic elements, which constitute a substantial proportion of many genomes, appear to be important in the functional reorganization of genomes. And, much more commonly than was once supposed, DNA from other organisms can be inserted into genomes.

It is interesting to reflect, in the light of some of these facts, on the surprise that is sometimes still expressed at the claim that human genomes are, say, 99.4 per cent identical to those of chimpanzees. One may well wonder what exactly this means, but without worrying about that, we can certainly wonder why we should care. No doubt if the genome were a blueprint this would be quite surprising. If the blueprints for two ships, say, are 99.4 per cent identical (without again worrying exactly what that means) we might expect two pretty similar vessels. But if we were told that they were made of an almost identical set of raw materials, or that they had identical engines, we would have no such preconceptions. The fact is that even if the genomes were 99.9 or 100 per cent identical, nothing much would follow as to the degree of similarity of the organisms of which they were the genomes. It *is* an interesting and important discovery that parts of genomes are very strongly conserved through very much longer periods of evolutionary time, and substantial proportions of our genomes are almost identical to parts of the genomes of worms and even bacteria. This does not tempt us to wonder whether we are really rather similar to worms or germs, though it does direct us to look for bits of chemical machinery that we may share with very different creatures. We may here recall some prescient remarks of François Jacob in 1977: 'What distinguishes a butterfly from a lion, a hen from a fly, or a worm from a whale is much less a difference in chemical constituents than in the organization and the distribution of these constituents' (1977: 1165).

In sum, genomic function is a very different matter from genetic sequence and it is genomic function that provides the differences on which natural selection can work. It is likely that creation of entirely new bits of sequence has been greatly overstated as a central process in evolution and that redeployment of existing genomic resources may be much more important in producing large evolutionary changes. At any rate, research in genomics is opening up a wide range of possibilities for thinking about evolutionary change, and we should certainly not be committed to seeing the evolutionary process solely in the terms developed over 50 years ago.

Let me now come at the topic from a rather different direction, the philosophical analysis of evolution itself. Much of this has been focused on the so-called units of selection problem, the question what exactly does natural selection select. Thirty years ago it was fairly uncontroversial that the primary objects of selection were individual organisms and perhaps also groups of organisms. Then came Richard Dawkins's notoriously successful popularization of the ideas of G. C. Williams, and a great many people were convinced that ultimately the only possible unit of selection was the gene, understood by Dawkins in a broadly Mendelian way as a difference in the DNA sequence that made a difference to the phenotype. The crucial premise for this move was the claim that only genes were inherited. Whereas organisms invariably perished in an evolutionarily trivial length of time DNA, in Dawkins's colourfully hyperbolic term, was immortal. The structure of DNA was passed on intact from parent to child and hence that structure was potentially immortal. Indeed, molecular biologists have been able to find large chunks of DNA more or less identical between ourselves and plants or even bacteria, and thus presumably preserved across the aeons of evolutionary time from our distant common ancestors.

But, impressive though this point may seem, the view that only DNA is inherited is quite unsustainable. And this is one of the reasons why most philosophers of biology, and some prominent evolutionists, have never been much convinced by the gene selection theory. It is easy to show that Dawkins's gene selection theory provides an inadequate model of evolutionary processes even if it is conceded that inheritance is solely mediated by DNA, and most philosophers concerned with these issues have accepted a pluralistic answer to the units of selection: selection acts on objects at a range of different scales, including genes, organisms, and very possibly groups of organisms. But we should not concede this view of inheritance. Broadening our understanding of inheritance suggests a much more radical rethinking of the units of selection problem that has been developed under the rubric of Developmental Systems Theory, or DST.[6] DST, I shall suggest, provides a context in which we can understand the significance for evolution of the recent advances in genomics.

The Central Dogma of evolutionary theory stated that the only transgenerational vehicle of inheritance is the genome. The negative phase of DST provides a fundamental critique of this dogma. In very brief summary, it asks the question whether there is anything unique about DNA that justifies its privileged status in evolutionary models, and offers a negative answer. It has been claimed that DNA is unique in its ability to replicate itself. But DNA requires a range of other structures and substances for replication and, with similar access to other resources, including DNA there is a wide range of structures that successfully replicate themselves in the course of development. The genome has been conceived as a privileged source of information. But it is easy to show that from an informational perspective the status of the genome is

---

[6] The clearest summary of DST I know of is Griffiths and Gray (1994). The classic statement is Oyama (1985).

symmetrical with other contextual resources through which information is conveyed. Just as the cellular environment provides a channel for conveying information about the genome, the genome provides a channel for conveying information about its environment. Though the issue is not uncontroversial, DST has placed a strong burden of argument on those who wish to show how the genome has a unique status in biological organization.

The positive claim of DST brings us back to the unit of selection. For DST this is the full life cycle of the organism. DST looks at the whole set of resources that are necessary for the reproduction of the life cycles of organisms, and the means by which parent organisms facilitate the availability of these resources for their offspring. This picture, of course, retains the basic Darwinian idea that evolutionary change is driven by the differential success that organisms have of launching, during their own life cycles, life cycles of organisms similar to themselves. By rejecting the picture of evolution as essentially no more than a sequence of gradually changing gene pools, this move makes room for the reintegration of development into evolutionary models.

It is plain that the restriction of inheritance to the genome cannot be right in the case of human reproduction. For a modern human in a developed modern society to successfully launch and sustain the life cycle of another modern human in that same society many other resources must be provided: maternal care in infancy, schools, hospitals, and much else. And these resources affect the course of development. Despite debates about the importance of innate underlying cognitive structures, it is impossible to deny that a human developing with access to the full range of such developmental resources will acquire a range of capacities—from reading and writing, to appropriate table manners and locally appropriate dress sense—not available to one denied these resources.

Such cultural developmental resources are not unique to humans. Birds must provide nests, termites must construct mounds, and beavers must build dams if they are to be successful in reproducing their respective kinds. No doubt there are greater or lesser innate dispositions displayed in these acts of provision—lesser for birds than for termites, for instance. But often the experience of exploiting the resource will also provide some of the information necessary for reproducing the resource when they become parents. Many species of birds, for instance, learn by imitation the songs necessary for attracting reproductive partners.

These developmental resources fully external to the bodies of the reproducing and reproduced life cycles are of obvious importance in human evolution as current human reproduction involves a vast infrastructure of resources that are maintained and improved upon by successive generations. What is less obvious is that the genome is not a unique bearer even of internal heritable information. A consequence of the critical work of DST has been to dismantle the conceptual firewall that some have tried to construct around the genetic to preserve its privileged place in evolutionary models. In reality, the minimum physical material passed on to an organism in reproduction is a single cell. The female egg contains a vast set of chemical materials. Though the production of many of these chemicals depends on genomic resources, the genome contains resources that could in principle produce an unimaginably large set of different chemical environments. The transmission of one particular set is potentially a transgenerational transmission of information of a complexity not incomparable with the transmission of the genome itself. And as I have mentioned, functionally important modifications of the genome itself, such as methylation, are also transmitted to an extent that remains unclear and, indicatively of its pivotal ideological role, highly controversial.

An obvious consequence of transmission outside the body is that this sort of inheritance is Lamarckian. By this I refer (with apologies to Lamarck[7]) to the inheritance of acquired characteristics. Schools, for instance, allow the acquisition of characteristics that can be transmitted to future generations. A consequence, or ideological function, of the dogma of only genetic inheritance is to emphasize the intellectual iniquity of Lamarckism. The phenomenon of methylation is one of a range of recent biological insights that threaten to open more fully the Lamarckian Pandora's Box. This is more controversial terrain than I have so far ventured into, but some recent results are suggestive.

I have already mentioned the research on maternal care in rats affecting gene expression in the brain of pups. Rats deprived of maternal care in infancy grow up more fearful and show stronger hormonal response to stress than normally nurtured rats. Closer to home, there are data showing that low birth weight of children born during the Dutch famine of 1944–5 not only had increased susceptibility to various later life illnesses, but passed this susceptibility on to their children, and epigenetic effects such as abnormal methylation patterns provide a plausible explanation. One other well-documented case is the effect of social status on the production of dopamine receptors in the brain. Higher ranking Macaque monkeys turn out to be less susceptible to cocaine addiction than monkeys that they socially outranked. This difference was traced to the fact that exposure to lower ranking monkeys, but not to higher ranking ones, effected changes in the expression of genes in the monkey's brains, specifically to the production of dopamine D2 receptors (Morgan et al. 2002). I don't know whether these changes are heritable, but certainly mechanisms exist whereby they could turn out to be.

Less controversial are effects of the environment that are not directed, but affect the rate of evolutionary change. There is work going back to Barbara McClintock that shows that the activity of retrotransposons, genetic elements that replicate themselves throughout the genome, is increased when plants experience stress. This will tend to cause genomic reorganization that can provide material for rapid evolutionary change. There is, at any rate, considerable evidence that these elements, which constitute a very substantial proportion of most genomes including ours, have important effects on gene

[7] Lamarck, of course, had a complex position in which this view was embedded and which I don't want to endorse; and many others, such as Darwin, were fully committed to the inheritance of acquired characteristics.

expression, and can have decisive effects in early embryogenesis, which is of course the point at which the largest effects on development can be expected.

Perhaps most intriguing of all are the small, non-protein-coding RNAs that are proving to be omnipresent in cells and to have vital, diverse, but very partially understood, functions. They appear able to bind to DNA, inhibiting its expression, they can control the activity of protein coding RNAs, and some can even bind to proteins, altering their behaviour. It would be impossible to begin to describe the intriguing findings that are beginning to emerge from this incipient research field, but it seems clear that it represents an entirely new level of cellular control. Small RNAs also have the ability to move between cells, and may prove to have important communicative functions between tissues. And, of course, a set of these RNA fragments is part of what is transmitted with the maternal ovum in reproduction.

A very radical and heretical view of evolution, most forcefully presented by Mary Jane West Eberhard (2003), suggests that adaptation is, in the first instance, a process of organismic response to the environment, facilitated by the developmental plasticity of organisms. Genomic adaptation follows. So far from selection among genes being the primary force behind adaptation, it is largely a consequence of phenotypic adaptation. Perhaps then we are well adapted to modern life, but our genomes are still catching up.

I have done no more than gesture at some of the extraordinary insights that are currently emerging in molecular biology. Why should we care? Recall the basic argument underlying the Stone Age origin of the human mind. Essentially the mind is a product of the genome. Behaviour, to be sure, responds differentially to environmental circumstances, but the basic structure of the mind is laid down in the genes. This is a thoroughly bottom-up picture. Genes, as the dogma has it, produce RNA, which produces proteins. Proteins provide the predetermined neurological structure that then interacts in a determinate way with environmental contingencies. We are not, as the Evolutionary Psychologists insist, exactly programmed to be rapists, but given the right set of stimuli in which our Stone Age minds calculate rape as the best reproductive strategy, rapists we become. And finally, the most fundamentally bottom-up part of the picture is the model of evolution that claims that evolutionary history is, in essence, no more than a sequence of genomes, each slightly modified to improve on its predecessor.

I have tried to indicate that this picture is entirely obsolete and unrelated both to contemporary molecular biology and the most plausible understanding of the evolutionary process. Certainly these bottom-up processes are important, but equally important are simultaneous top-down processes. The environment does not just shape the human mind in the uncontroversial sense of filling in gaps in a pre-existing structure—speaking English rather than French, or knowing which social rules to monitor for cheats. As shown by the high status monkeys who just say No to drugs, social factors can influence the expression of genes in the brain, and basic brain chemistry. Gross morphology can affect the shape of cells which can affect the chemical functions within cells, a process that has been found to be very significant in early cell differentiation (Folkman and Moscona 1978). DNA produces RNA, but while some of that RNA contributes, in very complex ways, to the coding for protein sequences, other bits feed back on the function of DNA or on the splicing and translation of coding RNAs. Proteins also feed back on the expression of DNA or contribute to the physical structure of DNA, also an essential determinant of gene expression.

Hence in reproduction it is not just a set of genes that is passed on to descendants, but an exquisitely complex and dynamic chemical system of which the genome is just one vital interacting part. And to the extent that organisms shape the environment in which their offspring are found either purposefully, as is carried to by far the highest level in our own species, or simply as a by-product of their characteristic behaviours, this will also affect the developmental sequence of chemical environments in the differentiating cell lines. How effective at tracking exogenous changes in the environment such a system will prove to be over evolutionary time is not to be settled by abstract calculations on the trajectories of naked DNA. Certainly there can be maladaptive time lags in evolutionary processes, but these are to be discovered empirically rather than proved a priori.

Let me summarize these conclusions by returning to the question of the universality and diversity of human nature. Evolutionary psychologists respond to the accusation that by seeing human nature in the genes they are reifying differences between people, by insisting that their primary concern is with the common genetic inheritance that we have all inherited from our Flintstone ancestors.[8] Still, where there are evident differences between people, as for example between homosexuals and heterosexuals, these must be located in the genes, and silly stories are made up about Stone Age homosexual shamans providing for their nephews and nieces. It is true that we are an unusually genetically homogeneous species, so perhaps these differences are not so great. An important exception, of course, is the difference between men and women. Since the genetic difference between a human male and a human female on some common measures exceeds that between a human male and a male chimpanzee it is not surprising that Evolutionary Psychologists have portrayed men and women almost as if they belonged to different species—perhaps even came from different planets.

The picture I have sketched is neutral on the uniformity of human nature, in large part because I am sceptical about the usefulness of the very concept of human nature. Of course there is a vast amount of human biology common to the human species, some of which we are just beginning to understand. The problem with the concept of human nature is that it tends to suggest fixity, indeed something like a traditional human essence. It is true that we have mostly learned to reject biological essences as incompatible with evolution but still, in the timescales that matter to us, this may seem a pedantic difference. And indeed, though they certainly admit that we have evolved and are probably evolving still, Evolutionary Psychologists perfectly illustrate this

---

[8] Thanks to Steven Rose for having drawn attention to this important popular precursor of Evolutionary Psychology.

effective essentialism with the claim that human nature is, as far as matters to us, stuck in the Stone Age.

I object that there is no reason to suppose that we are stuck in the Stone Age, and indeed that we are very likely quite well adapted to the twenty-first century. And it is possible that we may soon be adapted to something quite different. Is this an assertion of the blank slate view of the human mind so violently denounced by Stephen Pinker (2002)? Not at all. Human development is a much more complex process than the crude genetic determinism supposed by Pinker, but its very complexity may make it difficult to change in predictable ways. My point is just that organisms in general, and ourselves in particular, are much more subtle and interesting than the antiquated biological picture I have criticized suggests. In evolutionary time there are many ways in which they may respond to changing environments: partly by changing their genomes, though probably the important changes to the genome amount to the redeployment of existing genomic resources, and perhaps more importantly still by changing environmental factors that elicit new employments of existing resources both genomic and more widely biological. Because for our species many changes in this last category can be purposefully effected, it is possible that significant evolution could have happened very rapidly. And the great differences in all except genomes between ourselves and our nearest relatives clearly point to the conclusion that it has. This evolutionary flexibility is, from a proper perspective, inextricably connected to developmental flexibility. It is not, as the accusation of blank slateism suggests, a trivial matter of producing whatever developmental outcomes we might like to imagine, but nor is the production of new developmental outcomes something ineluctably barred by genetic fate. Human differences, the diversity, however much there may be, in actual human developmental outcomes is our best clue as to the diversity of outcomes that might be achieved with a will and a better understanding of human development.

No doubt there are many kinds of time lags. Much of our genomic machinery is inherited from simple organisms billions of years ago, though it seems to be rather adaptable to new uses. At the opposite extreme rapid social change produces developmental obsolescence. People of my generation are surely less well adapted to the age of information technology than will be today's teenagers. And it may be that we have deeply engrained tendencies of behavioural development that stem from exigencies of some part of our evolutionary history. But if so this needs to be empirically demonstrated in detail, not proved by a priori argument. And even if such atavistic defects are demonstrated, there is no reason to suppose that they are somehow immutable.

One message of this chapter is a sceptical one. The more we understand of contemporary biology, the more we see how much we don't know. We still understand very little of the development of the simplest organisms, let alone the most complex. How, as our knowledge of molecular biology and ontogeny develop, these will bear on more refined understandings of the process and tempo of evolution is perhaps even more difficult to discern. The biology underlying Evolutionary Psychology, at least, can be confidently rejected as based on assumptions that have unravelled in the last couple of decades.

Conclusions drawn from so rickety a base about a matter as important to us as Human Nature should be rejected not only for the epistemological worthlessness, but because groundless guesses in this area can be extremely dangerous.

Let me end where I began, with the Krispy Kreme doughnuts. There can of course be no doubt that biological facts about humans, a part even of human nature, are engaged in the attraction of many of us to fat and sugar. These are, after all, good sources of energy that we are physiologically equipped to exploit. But what is interesting about the case is the diversity of human responses to the omnipresence of these resources. Obesity is not an inevitable response to the overabundance of cheap calories. In fact, and unsurprisingly, obesity seems to arise most strongly where overabundance intersects with poverty, that is among poor people in rich countries. Unfortunately such observations do not differentiate between the hypothesis of a fixed psychology responding to varying circumstances and a variable psychology developing in response to varying environments.

So how do we choose between these alternatives? Are we genetically programmed fat-guzzlers sucked inexorably towards the doughnuts, or are we blank slates, haphazardly imprinted with the culture of Mars bars or a healthy bourgeois love of broccoli? Of course we are neither. The way to break down the dichotomy between these equally hopeless alternatives is to begin to appreciate the intricate hierarchy of upward and downward interactions between objects and structures at all levels of the biological hierarchy. In doing so we dispense with the stultifying dogmas I have mentioned in this chapter, and we see the importance of a perspective on evolution that encompasses the diversity of processes susceptible to selective change. And, finally, we can begin to understand the vast changes in human behaviour that have occurred over the last few thousand years without seeing ourselves either as formless lumps of psychoplasm or atavistic relics from the mists of prehistory.

15

What Genes Are, and Why There Are No 'Genes for Race'[1]

Talk of the genetic basis of race has resurfaced in the aftermath of spectacular progress in the development of genetic technologies, most especially technologies that provide genetic tests, allegedly 'for race', that coincide quite closely with self-reported racial identities. In a much discussed Op-Ed piece for the *New York Times* (14 March 2005), Armand Leroi, an evolutionary biologist at Imperial College, argued that the classic claim by Richard Lewontin (1972) that human variation was overwhelmingly within rather than between races, so that traditional racial categories could be seen as socio-economic constructs, was based on an elementary statistical error.[2] Whereas, taking all genes separately, Lewontin's claim was true, the clustering together of genes characteristic of particular groups was able to show a distinctive genetic inheritance to traditional racial groups. These developments carry a significant danger of lending new respectability to controversial speculations about racial differences in such politically charged characteristics as IQ.

Unfortunately, the further these discussions move away from the technical contexts in which these genetic tests originate, the more misunderstandings appear. Others (Bolnick 2008; Fullwiley 2008) have explained in detail the statistical procedures to which Leroi is referring and the limitations of conclusions drawn from them. As I shall try to show in this chapter, insights from recent genomic science have helped clarify and highlight the ambiguities and misunderstandings that threaten incautious interpretations of genetic data and even of the very concept of a gene. I shall also show how misunderstandings of these concepts can and do lead to spurious conclusions of the kind just outlined with regard to race and that, in fact, the kinds of tests just referred to do nothing to underwrite traditional racial categories.

The topic can be approached by considering the apparently quite straightforward claim that there are genes for race. This may seem banal and obvious; physical characteristics such as skin colour presumably have a genetic basis. Far from being

---

[1] The chapter has benefited greatly from comments on an earlier draft by various colleagues in Egenis, especially Christine Hauskeller, Staffan Muller-Wille, and Maureen O'Malley. I also received very helpful comments from Sarah Richardson. Finally, my understanding of the topic was much improved by attendance at the Authors' Conference in Stanford in January 2006.

[2] Leroi attributes the statistical insight to Edwards (2003). Robust replies can be found in Graves (2005).

banal, the claim just mentioned is so difficult to interpret as to be close to unintelligible. This is because of the great difficulty of making clear sense of its main terms. The difficulty in defining race is familiar: although it is quite widely accepted by relevant experts that race is primarily a socially constructed concept, there are still many who think of the concept as fundamentally biological. And whether it is a social or biological concept, there is no agreement as to how many races there are.[3] Difficulties with the concept of gene may be less familiar. The first task of this chapter will be to explain why this concept is so problematic. This will make possible the differentiation of various interpretations of the claim under consideration about genes and race and examination of the plausibility and implications of each of them.

Are there genes?

To explain the complexities of the various uses of the term 'gene', it will be helpful to provide a very condensed and somewhat Whiggish history. Genetics is generally thought of as starting with the work of the Austrian monk Gregor Mendel. Mendel's famous experiments in the 1860s involved crossing varieties of peas which, after generations of work by plant breeders, consistently produced plants with known phenotypes. For example, he interbred lines of peas with yellow- and green-coloured seeds. The first generation produced by this crossing was found to have uniformly yellow seeds. When these hybrid peas were crossed with one another, the second generation was found to have 75 per cent of plants with yellow seeds and 25 per cent with green. Similar results were claimed with other observable features.

The (mildly Whiggish) interpretation of these results goes as follows. Plants are assumed to contain factors that produce the observable traits. Each plant contains two such factors, one of which is passed on to the offspring. If we call the factor that produces yellow seeds Y and the other G, we assume that the true-breeding lines have two copies of the same factor, and we refer to the lines as YY and GG. The first-generation hybrids will therefore have one factor of each kind, which we refer to as YG. The observation that the first-generation hybrids are all yellow, hence that YG plants have yellow seed, is interpreted as showing that the Y factor dominates. In the second generation, assuming that parents are equally likely to pass on either of their factors, the plants will be divided between YYs, GGs, and YGs in the ratio 1:1:2. Since only the quarter that are GG will produce green seeds, this explains the quantities found in the classic experiments.

Also famously, Mendel's results were largely ignored until they were taken up by several scientists independently at the beginning of the twentieth century. In 1909 the Danish biologist Wilhelm Johannsen named these factors *genes*. And in the first few decades of the twentieth century a highly successful research programme, now referred

---

[3] For these issues, see several papers in Koenig et al. (2008), the volume in which the present chapter was originally published.

to as Mendelian genetics, greatly expanded empirical knowledge of the transmission of traits from organisms to their descendants. The most influential embodiment of this programme was the work of Thomas Hunt Morgan and his students and collaborators on the fruit fly *Drosophila melanogaster*.

The most crucial thing to note about Mendelian genes, the objects of study in this episode of scientific history, is that they were causes of differences. No difference, no genes. In this strict Mendelian sense, there are no genes for traits that are universal in a population. This is a concept suited to evolutionary theory, where selection can only work on differences, and one that remains prominent in medical genetics, since medicine is centrally concerned with deviations from the norm—and hence with genetic peculiarities that cause differences. In light of this general point we can easily see that the idea of genes for race is highly problematic.

Even supposing, for the sake of argument, that belonging to a particular race is a biological trait at all, it is certainly not the kind of trait that could be the subject of a Mendelian experiment. We could not, for example, examine the offspring of two black people, or a black person and a white person, and decide how many of them were black. Whatever the phenotypic criteria are for these categories, they are not immediately accessible to inspection. This simple observation, incidentally, is already enough to throw serious doubt on the idea that race might be a biological trait, but we will continue for the moment with the counterfactual assumption that it is. Minimally, the problem is that race, even if it were a biological trait, would be far too complex a trait. Perhaps there are Mendelian genes for dark skin, hair texture, the shape of facial features, and so on, but race is at least a matter of there being many such traits. Some people have some but not all of the relevant set of traits. There is quite certainly no gene that makes the difference between being black or white, even ignoring, for the moment, the fact that there is an important social aspect to many racial categorizations.

The simple preceding point is important because a lot of talk of genes is still firmly embedded in the tradition of Mendelian genetics, a tradition that, first, explicitly licenses the idea that genes are *for* a phenotypic trait, the trait to which they make a difference, and, second, almost inescapably suggests the erroneous inference that the trait is caused by the gene. Medical genetics, as just noted, is still very largely Mendelian, in that its traditional and continuing central concern is with genes that make a specific difference, resulting in sometimes devastating pathologies. Medical genetics, it is true, is now moving rapidly towards a concern with predisposing genes, specific alleles that make it more probable that a particular pathology will develop. Good examples are the BRCA-1 and BRCA-2 genes, which strongly predispose women to developing breast cancer. But this hardly brings us nearer to a promising line for understanding genes for race. The idea of an allele that increases the probability of belonging to a particular racial group is a nonsensical idea. Many of the hypothesized genes for complex properties—intelligence, sexual orientation, violence, and so on— that appear regularly in the popular press raise a similar problem of spurious assimilation to traditions of Mendelian research on heredity, a further reason to emphasize the conceptual pitfalls. But to get to some slightly more plausible lines of thought, we should briefly move to some more recent history.

From the early stages of this programme it was widely, but by no means universally, assumed that Mendelian factors, or genes, would eventually turn out to be specific material entities. Quite quickly, a consensus emerged that these were located on chromosomes—threadlike structures that were observable with the microscopes of the time. This consensus was reinforced as techniques were developed that enabled the order of the genes along the chromosomes to be ascertained, techniques which also made the hypothesis that genes were physical entities increasingly hard to resist.

A turning point in the attempt to convert hypothetical Mendelian genes into something solidly material was the determination of the structure of DNA by James Watson, Francis Crick, and others in 1953. The molecule had a number of features that seemed essential for anything that could be the bearer of Mendelian genes. The very long sequence of varying components making up a DNA molecule could be seen to have the information-carrying capacity, by varying the order of its nucleotide components, to specify the many traits for which there were genes. The double helical structure, allowing the possibility of separating into two strands, each of which could provide the template for a new double helix, provided a mechanism for the indefinite transmission of this information. And DNA was also a sufficiently stable molecule to maintain with some reliability the information it carried. It was naturally hoped that Mendelian genes would turn out to be specific sequences of nucleotides in the DNA that made up the chromosomes. When, a few years later, the 'code' through which triplets of nucleotides 'represented' specific amino acids was discovered, revealing the way in which sequences of DNA could provide information for the production of functional proteins, such a hope seemed to some even closer to realization.

However, the last 50 years of molecular genetics can also be seen as a gradual unravelling of this attractive vision. First of all, functional proteins correlate very poorly indeed with the phenotypic traits that are of interest to the student of whole organism inheritance. Even quite simple traits turn out to be the result of developmental processes involving many different protein products and much else besides, and for more complex traits the number of proteins involved might be hundreds or thousands. Hence, a particular gene, conceived as the sequence of DNA coding for a particular protein, would typically have no very specific phenotypic upshot. Not only did traits turn out to have many genetic causes, but a particular gene would generally contribute to the development of many traits.[4]

More recently, the situation has proved to be far more complex still. First, the assumption that identifiable bits of DNA sequence are even 'genes for' particular proteins has turned out not to be generally true. Alternative splicing of fragments of particular sequences, alternative reading frames, and post-transcriptional editing—

---

[4] Not to exaggerate the Whiggishness of this history, I should note that the many/many relations between genes and traits were not unfamiliar to geneticists in the first half of the twentieth century.

some of the things that happen between the transcription of DNA and the formation of a final protein product—are among the processes the discovery of which has led to a radically different view of the genome. The relationship between stretches of DNA and protein products is already many/many. Coding sequences in the genome are therefore better seen as resources that are used in diverse ways in a variety of molecular processes and that can be involved in the production of many different cellular molecules than as some kind of representation of even a molecular outcome, let alone a phenotypic one.

Moreover, most of the genome doesn't code for anything. When it was still assumed that the function of the genome was to code for proteins, this non-coding sequence came to be known as 'junk DNA'. As a more complex view of the genome is emerging, it has become an increasingly more plausible project to look for different functions of this material. It is understood that most of this non-coding DNA is nevertheless transcribed into RNA, and the list of identified functions of these RNA molecules is growing rapidly. DNA sequences at other sites attach to various chemicals in the cell which in turn affect the rate of transcription at related loci. And it is plausible that even parts of the chromosomes that do not have specific chemical functions may have structural importance. The structural configuration of the chromosomes will affect, for instance, which parts are accessible to the transcription machinery. The assumption that the genome merely stores information is becoming untenable, and it now appears rather as an object in constant dynamic interaction with other constituents of the cell.

A problem that emerges from all this and that is exercising a growing number of philosophers of biology is whether any coherent interpretation of the concept 'gene' can be recovered from these complexities. The first part of an answer is well captured by the proposal of Lenny Moss (2003) to distinguish two kinds of usage, which he calls genes-P and genes-D. Genes-P are related to phenotypes and the biological tradition of preformation. They are most obviously the genes for this or that phenotypic trait found in the Mendelian tradition. Genes for cystic fibrosis or Huntington's disease are genes-P, as are the red-eye or wingless genes in *Drosophila*. Mendelian genes retain important, but highly circumscribed, uses, but they are quite unsuited to general characterization of the genome. Genes-D are understood in relation to development. Genes-D are defined not by their phenotypic outcome but by their molecular sequence. Though they are often referred to by means of a protein for which they 'code', it is important to be aware that the sequence is fundamental. So, for instance, the N-CAM gene, named for the neural cell adhesion molecule for which it codes, can actually produce perhaps 100 different isoforms of this molecule in different tissues and at different developmental stages. Genes-D are the functional constituents of the genome as these constituents need to be distinguished in order to understand molecular function and, more specifically, the way genomes contribute to the differentiation of cell function in development. The philosophical issue here is whether there is any canonical division of the genome into genes-D or whether, as I and a number of commentators suspect, this is just a name for any sequence of nucleotides that may, for a particular investigation, be of interest to a particular group of researchers.[5] But what is clear is that genes-D cannot, in general, be identified with relation to their outputs even at the level of functional proteins, let alone at the level of phenotypic traits.

Kinds of genes

To get a better sense of the diversity of contexts and uses in which the term 'gene' appears, I will now summarize some of the more prominent uses of the term with a view to exploring what relevance, if any, they might have to our understanding of human race. I don't claim that this is a complete list of such uses or the only way in which this concept could be conceptually divided. I hope this list will, however, illustrate the diversity of uses across the Mendelian/molecular and gene-D/gene-P divides and, consequently, make it clear how hazardous it is to talk about genes, or the genetic, without a good deal of clarity about what is intended.

1. *The hypothetical cause of a phenotypic difference.* This is the original meaning of the term 'gene' in classical Mendelian genetics and the standard gene-P. I mentioned, as an example, the gene for cystic fibrosis.

2. *The physical cause of a difference.* There are, of course, physical features of the genome corresponding to traits with Mendelian inheritance patterns. These may be point mutations, deletions, insertions, inversions, duplications, and so on. They are of continuing interest mainly in the study of genomically based pathologies and also in the application of genomic knowledge to the improvement of techniques for plant and animal breeding. But they are not the objects of primary interest in the study of normal development. The 'gene for cystic fibrosis' actually refers to a large number of possible mutations in a particular part of the genome, so the relation of this concept to the previous one is not straightforward correspondence.

3. *The physical cause of a trait (the gene for X).* Such a thing can only be assumed to exist at all in so far as there is a Mendelian trait; in which case it will be the kind of thing described in 2. For most X, there is no genomic feature or set of features that can be distinguished as the gene or genes for X. This is very likely the case for most of the Xs for which genes are regularly announced in the popular press, contributing to massive popular confusion on the general nature of genetics. The expression 'gene for X' is, according to Moss (2003), the canonical expression of the failure to distinguish genes-P from genes-D and very often signals a conflation of these two concepts.

---

[5]   See various discussions in Beurton et al. (2000). For empirical evidence that biologists do not have a clear consensus on the meaning of 'gene', see Stotz and Griffiths (2004).

4. *Quantitative trait loci.* This category constitutes a technical qualification of the negative remark at the end of the last category. Breeders interested in, for example, leaner cattle or bigger cabbages, can locate particular genomic loci that have particular relevance to such features and use this information to improve breeding programmes. These are 'genes for big cabbages' only in the sense that changes in these loci tend to affect the size of cabbages more than other loci. These loci may have countless effects, and their effects on cabbages may be dependent on interactions with numerous further genetic and environmental factors. Quantitative trait loci can be identified for breeding purposes through genetic markers (see below).

5. *An open reading frame (ORF).* This is the sense of 'gene' intended, more or less, when we are told that there are only 23,000 genes in the human genome; its beginning and end are marked by 'start' and 'stop' codons—signals to begin and end transcription. It is a bit of sequence that is sometimes transcribed as a block into RNA. These transcripts are then subject to the processes of alternative splicing, editing, and so on that result in the much larger number of proteins, probably at least 10 times the number of ORFs, though still a very speculative quantity. This is a concept located firmly in genomic studies, and it is unlikely to be confused with a 'gene for' a specific trait.

6. *A functional part of the genome.* This very loosely defined concept may well be the way the concept is heading in technical molecular biology. This, I take it, is the paradigmatic gene-D. I gave the example above of the N–CAM gene, which plays a central role in the production of a set of closely related proteins. These can coincide with the ORF or be much smaller genomic elements.

7. *An error in the genome.* This is the molecularized Mendelian concept in much of medical genetics. It refers to any peculiarities of the genome with pathological consequences. For example, the cause of a particular case of cystic fibrosis is a particular error, one of the set of mutations corresponding to the Mendelian gene.

8. *A genetic marker.* These will be discussed below. Whether they are properly referred to as genes is questionable, but this is the concept that underlies many recent claims about genetics and race. A genetic marker is a specific bit of sequence that need not have any function or correspond to any natural unit, but which is used to locate a part of the genome, generally because it is close to some functionally interesting part, for which it can serve as the marker.

Race and genes

I have already noted that the first category on the list has no possible relevance to race and, by implication, neither has the second. Whatever races are, they are not Mendelian traits, traits that are present or absent in any individual and are transmitted in specific ratios across generations. Number 3 deserves a little more discussion. If there are biological features that constitute belonging to a particular race (and I continue to assume this for purposes of the argument), then surely there is some set of factors that causes those features. In considering this suggestion, it is worth recalling that all or most biological kinds encompass a substantial amount of variation.[6] Hence, if there are causes that make something a member of a kind, these causes are themselves likely to be diverse. Anyone who thinks that particular races are objective biological kinds must admit that they are variable kinds with diverse memberships.

It must be admitted that the variability of a trait does not in general prevent genetic analysis of the processes involved in its ontogeny. Examples are the complex diseases—diabetes, Alzheimer's, cardiovascular disease—that are currently undergoing this kind of investigation. Two further points distinguish such cases from the case of race, however. First, it is important to distinguish the basis of correct function from the basis of dysfunction. Various blows on the head can disrupt proper brain function, but there is no correct blow on the head that explains normal brain development. In the case of diabetes, for example, part of the genetic project is distinguishing different classes of genetic failure which call for quite different therapeutic responses. None of these failures is in 'the gene for correct blood sugar regulation'. But second, and most importantly, the fundamental reply is that race is not a biological trait at all, it is a social classification. So there is no candidate subject for genetic explanation. In some cases this is self-evident, as in the category of 'black' defined by the one-drop rule in the United States. More generally, this conclusion is entailed by the failure of races to constitute credible biological kinds. What I mean by this, and the reasons for it, should become clearer in the later stages of this chapter.

Of course, even though races are not kinds, and even if race is not a biological trait, several phenotypic features strongly associated with conceptions of race are traits of some kind. Most obvious is skin colour. Skin colour is not, however, a Mendelian trait but, like height, the kind of continuously variable character associated with many genetic loci. If one were interested in breeding people with darker or lighter skins, one could probably discover QTLs (qualitative trait loci) that would facilitate a sophisticated breeding programme of this kind. But particular alleles contributing to skin colour are quite certainly not located exclusively in members of one (socially defined) race. Apart from the fact that people of different races do, of course, interbreed, this is evident in the continuous variability of skin colour.[7] And again, of course, there are people who count socially as white who have darker skins than some people who count as black (a well-tanned person from Southern Europe, say, versus an American with one grandparent of African descent).

---

[6] Issues about biological kinds are discussed in several essays in Dupré (2002).

[7] I understand that in parts of Latin America social discrimination is based on continuous variations of skin colour. Though this may no doubt ground a form of racism, it does not assume the dichotomous view of distinct races that is the present subject of discussion.

These elementary observations about variability of race-indicative phenotypes and the regular interbreeding between people socially defined as belonging to different races is enough to show that there is no interesting work to be done in this area by genes-D. One could investigate the developmental processes by which skin colour, say, is determined, and this would include investigation of various genes in the sense of number 6, above (genes-D). It is almost certain that a variety of developmental processes might equally be found to lead to the same skin colour, and certain that no such precise developmental sequence would be exactly correlated with any particular socially defined race.

Development, the more or less species typical physiological trajectory of an organism, depends on a great variety of factors. Though the genome is, of course, an essential factor, what genomic resources are deployed at any point in the developmental cycle depends on many other factors of many kinds. Familiar metaphors for the genome—blueprint, recipe, program, and so on—suggesting that the genome alone determines the development of an organism, are entirely misleading. The resources required for development are sometimes divided between the genetic and environmental, but within such a division environmental resources will range from parental care and social context to the set of extragenomic factors passed from mother to offspring in the egg cytoplasm. Any of these 'environmental factors', up to and including the social, can affect the chemical and physical structure of the genome in ways that will contribute to the determination of which genomic resources are exploited by the developing organism. These complexities of development and of genomic function explain more deeply the point stressed earlier in this chapter: that it is in general quite mistaken to think of bits of the genome having specific functions defined in terms of phenotypic outcomes. So, finally, there is no reason to expect a particular set of genomic features to provide a complete causal explanation of a feature such as skin colour. Like other features, we should expect skin colour to be the final outcome of various possible developmental pathways, exploiting a range of genomic and other developmental resources.

Genetic testing for race

With this background we can address the concerns that have been raised by recent claims that race can reliably be discerned with genetic tests. First, we need to relate the relevant categories of genetic test to the various gene concepts that have been distinguished earlier in this chapter. These genetic tests depend on large numbers of genetic markers; and though these are genes-D, in the sense that they are specific bits of DNA sequence, there is no necessary, or even expected, connection to functional units of DNA; the tests in question are essentially similar to the technologies used in criminal forensic genetics and paternity testing.[8] Human genomes differ considerably in detail,

---

[8] Individual genetic fingerprinting is most often based on measuring repeated sequences of variable length in non-coding DNA. This is a somewhat simpler technology, but the basic point is the same, namely, an inventory of variable aspects of the genome. It is also worth noting that most tests of this kind use loci on

and one current successor project to the Human Genome Project is the cataloguing of single nucleotide polymorphisms, SNPs, particular points in the genome at which different individuals are found to have differing nucleotides. A particular SNP can provide the basis for a specific genetic marker. If one tests an individual for a sufficient number of common SNPs, one can find a profile that becomes more unique the larger the number of SNPs for which one looks.[9] An important thing to note about all these technologies for genetic profiling is that the variation studied is preferentially drawn from parts of the genome that do not have coding functions. This is for the simple reason that the less functionally critical the sequence is, the more variation will accumulate in it. Variation in sequence with important coding function is likely to have deleterious effects on the organism, so that integrity of sequence is maintained both by internal editing processes and, failing that, natural selection.

As well as being clear about the relevant concept of gene here, we need to consider how race is being conceived. The simple answer is that race is being identified with geographical ancestry. Certainly, the concepts are related. Traditional broad racial categories were assumed to coincide with origins in the world's major continents, and more local racial concepts are often identical to concepts based on ancestry: African American means, more or less, having ancestors from (West) Africa. In the United States 'black' is often understood as synonymous with 'African American', though in the United Kingdom the former term is used much more widely to include more or less anyone with naturally (i.e., not environmentally induced) dark skin and can include South Asians or aboriginals from Australia or New Zealand.

The reason why SNPs or other indicators of genetic variation can track ancestry is clear enough. Since SNPs appear by random mutations and are passed on to offspring, they will, if they are fortunate, diffuse slowly around and away from the populations in which they originally occurred. SNPs will for a considerable time be most common in those areas in which they have originated. Thus, and especially where populations are less mobile, particular SNPs will tend to characterize particular geographical regions. We are now in a position, finally, to consider the significance of recent reports (Rosenberg et al. 2002; Bamshad et al. 2003), apparently disturbing to some, that a genetic test can give results that very reliably predict whether Americans categorize themselves as African American, white (or European American), or Asian. Such tests indicate whether a person has ancestors who came from a particular geographic area in which particular SNPs originated. Thus, the ability to distinguish, through these tests, those Americans who identify as African American shows that those who so identify tend to have more ancestors from a particular region, presumably in this case West Africa. This should hardly surprise

either the Y chromosome or the mitochondria, thereby restricting their relevance to only male or female ancestors. This has the rather striking consequence that claims to ancestry at, say, 10 generations in the past will actually be based on the genome of just one of the 1,024 ancestors in that generation.

[9] Actual testing, both for ancestry and for individual identity, in fact uses a range of variable genomic features. I mention SNPs in large part because they are the easiest to explain. The differences between these features are not important for the present discussion.

anyone. It should be stressed, though, that all of these SNPs will be found in many people who don't identify as black since racial interbreeding will ensure that they are gradually spreading through the wider population.

If one is trying to specify an individual from among a generally interbreeding population, the desirability of tracking non-functional, or minimally functional, parts of the genome is clear. The more functional the locus, the more selection will work to reduce variability. This argument is less straightforward in the case of testing for geographical ancestry. If populations have adapted to local conditions, then genes involved in such adaptations will be the most reliable indicators of that ancestry. Markers linked to such selected genes will be the ones that will tend to spread through the population. As a matter of fact it is unclear whether such functional genes are available.[10] The majority of loci actually applied are variations that occur in all populations and differ only in their frequencies within populations. The best correlation with West African ancestry in US populations is exhibited by the so-called Duffy null gene, a variation that confers almost complete protection against the malaria *Plasmodium vivax*. This gene occurs in a very large majority of sub-Saharan Africans but is rare in Europeans. This degree of bifurcation is not currently known for any other locus. The extent to which the genetic variation characteristic of geographical locations is due to such adaptive histories is unknown, though for reasons that will be briefly discussed in a moment, much of this variation is likely to be a great deal more local than even the geographically restricted racial categories currently being considered. In what follows I shall mainly consider non-functional SNPs.

It should be noted that the same process of geographic origin of genetic mutations does explain the concentration of genetic disease in people of particular geographic origins. Many genetic diseases are consequences of simple point mutations in coding sequences that lead to a pathological defect in a functional protein. Echoing the preceding discussion, selective processes such as linkage to a selected locus or heterozygote superiority (as in sickle cell disease) will most effectively spread the deleterious genes, though drift may also be a sufficient explanation. Such localized genetic diseases lend some support to the idea that racial identification may have a use in targeting of genetic medicine, though the weaknesses of such a strategy should also by now be clear. First, ancestry and race are not identical concepts, and only ancestry has any relevance to the incidence of genetic disease. Racial self-identification is at best a rough proxy for a specific ancestry. Second, as with the functionally neutral mutations that are typically sought in ancestry testing, deleterious mutations will not be confined to people who identify as having ancestors in the relevant geographic region of their origin. There is distinct danger that exaggerated correlations will be assumed, and disease will be overlooked in people who do not identify with the right racial proxy

---

[10] Some commercial providers of ancestry tests do claim to use mainly or entirely functional loci, and I don't wish to query (or endorse) their claims. One reason it is difficult to do so is that most of these loci are proprietary information. As noted in the text, I don't think that anything fundamental is at stake.

groups. We may hope that as genetic tests become rapidly cheaper, diagnosis by racial classification will be a very temporary transition towards more general disease screening or individual genetic testing.

Are there biologically distinguishable human kinds?

There is a long philosophical tradition of asking whether the categories into which we sort things are in some way given to us by nature or rather imposed on the world. Categories such as the chemical elements or, though more controversially, biological species are often assumed to be given by nature and discovered by us, and these are often referred to as 'natural kinds'.[11] Pencils, penitentiaries, and philosophy professors, on the other hand, are clearly humanly created categories. A formulation of the question about race that has been partially resurrected by recent genetics is whether human races are natural kinds (see Haslanger 2008). I take the answer to be an unequivocal no. The procedures just discussed for testing for geographic ancestry in America are effective because the large majority of African Americans have ancestors in a relatively specific geographic region. This does nothing whatever to support broader racial categories. As noted above, in the United Kingdom people are classified as black if they are non-white and, hence, experience discrimination. It includes people with Asian origins as well as those of African and Afro-Caribbean descent. It would also include (no doubt a very small number of) native Australians or New Zealanders. This is about as heterogeneous a group of humans in terms of origin, and therefore genetic profiles, as it would be possible to construct. In the United States the category is more narrowly circumscribed in terms of descent from black populations in West Africa, though to the extent that the one-drop rule is taken seriously this could make the category even more heterogeneous in terms of origin and genetics than the British version. At any rate, no serious scientist thinks these categories, even if the American category is interpreted in terms of some greater predominance of African ancestors, have any biological grounding that could justify any claim to the status of a natural kind.

However, as previously remarked, the human population does have some geographic structure. Relatively isolated populations of humans, as with most species, make minor but specific evolutionary adaptations to their environments quite quickly. Skin colour has been mentioned as one superficial characteristic which is notably fluid in human micro-evolutionary history. Following Kaplan and Pigliucci (2003), I have discussed elsewhere (Dupré 2003: ch. 7) the relevance of local human 'ecotypes' to discussions of race. The point is just that while there have been and continue to be numerous very local human types adapted to specific local conditions, this is a vastly finer-grained classification than any standard racial category. This phenomenon very possibly explains such observations as the dominance of Kenyans among international

---

[11] For general discussion, see Dupré (2002), especially chs. 1, 2, and 8.

marathon runners as the consequence of local adaptation to a culture involving extensive running at high altitude. (It has also been pointed out, though, that this tradition of success promotes a culture of aspiration in this particular direction.)

Broad racial categories, at any rate, comprise large numbers of ecotypes that are likely to differ in most respects of local adaptation. All equatorial peoples share dark skins and related adaptations to high temperature and strong solar irradiation, but there is little reason to suppose that they share any other adaptations not specifically responsive to climate. This is, of course, why claims such as a correlation between race and IQ are so biologically implausible. Though it is possible that local adaptation may promote subtle differences in cognitive skill sets, no good reason has been offered why these differences should be common to all ecotypes in low latitudes. Given that race broadly defined is a social kind with no interesting biological grounding, it is overwhelmingly plausible that familiar social explanations—less educationally enriched environments, subtly culturally biased questions, and so on—will be more relevant to explaining prima facie data of this kind.

Why does all this matter?

It is sometimes remarked that it is misguided and even dangerous to engage in debates about the biological reality of race. By doing so, it is said, one is offering quite unnecessary hostages to fortune. What if races did turn out to be biologically significant categories? We would still have no reason to discriminate against people because they fell into a different biological kind from our own. After all, male and female are indisputably significant biological categories, but this provides no justification for treating women (or men) as inferior. This is all true enough, and it is certainly important, if only because of the last point about sexual kinds, to insist that biological difference is no simple justification for social discrimination.

However, I think it is important to engage with the biological issue. First of all, there is no very serious hostage to fortune involved. We know enough about race to be quite confident that races will not turn out to be significant biological kinds, and it is at least worth explaining recent developments in genetics which are liable to be interpreted as underwriting biological interpretations of race.

Second, and more importantly, although we should not (of course) unjustly discriminate against people on the grounds of difference, real differences can and do provide reasons for different treatment. The political consequences of sexual difference are a much more complex issue than those of racial difference. Minimally, the fact that most women bear children is a reality that cannot simply be ignored in a just society. The problem, or one problem, is to make sure that this fact does not lead to systematic and unjust disadvantaging of women. In the case of race, by contrast, there is no such difference and therefore no such problem. If, as some racists may once have thought, black people were an evolutionary experiment somewhere on the step from apes to white people, there would be a real question as to what differences in treatment, if any, this justified (as, indeed, there is beginning to be a debate as to whether we are morally justified in treating apes in quite different ways from humans). But such a sharp distinction between human races is, needless to say, biological nonsense. Racial categories group together highly diverse groups of people on the basis of multiply evolved and trivial surface characteristics, and it would be miraculous if there turned out to be systematic biological differences dividing members of socially distinguished racial groups. So there is no question of what differences there should be in the treatment of people of different races: there should be none. The only question is the political one of how we move from racially divided societies practising racial discrimination to a situation in which race ceases to be a concept of any interest to anyone. Addressing biological misunderstandings doesn't do much to get us there, but it provides a small part of the necessary groundwork.

Conclusion

Contrary to some popular misunderstandings, there are no 'genes for' race in any of the various senses of the word 'gene'. There is a lot of local variation within the human species, as there is for almost any widely distributed species; but as migration, easier travel, and so on make the species increasingly panmictic, this variation is likely to become ever more dispersed. This variation, moreover, provides no grounding what-ever for the much coarser classifications that make up traditional racial categories or, indeed, any other comparable higher-level categories. The human species is an unusu-ally genetically homogeneous one, and there are no important natural kinds distin-guishable within it. As I have also discussed, genetic techniques make it possible to identify the geographic origins of some of the ancestors of individuals. But this reflects random and insignificant changes that occur in local human populations, or perhaps superficial adaptations to very local conditions, not the discrimination of significantly different kinds. Recent biology has confirmed the conviction of those who have long insisted that racial kinds were social kinds, and it has undermined any possible argument for placing these kinds in the realm of the biological. In its broadest and most common understanding, the concept of race remains little more than the reified residue of racism.

16

Causality and Human Nature
in the Social Sciences

Introduction

It would be hard to find a more fundamental concept for the social sciences than human nature. The social sciences are, after all, about *human* societies, so they had better have some idea what the constituents of such societies are like. But the issue central to the present chapter is whether human nature is something that the social sciences presuppose, an exogenous input from some other part of the intellectual map, or whether it is rather the subject matter of the social sciences, something that the social sciences aim to illuminate. Or, and here is where I shall suggest the truth lies, perhaps it is not quite either, but human nature is a concept that can only adequately be understood from multiple perspectives, some, but not all, of which form parts of the social sciences. The other topic of this chapter, causality, is fundamental to explaining this last point, as will emerge, I hope, as the chapter develops.

The reason that there has been a question about the role of human nature in recent years is that there has been an active and influential movement to insist that this was a question entirely, or almost entirely, outside the social sciences, somewhere on the boundary between biology and psychology. A natural, if ultimately arbitrary, point to date the beginning of this movement is with the publication in 1975 of E. O. Wilson's *Sociobiology*, and the heated controversy that followed this event. Wilson famously suggested in this work that the extension of evolutionary biology he was advocating would lead to the 'cannibalization' of the social sciences and ethics, as human behaviour, both social and individual, was increasingly understood as an elaborate set of fitness-maximizing devices.

This reductive vision was rightly subject to severe criticism (Lewontin et al. 1984; Kitcher 1985), firstly because of its scientific inadequacy, but also because of its unsavoury potential social and political implications. But for two reasons this is hardly the end of the story. First, as I shall describe in a moment, the same basic ideas emerged soon after in a slightly different guise. But second, the extreme reaction to the sociobiological picture, reductive environmentalism, is no improvement. Indeed the latter may be the position with the more disastrous potential implications. Biological determinism suggests political nihilism, as attempts to alter the natural biological state of human life must ultimately be futile. But environmental determinism suggests a plasticity of human nature that may legitimate any political system, however repellent it may seem to us, now. Worker bees, one assumes, do not yearn for the freedom to choose their way of life and nor would we if our upbringing and social milieu had properly conditioned us to the lives of slaves.

The remainder of this chapter will take on three tasks. The first will be to describe the successor project to sociobiology and briefly point out some if its major weaknesses. The second will be to sketch a more adequate view of the relation between biology and society in the development of human nature. And finally I shall say something more contentious about the way this positive view presents a possible view of human freedom. This will also make clearer the vision of causality that, I believe, makes most sense of the problem addressed in the second part.

From sociobiology to Evolutionary Psychology

As mentioned above, sociobiology slipped out of view during the early 1980s, in part in response to some severe criticism. However, something similar re-emerged in the latter half of that decade, rebranded as Evolutionary Psychology.[1] There is considerable debate as to how much this scientific venture differed from its predecessor. The official story is that sociobiology had ignored a crucial link between evolution and behaviour, the cognitive mechanisms that had evolved to produce appropriate behaviour in response to environmental information (Cosmides and Tooby 1987). It seems unlikely that Wilson had been unaware of the necessity of some kind of cognitive mechanism or, to put it differently, of the distinction between proximate (neurological) and ultimate (evolutionary) causes.[2] However, there is no doubt that Evolutionary Psychologists devoted more attention to this intervening entity, and this led to an aspect of their account of the mind that I want to stress, what I refer to as its atavistic character.[3]

Evolved cognitive mechanisms are devices evolved to respond to problems organisms face in surviving and reproducing. But exactly which problems will these be? Clearly they will not necessarily be the problems that the organisms are currently facing: evolution is not an instantaneous process. In fact, one of the most distinctive features of Evolutionary Psychology was the quite specific answer it gave to this question: human cognitive mechanisms evolved in the Pleistocene, the period from

---

[1] Following Buller 2005, I capitalize Evolutionary Psychology to refer to the specific and influential school discussed here, and associated especially with John Tooby, Leda Cosmides, and David Buss. Classic statements are Barkow et al. (1992) and Buss (1999). In lower case, I mean by evolutionary psychology any attempt to understand how it is that humans came to have (evolved) the mental capacities they now exhibit. Provided the latter project does not assume a specific and controversial understanding of evolution, it is of course unexceptionable.

[2] This distinction was made famous by Ernst Mayr (1961).

[3] I shall concentrate my criticism of Evolutionary Psychology on this point. This is far from exhausting the difficulties the position faces. For more comprehensive criticism see Dupré (2001); Buller (2005). I explain the present objection in more detail in Chapter 14.

about 2 million years ago, to about 10,000 years ago, the end of the last Ice Age. Motivating this choice is the thought that substantial periods of time are required for significant evolutionary change, and the Pleistocene is conceived of as a sufficiently extended period with reasonably constant conditions to which human life could adapt. It is also the most recent such period, and therefore an appropriate era during which to look for characteristics that distinguish humans from other lineages from which they have diverged, most recently the great apes. Much of Evolutionary Psychology has consisted of reflection on the conditions that might have obtained during this period, and on the behaviours that would have been most favoured by natural selection given those conditions. This has been more or less supplemented by empirical investigations aiming to show that the appropriate behaviours have, indeed, evolved.

There are, unfortunately, many problems with this line of thought. To begin with, knowledge of the conditions in the Pleistocene is a lot less certain than one might wish and, more importantly, those conditions were probably far from stable. It has been argued that the safest inference from the Evolutionary Psychologists' assumptions would be that human psychology should be enormously flexible to take account of this variability. But even if we did know as much as we could wish about the Pleistocene, including that the relevant conditions there were highly stable, the procedure in question would be highly dubious. First of all, a lot of human behaviour has roots that are far more ancient, and that are shared with many of our not even very close relatives. Sociability, for instance, is not a uniquely human attribute, though its detailed implications may be different in humans than in other animals. But then, second, the assumption that significant evolutionary change must have taken at least hundreds of thousands, perhaps millions of years, is also questionable. This latter assumption is based on a model of evolution as change in gene frequency resulting from selection of advantageous alleles. But significant changes in the nature of human sociality are evident over historical periods of tens or hundreds of years, presumably because they are due to cultural, or possibly epigenetic, processes. Why should similar processes not also facilitate the evolutionary divergence between humans and non-human relatives?[4]

I mentioned that Evolutionary Psychologists attempt with varying degrees of commitment to provide empirical backing for the hypotheses derived from reflections on the Pleistocene. It should be stressed that empirical support is being sought for universal claims about human psychology. There is some room for explanation of diversity in human behaviour through appeal to different environments in which people grow up, and specific differences in the experiences of individuals. But the object of interest is what is common to all humans: human nature. There are, certainly, worthy motivations for a concern with universal human nature, for example it may serve as a ground for rejecting racist views that claim deep differences between groups

---

[4] Limitations to the neo-Darwinist view of evolution assumed by Evolutionary Psychology are discussed more generally in Chapter 9.

of humans. On the other hand, Evolutionary Psychologists do make a lot of the differences between the sexes; from an evolutionary perspective it is certainly a highly salient one. The historical message seems to be that with sufficient ingenuity views about human nature can be deployed on either side of most political issues.[5] However, the Evolutionary Psychologists' treatment of sexual difference does point to deep theoretical difficulties with their general position.

The notion that there is no difference at all between the human sexes except what local conventions of gender dictate has largely been abandoned, and this is probably a good thing. It is an unhelpful view because it represents exactly the veering to reductive environmentalism that I mentioned above. There are, of course, biological differences between men and women. The trouble is that although Evolutionary Psychologists claim that their theories are interactive—the psychological modules we all share determine behaviour in ways responsive to and hence appropriate for environmental circumstances—their evolutionary arguments are presented in terms of universally optimal behaviour for humans, for males or for females. Moreover, the dispositions that humans develop through their lives are universal. If humans universally have a tendency to reciprocate cooperative behaviour, let us say, and to punish selfish behaviour, the interaction is only at the point of detecting an instance of cooperation or selfishness and then behaving appropriately. Development, the process of becoming a mature human with a particular set of responses to contingencies in the world, turns out to be irrelevant. A proper interactionism, on the other hand, does not merely involve appropriate interaction with various environmental contingencies, something that probably characterizes every life-form on the planet, but rather refers to development that produces different mature phenotypes in response to different environments. This much is also true of many organisms, perhaps most strikingly plants. What is developed to a unique degree in humans is the ability to develop a cognitive phenotype, a set of cognitive mechanisms, if you like, that is adjusted to the environment in which it matures. And this is something that the evolved cognitive mechanisms of the Evolutionary Psychologists are wholly unable to comprehend. So I now turn to a view of evolution that is better fitted to this task.

From Evolutionary Psychology to developmental systems theory

Evolutionary Psychology, as I have tried to explain, is ultimately committed to a view of development that sees the basic parameters of cognitive systems as somehow

---

[5] The political versatility of scientific findings is illustrated in some detail in the second half of Barnes and Dupré (2008) with respect to genetics and genomics. What we describe there as 'astrological genetics' the vulgar view that sees details of human behaviour ineluctably inscribed in genes, would be difficult to deploy in a politically progressive way. There is probably no reputable scientist who believes the extreme vulgar view, though it is easily read into a lot of popular writing, not least by Evolutionary Psychologists, and it is often implicit in casual statements by scientists extolling the importance of their fields.

inscribed in our DNA. One reason that it does this, to which I have already alluded, is that it is still very much mired in the assumption central to neo-Darwinist thinking, that the products of the evolutionary process could only be preserved in the long term if they were entrusted to the care of the genome, to Dawkins's 'immortal coils' (1976). This assumption has little to be said for it, however. Genes are by no means the only vehicles by which information about development can be passed down from one generation to the next, and it is far from clear what degree of stability—immortality—is required for such a mechanism of heredity to function in an evolutionary process. Three generally interconnected processes that have come under recent investigation and that illustrate the limitations of traditional gene-centred neo-Darwinism are epigenetic inheritance, transgenerational niche construction, and cultural evolution. I shall next say a little about each of these.

Epigenetics

Epigenetics embodies a fundamental re-evaluation of the ways that genes work. Genomes are constantly undergoing chemical modifications through interactions with the cellular environment. The most well known of these is methylation, the alteration of cytosine, one of the bases that make up the famous genetic code, by the addition of a methyl (CH_3) group. Other epigenetic processes modify the protein core that forms part of the structure of the chromosome. Methylation generally reduces the probability that the sequence of DNA in which it occurs will be transcribed, thus changing the overall output of RNA transcripts from the genome. Processes of this kind help to explain the different behaviour of genetically identical cells in the different parts of the bodies of multicellular organisms. The crucial implication of the expanding understanding of epigenetic phenomena is that it finally lays to rest the idea that the nature and behaviour of an organism was somehow inscribed in the sequence of nucleotides in its nuclear DNA. It is now clear that this sequence provides no more than a (vast) set of chemical possibilities; what is actually done even in terms of the transcription of RNA molecules, depends on a further level of chemical modification, and one that is far more transitory than DNA sequence.

Contrary to an earlier belief that at least only DNA sequence was passed on to subsequent generations, it is increasingly clear that some epigenetic changes can be inherited too. Striking illustrations of this kind have emerged from the UK Avon Longitudinal Study of Parents and Children (ALSPAC), a project involving 14,000 mothers enrolled during pregnancies in 1991 and 1992. The findings of this project have included a correlation between smoking by men prior to puberty and obesity in their male offspring, and—bizarrely—an inverse correlation between the availability of food for men in childhood, and the longevity of their grandsons (but not granddaughters) (Pembrey et al. 2005). Although it is difficult to assemble conclusive evidence, such results add to the plausibility of the long-held suspicion that descendants of victims of the Dutch hunger winter of 1944–5 showed symptoms such as low birth weight, and that these were the consequence of epigenetic inheritance.

It is also important that epigenetic inheritance need not involve the direct transfer of molecules between generations. A fascinating illustration of this point can be found in the research on maternal behaviour in rats by Michael Meaney and colleagues (Champagne and Meaney 2006). It appears that attentive mothering by rats, involving a lot of licking of rat pups, produces calmer, less nervous adult rats, and that this is a consequence of epigenetic effects in the developing rat brains initiated by maternal care. These calmer adults, if female, are likely to lick their pups more. Hence the epigenetic changes to the rat's brain can be passed on by means of a process involving behaviour alterations between parent and offspring.

Another important point about this example is that it illustrates the fact that environmental influences on the organism can produce epigenetic changes, another crucial idea in articulating a picture of development that goes beyond simplistic genetic determinism. A disturbing example of this point is provided by the growing evidence that assisted reproductive technologies, by providing an abnormal developmental environment at a crucial point in embryonic development, can have epigenetic effects that may produce disease. These certainly include rare disorders known to be epigenetic, and it is increasingly suspected that these technologies substantially increase the risks of diabetes and obesity in later life (Pembrey 2010). More speculative is the thought that the realization that the environment can affect the behaviour of genes and can do so in ways that may be heritable, raises the spectre of Lamarckian processes in evolution. This is an issue I shall not pursue here however (but see Jablonka and Lamb 1995).

Niche construction

It is still often supposed that there exist niches in nature, and organisms evolve to occupy them. On the other hand it has been known, at least since Charles Darwin's extensive and classic investigations of earthworms (Darwin 1881), that organisms can have a profound influence on their environments, and can do so in ways that are beneficial or essential for their ways of living. Of particular importance is the fact that the niche that the organisms construct is the environment in which subsequent generations develop. Thus, as opposed to the niche being a pre-existing space to which natural selection adapts a group of organisms, the organisms come to be adapted to the environment that its members have constructed, in part because that environment provides some of the conditions that enable them to develop in an appropriately adapted way. I shall therefore sometimes refer to the constructed niche as a developmental niche.[6] Classic examples of niches both constructed and developed are provided by the beaver, the entire life of which focuses on the resources provided by the dam that it itself constructs, and the termite, whose mounds are remarkable achievements in climate control and much else. But these are only extreme examples. It is increasingly

---

[6] I learned this very useful term from Karola Stotz.

acknowledged that niches are not pre-existing givens, but rather co-evolve with the organisms that inhabit them (Odling-Smee et al. 2003). And surely the organism that has taken this phenomenon to the highest level is *Homo sapiens*.

From certain perspectives one may admire the climate regulation system of a termite mound more than the energy-guzzling air conditioning systems that keep the inhabitants of Los Angeles or Hong Kong comfortable on hot days, but it would be hard to deny that the latter constitute even more complex systems, and ones that would not have been possible without the unique cognitive endowments of the human species. More fundamental to human development, on the other hand, are the hospitals in which most of us are now born, and which contribute to the extensions of our lifespans, and the schools that provide us, over many years, with the skills necessary to negotiate successfully the enormously complex material and social environments we construct. No one could be tempted to imagine that a human infant raised in the wild by non-humans would acquire these skills by sheer force of genome.

One way of thinking about these phenomena is through Richard Dawkins's (1982) notion of the extended phenotype. On Dawkins's view, a termite's genes don't just build termite bodies, they build termite mounds by determining the behaviour of termites that results in the building and maintenance of mounds. It should be noted that this provides a very different causal path between the generations from the familiar idea of a genome directing the development of an organism. For one thing, it is evidently impossible for a termite to build a mound by itself, so that the termite genome is at best only part of a much larger system that in its entirety provides the conditions for the production of new termites. My own view is that the differences are greater than the similarities, and Dawkins's way of describing things is likely to mislead more than it enlightens. But I don't need to pursue that argument here, since the focus will remain on the human case. And no one could suppose that the environment that humans create for, among other things, the production of new humans, is simply a consequence of genetically determined human behaviour. The point is probably too obvious to require argument. It is sufficiently established, for example, by the diversity of human environments. Of greatest interest here, and one of the central explanations for that diversity, for the particular ways in which particular groups of humans shape their environments, is cultural evolution. To this I now turn.

Cultural evolution

That culture can generate processes similar to biological evolution has been a familiar idea for a long time. Recent discussions generally date from the sometimes rather technical analyses of Cavalli-Sforza and Feldman (1981) and Boyd and Richerson (1985). The basic idea is that elements of culture are transmitted from one human to another, and if the cultural item is beneficial to its possessors it will tend to be passed on more often and become more common. This deliberately vague summary covers many possibilities. Transmission may be from parents to offspring, but it certainly need not be: transmission from teachers to students or between peers is perhaps equally or more common. 'Beneficial' could be interpreted in a way analogous to biological evolution as promoting survival and reproduction, but it also need not be. It might just mean something the possessor enjoys, or it may be pleasurable or otherwise advantageous to transmit it. Cocaine use probably doesn't increase reproductive success, but the habit appears to be easily picked up, and the economic context of many contemporary societies tends to generate a subset of users with a strong interest in finding new recruits to the practice. The sources cited above offer a range of different plausible and even empirically supported dynamics for the evolution of various cultural items.

Another approach that has received a good deal of attention starts rather from the perspective of the cultural element itself. I refer to so-called memetics (Blackmore 1999, following Dawkins 1976). Here the idea is that there are certain cultural items, 'memes', that are very good at getting themselves transmitted from one human mind to another, and human minds thus end up being colonized by the most successful such memes. Although this perspective can provide some illumination in particular cases, as a general approach to cultural evolution it is highly simplistic, and not surprisingly it shares many of the defects of simplistic gene-centred approaches to biological evolution. For example, it has become increasingly clear that the division of genomes into a specific number of distinct genes is a human imposition rather than a reflection of the nature of things (Barnes and Dupré 2008). That culture does not exist as an objectively determined set of discrete elements is far more obvious.

The last remark points to some very serious issues that I have glossed over. My talk of cultural elements or items above is no more justified than the assumption that culture can be divided into memes. Indeed, and worse, I have written as if it was unproblematic what the word 'culture' refers to, and certainly this is not the case. Fortunately, I do not think it is necessary to go into any of these difficult questions here. All I want to insist on now is that a wide range of behaviour transmitted between human individuals, including from more mature to juvenile individuals, is part of the set of resources involved in the successful development of human individuals. I have wanted to indicate that there are interesting questions to be asked about the processes by which this behavioural repertoire changes over time, though I certainly do not want to commit myself to the view that this is best studied in terms of formal models, or indeed that all such phenomena are amenable to such study at all. Given only this very general assertion, it is possible to see how far the human developmental system differs from that implicitly assumed by evolutionary models limited to an obsessive focus on the genetic.

Developmental systems[7]

The point I have been making is in many ways an obvious one: the successful development of a human takes the confluence of a considerable variety of resources.

---

[7] My use of this concept is indebted to Developmental Systems Theory as developed in Oyama (1985), Griffiths and Gray (1994), and Oyama et al. (2001).

These include a great deal that is provided by other humans, some through direct interaction, many more through the construction of the environment in which contemporary human life is possible. There are also, of course, many biological conditions. Although one may say that first among those is a zygote with an appropriate and not fatally corrupted genome there is much more. The zygote and the developing embryo and foetus undergo a series of interactions with the environment provided by the mother's body, and the influence of this environment is to some extent affected by the wider environment in which the mother herself is placed. All of this makes nonsense of the idea that somehow the future adult human is inscribed in the zygotic genome, if only we had the ability to read it. Although few contemporary theorists assert so crassly the preformation of the adult in the genome, many implicitly or explicitly assume more of this picture than is defensible.

The appreciation that evolution can act on many different aspects of the developmental system is another way of seeing the inadequacy of Evolutionary Psychology. Most obviously this is illustrated by cultural evolution—the clue, after all, is in the name. Cultural evolution has surely had a great deal to do with the very different phenotypes (behavioural, at any rate) exhibited by contemporary humans and their ancestors a few centuries ago, and indeed between those exhibited in (say) New York City, rural England, and the forests of New Guinea. Genetically minded evolutionists are inclined to respond that cultural differences are easily mutable, and hence superficial. And it is true that an infant born in rural England or even New Guinea and transplanted to New York City might grow up as a typical New Yorker. But even assuming this is true, it of course begs the question by assuming that all that really matters is the 'deep' biology. This, and the argument that deep biology (genetics) takes a very long time to change significantly (a premise increasingly questionable in the light of epigenetics), are what underlie the argument for Evolutionary Psychology that I have been particularly concerned to refute.

One way to see the power of cultural evolution, on the other hand, is to stress its role in the reconstruction of the human niche. Let us focus on a very small episode of cultural evolution, say that which has occurred in Europe over the last two centuries. Human behaviour is, I suppose, significantly different between the ends of this period. At the beginning of the period a much higher proportion of people were occupied with agricultural work of some kind, and the kind of agricultural work was mainly different from anything available today. The affluent travelled in horse-drawn vehicles, the rest on foot; most people stayed much closer to home than they do today. No one watched television or played video games. Generally people did different kinds of work and entertained themselves in different ways.

The biologically inclined will tend to acknowledge these differences, but stress that both then and now people had sex, raised children, competed with one another for status, and so on; in these fundamental ways nothing changed. But as these activities do not even distinguish us from apes, or indeed most other animals, it is clear that a rather finer grain of description is relevant. No doubt there are finer grains of description than these that will count the populations in question as similar. One of the deeper problems in this area is that between any two groups of organisms there will be similarities and differences. As a population evolves new differences will appear and old similarities will disappear. What constitutes significant, interesting differences that should be marked by the term 'evolutionary change'? I do not see how any answer to this could be given by nature; it is up to us how we use this term. We might decide by fiat to apply it only to genetic changes, but if we did we should be careful not to infer anything from this about the importance of different kinds of change in nature. My point is just that in terms of changes that are of interest to us, very considerable differences occur to humans in relatively short periods of time, and whether or not these involve genetic differences may be an interesting question in its own right, but has little bearing on how significant the changes may be.

But to return to the main thread, I wished to emphasize particularly the ability of cultural evolution to transform the developmental niche. And here, at least in contemporary developed countries, it seems clear that humans have learned in quite recent time to construct a remarkably novel environment for the development of their young. Our homes are heated, plumbed with incoming water and outgoing waste, and provided with electricity. Entertainment arrives through the air or in subterranean cables at specially made receivers that project images of musicians, actors, etc. Our food comes from supermarkets, sometimes in cans or ready-frozen meals. If our health is threatened we are moved to special facilities where specialists intervene to restore our proper functioning. Massive infrastructures facilitate our movement through space and our communications with one another independent of physical proximity. And most importantly of all in the present context, other locations house specialists who impart to the young some of the vast body of information necessary to thrive in these very complex environments. All of this is entirely banal. What is curiously often overlooked, however, is that these prodigious changes to the human environment, concretizations of our rapidly evolving culture, profoundly affect the developmental resources available to growing humans. For that reason their introduction should be seen as representing major evolutionary change.

One simple example may further illustrate the point. The mobile phone did not exist when I was a child. In fact it is for hardly more than a decade that it has been omnipresent, a mandatory accoutrement for everyday life in developed countries. And whereas it may seem only more or less mandatory for people of my generation, for those aged, say 10–20, it is as unthinkable to be deprived of one's phone as to wander the streets stark naked. Most teenagers move through the world, by virtue of this technology, in a continuous dialogue with a group of friends who need not be in any physical proximity. In fact the virtual community seems far more salient than the contingency of physical proximity, very probably the cause of considerable conflict in spaces such as train carriages, in which an older generation continues to see physical proximity as a decisive basis for at least polite interaction. It is not, therefore, merely behaviour that has changed for those who have grown up with the mobile phone, but the entire experience of social space, transformed from a direct function of physical space, to a virtual space within the voluntary control of the individual. Needless to say, the rate of such evolutionary change is entirely different from the genetic change so beloved of neo-Darwinists.

Human nature

It is now possible to see why I want to deny that there is any such thing as human nature, when this is understood as something constant through the history of the species and across members of the species. By human nature, therefore, I shall in what follows mean only the nature of a particular human, or the nature typical of, or average for, a particular group of humans. Human nature as a population average can evolve rapidly over time; and individual human nature can vary considerably within a population at a time. The reason for this is not, as Evolutionary Psychologists imagine to be asserted by the 'Standard Social Sciences Model' (Barkow et al. 1992), that human nature is something superficial and trivial that can be written on the blank slate of the human mind by any ambient culture. On the contrary, it is a consequence of the complexity of the way human nature develops, the multiple causal factors involved in the progression from zygote to mature human with a relatively settled set of behavioural dispositions.[8] The complexity of the process and the number of factors that influence it explain both these dimensions of diversity. Evolution can change the characteristic or typical behaviour of a population through the accumulation of (at least) genetic, epigenetic, and cultural changes. It is safe to say that in recent human history the last-mentioned has been the leading driver, as cultural evolution has drastically altered the species-typical developmental niche. It may well be that some of these changes have become more firmly entrenched through parallel epigenetic or even genetic changes.

It is equally clear that recognition of the variety of factors involved in development makes possible a diversity of individual outcomes within even quite narrowly defined populations. Everyone recognizes that there is genetic diversity within most populations and specifically among humans. A great and currently increasing quantity of work goes into correlating these genetic differences with phenotypic differences. A major form of contemporary biomedical research is the Genome Wide Association Study (GWAS), which uses the very large volume of genomic data we now have about human populations to find correlations with medical outcomes—physiological and psychological disease. I don't mean to raise an objection to such studies, which may well succeed in usefully identifying causal factors involved in pathological processes. However, as everyone involved in such research is aware, this is a hardly a search for

---

[8] I say *relatively* settled. In fact human development should be seen as a process that continues from fertilization of the egg until death. It is probably safe to say, however, that dispositions are a good deal more fixed in the last few decades of this process than in the first.

sufficient causes. GWAS will at best provide clues to the detailed causal processes involved in pathology.

A good indication of the difficulty can be gained by reflecting briefly on by far the strongest known correlation between a genomic factor and a psychological pathology, a correlation far too well known to require anyone to launch a GWAS, namely the genetic cause of violence. The cause in question is, of course, the Y chromosome. Possession of this genomic feature increases the probability that a person will commit a violent crime by a factor of 5 to 10, the sort of finding which would be likely to achieve considerable publicity if it related to schizophrenia or cancer, say. The example can usefully highlight a number of quite general, mainly fairly obvious, points.

To begin with the most obvious point, a genetic cause is not generally a sufficient cause. Most men do not commit violent crimes. And it is not a necessary cause. Ten to 20 per cent of violent crimes are committed by women. Like any other human trait, the disposition to violence develops in interaction with a range of other factors, for example those explored by social scientists interested in the causes of violence. The variation in these factors, presumably, explains the wide differences in the prevalence of the trait in different social contexts.[9] But saying all that is not to deny that the genetic difference plays a role. This might mean that in all actual and most imaginable social contexts there would be a predominance of male over female violence. It is easy enough to imagine differences in hormone levels, the autonomic nervous system, or even more specific cognitive biases, that could result in such an enhanced disposition. And these differences may even be explained, in part, by the evolutionary scenarios offered by Evolutionary Psychologists.

But the point I want to emphasize most strongly with this example is that even with such a robust phenomenon and a well-grounded belief in causal relevance, the usefulness of this genetic information is very limited. No one seriously advocates addressing the social problem of violence by universal incarceration, elimination, or selective abortion of foetuses with Y chromosomes. This is a relevant factor in that causes of male and female violence may well be significantly different, and because it alerts us to the greater importance of focusing on the causes of male, rather than female, violence. But any practical impact on the social problem will require understanding in real depth and detail the processes that lead some people with Y chromosomes (and a smaller number without) to end up as adult humans with an atypical tendency to resort to violence.

One final point should be added with respect to the causally complex situation just described. There is a widespread if inchoate intuition that there is something specially deep and important about genetic causes. One thing that may contribute to this is a sense of their immutability: apart from some very recent and still quite unreliable technologies, there is nothing much we can do about genetic causes. But for the

---

[9] For an analysis of some factors affecting the prevalence of domestic violence, for example, see Archer (2006).

multicausal situations I have been considering, this is a reason for inferring the lesser importance of these causes. A long tradition of philosophical analysis has considered the question how we pragmatically distinguish a particular factor as 'the cause' from a complex causal nexus (Mackie 1974). A central conclusion is that we distinguish a fixed background (standing conditions) from the distinguishing and not necessarily expected factor. Thus, in one classic example, an electrical short-circuit rather than the presence of oxygen is offered as the cause (and, more obviously, the explanation) of a fire in the hay barn. The short-circuit is the 'difference-maker'; the oxygen is present just as it is in countless other non-burning barns.[10]

The preceding idea alerts us to the importance of being very clear about the scope of the questions we are considering. If we are interested in the general phenomenon, why men are more disposed than women to violence across a whole range of social contexts, then it may be that some physiological upshot of the Y chromosome is what makes the difference. But for most explorations of violent human behaviour being male is a background condition, and we are interested in causes that make the difference between violent and non-violent men. Similarly when we are interested in cross-cultural differences we will look at the differences between cultures, and the distribution of XX and XY karyotypes will be a background condition. As with almost any variable human trait, there are likely to be other genetic differences that affect the trait to some degree. Experience so far, however, suggests that it is most unlikely that there will be anything with an effect comparable in size to that of the Y chromosome.

Human autonomy

I have said a good bit about the genetic determinism which is still such a regrettable concomitant of much thinking about genetics. I want to finish on a rather different topic, determinism in general and the worries that this has long engendered about human autonomy, or free will. Space will not permit a detailed defence of my rejection of the still widely endorsed deterministic perspective.[11] What I would like to argue is that, contrary to a common philosophical assumption, rejection of the deterministic world-view does in fact have significant consequences for our view of what it is to be human.

Outside the philosophy of science it is still widely assumed that a commitment to determinism is an inescapable concomitant of taking scientific knowledge seriously at all. However, it is a quite different story among philosophers who have attempted to engage seriously with the contents of scientific belief. Philosophers of physics have, of course, given up on classical ideas about determinism since the general acceptance of quantum theory, though it is still often supposed that determinism can somehow reappear unharmed

---

[10] A sophisticated development of a similar idea, but based on the idea of the potential manipulability of a cause, has been developed by James Woodward (2003). However, for present purposes the simple idea outlined in the text will be sufficient.

[11] For this see Dupré (1993: part 3).

at the macroscopic level. To this I comment only that such containment of indeterminism seems incredible. Schrodinger's cat may or may not be around to kill the mouse that would have moved the nail that stuck in the shoe of the horse that would have . . . The fact that there are deterministic processes that emerge at the macroscopic level cannot exclude the amplified effects of microscopic events that are not deterministic from interfering with the orderliness of the macroscopic sphere.

Philosophers of biology are perhaps not typically much exercised by this question since, on the whole, they have now given up on the reductionism that, it was once imagined, might import determinism from the microscopic sphere. On the face of it the regularities that biologists discern or the models that they construct look anything but deterministic. Biologists, it is true, do tend to assert their commitment to determinism and reductionism, but it generally turns out that these doctrines are understood as methodological commitments rather than metaphysical doctrines. As such—assume that phenomena of interest have causal explanations; look for underlying mechanisms—these commitments are surely unexceptionable. On the other hand, the rise of systems biology in the last five years or so has brought a good deal of discussion of holism, emergence, and related ideas to the forefront of theoretical biological thought (Boogerd et al. 2007; O'Malley and Dupré 2005). Picking up on an idea promoted long ago by Donald Campbell (1974), biologists and philosophers have even started to consider seriously the idea of downward causation, the causation of the behaviour of parts by the whole.

But here I don't propose to review the arguments for or against these positions, but want only to consider whether the rejection of determinism and physicalist reductionism, together with the acceptance of emergent properties or downward causation would make any significant difference to the way we should think about the nature of the human. In particular, can these ideas begin to make sense of human autonomy, or freedom of the will? I want to argue against the still orthodox assumption that such issues are irrelevant to the issue of free will.[12]

The reason why these issues are generally thought to have little relevance to the question of free will is straightforward. It is naturally supposed that the alternative to determinism is indeterminism, lack of causality, or randomness. But the concerns that people have about determinism, that it may seem to imply lack of control over or responsibility for, one's actions, are hardly ameliorated by the thought that they are randomly generated. As philosophers since Hume have observed, it is a rather more attractive thought that they are caused by one's beliefs and desires.

That propositional attitudes such as beliefs and desires explain actions is largely uncontroversial, and most philosophers now hold that they do this because they cause actions. But what does this mean? One common picture is that beliefs and desires are states of the brain, and that these initiate signals down nerves which, in turn, cause the motions of parts of the body that constitute actions. But this, of course, is a picture

---

[12] An ancestor of this argument was presented in Dupré (2001: ch. 7).

that fits naturally with the philosophical vision of microscopic causal transactions to which the apparent actions of macroscopic agents are mere epiphenomena. A quite different picture begins with the rejection of the assimilation of beliefs, desires, and so on, to states of the brain. This rejection is often motivated nowadays by externalism, the view that a belief, for example, depends for its identity on things in the world beyond its human possessor. The alternative position, however motivated, is that believing that p, say, is a property of a whole human, and that the reification of a belief required in locating it in the brain is wholly unwarranted. If a belief, or an instance of believing, is indeed a property of a whole human, then its causing of the movement of a part may be seen as a case of downward causation, the influence of the whole on one of is parts. If this seems metaphysically extravagant, note that the familiar philosophical example 'I raise my arm', unless the I is a Cartesian ego or its current neurophysiological analogue, is an example of a whole ('I') acting on a part ('my arm'). So the rehabilitation of downward causation is an important step in beginning to make sense of the human agent as something causally efficacious, capable of making things happen, rather than merely an epiphenomenon of constituent microscopic happenings.

This will all continue to seem to most philosophers metaphysically extravagant in comparison to the alternative story at the microphysiological level in which a complex array of physical particles in my brain acts on another such array in my arm. If a belief is more than an array of stuff in my brain, then it may still only be that part of the belief that does the actual neurophysiological causing. Again, the description of all this in terms of whole person agency may seem epiphenomenal.

But why does this alternative picture looks so much more philosophically plausible (if it does)? The answer, I think, is that many of us are still captivated by a neo-Laplacean picture in which everything really happens at the microphysical level, which is causally closed and complete. And this picture cannot escape the implication that *everything* above the microphysical level is merely epiphenomenal. If the parts of a thing have their behaviour determined by microphysics then so must the behaviour of the composite thing be determined. Any appearance that it has casual powers of its own is illusory. It is no more or less necessary to appeal for causal explanation to the properties of my mental states than it is to the liquidity of water or the motion of tectonic plates. To a Laplacean calculator all are just the upshots of countless microscopic movements.

The resolution of this problem, in my view, lies with abandoning the assumption of the causal completeness of the physical. Although I cannot offer detailed arguments here against this assumption, I shall try to give some sense of why I think it can safely be abandoned. The microphysical determination of everyday events is, at least, hardly something open to casual inspection. It is, on the contrary, a metaphysical assumption, and once open to serious consideration it is, it seems to me, a highly implausible one.

Abandoning the assumption of causal completeness is giving up the idea of the universal reign of law, the assumption that everything happens in accordance with some universal causal regularity. Speculatively (though the speculation is of course not original), I suggest that this is an idea grounded in the prescientific conception of law as the edict of a supreme lawgiver. Certainly God should be capable of regulating every event, however minute; whether nature could or should be expected to accomplish the same feat is another matter. Reflection on biology, on the other hand, should make such universal regularity quite implausible. Not only are life processes constantly beset by at least the appearance of irregularity and unpredictability but, more significantly, regularity is won with great difficulty and ingenuity. The mechanisms that make possible the regularities that constitute the persistence of living things are more astonishing the better we come to understand them.

Of course, this will seem entirely beside the point to someone convinced that universal law reigns at the microphysical level. My point so far, however, is not to show that biology refutes microphysical determinism, but that it is incumbent on the determinist to offer an account of the relation between physical and biological phenomena. This account will be reductionist, but not in the sense of explaining biological laws, since in the determinist's sense there are none, but in the sense of explaining in principle every specific biological event. Irregularity is then an expected consequence of the microphysical heterogeneity of biological entities and processes. But then it appears that the determinist has explained too much; for biological regularities, the regularities that make possible the persistence of biological processes, while far from universal, are highly impressive and certainly in need of explanation.

I will not attempt to show that the determinist can't meet this challenge, but rather suggest that this is a point in the dialectic at which an entirely different perspective begins to look much more attractive. This is the idea that causal regularity is in fact a rare and precious thing, bought at great cost in energy or ingenuity. Biology, from this point of view, is not so much about tracing out how the causal regularities at the microphysical lead deductively to the (partial) regularities at the biological level, but rather is a matter of seeing how the causal properties of physical entities are employed to constrain events and maintain the persistence of complex systems. New properties, put to such purposes, are constantly emerging as more complex entities come into being. The complex macromolecules employed by living systems have properties—catalysing other reactions, forming structures with strength, elasticity, etc., neutralizing alien biological entities, and so on—that are a result of their particular complex structures. The combinations of these new causal capacities in turn create systems with entirely new (emergent) capacities—the abilities to fix atmospheric nitrogen, say, or run down and consume prey—capacities that contribute to the persistence of the highly complex systems of which they are part.

In this light, now consider the human developmental system, surely the most complex system in our experience. This deploys the causal capacities of humans and the countless artefacts they create, and perpetuates the survival of the human lineage and the structures that serve that survival. Central to this system is the human mind, an abstraction that I take to refer to the densest concentration of causal capacities in our experience, the capacities exercised in human intelligence, and without which it would be inconceivable that the human developmental niche could be maintained and indeed give rise to ever larger numbers of humans, in turn creating a set of problems that human intelligence may or may not ultimately succeed in solving.

This then, to summarize, is the major step towards an understanding of human autonomy made possible by the rejection of determinism, and indeed leads to a far more satisfactory metaphysics of human nature. Causal order is not something found saturating every part of the universe. On the contrary it is something quite rare and specific in its locations. It is found in the simplicity of massive physical processes such as are studied by astronomers; it is created with great difficulty in the complex, elaborately controlled and isolated machines built by physical scientists; and most spectacularly, though very differently in form, it is found in living beings.

If there is a scale of nature, it is an increase in the causal powers, the construction of causal order and regularity. One respect in which the human mind constitutes a further step in this scale is because it involves a new level of capacity to transform the world beyond the organism. Humans, in my view, are the densest concentrations of causal capacities, or causal power, in our experience. The niches we have constructed for ourselves—warm and sheltered housing, landscapes dominated by edible plants and docile and tasty animals, roads and machinery for moving ourselves about, and so on— are remarkable testimony to our causal potency. But still, it may be asked, does this amount to real autonomy?

How much autonomy do we want? As I have already mentioned, we don't want to conceptualize ourselves as random action generators. And we do want our actions to be properly related to our mental states, our beliefs and desires. Is there any sense that we can be said to choose our beliefs and desires? Or if we cannot be said to choose them, can we at least in some sense own them? It seems to me that we can do so to the extent that we organize our lives in pursuit of consistent goals or principles. If I simply act in pursuit of whatever passing whim is uppermost at the moment I exhibit no more causal power than any other animal. If I choose to build a bridge, write a book, or cook dinner, and subordinate my choice of actions to this decision, I exercise to a greater or lesser degree a distinctively human ability to shape the world.[13] In the social realm, the ability to conform to principle, above all moral principle, is the kind of regimentation of behaviour that constitutes a uniquely human achievement. And in the terms I have just been employing, it is through such plans or principles that human minds are able to impose regularity on the world. Clearly some acknowledgement of Kant is in order here, though the view I am proposing is a lot less arduous than Kant's in its account of the kind of principle that might constitute freedom. Rather than one rationally grounded canon of morality that constitutes an action as free or unfree, I would rather suggest a spectrum of degrees of causal efficacy, ranging from the person described by

---

[13] I take it that this has a lot to do with the importance that many thinkers, perhaps most famously Marx, have attached to the autonomy exhibited in labour. John Ruskin's view of the Gothic cathedral is a powerful if romantic expression of the point.

Harry Frankfurt (1988) as the wanton, to those most efficacious in affecting the world through the subordination of their immediate desire to goals and principles.[14]

There are of course many big questions unanswered. Can we choose what kind of person we will be, and if so when and how? Is it better to be causally efficacious than merely content (Socrates or a satisfied pig)? And no doubt many more. My point is only that inverting the familiar question about human freedom, might humans be an exception to the otherwise universal rule of law to the almost diametrically opposite question, might humans be an extreme exception to an otherwise largely disordered and unruly universe, opens up a quite different, and perhaps more productive, set of questions.

Conclusion

This essay has had more to say about what human nature is not than what it is. But this is no accident. Ultimately my central contention is that human nature is open. Humans have powers to shape the world and themselves which, while no doubt not without limits, have surely not yet encountered those limits. Hence I started this essay with my opposition to an influential perspective that not only insists on the importance of human nature, but offers us a methodology for determining exactly what it is. Unfortunately this methodology is grounded in an obsolete and simple-minded view of evolution. Or perhaps I should say, 'fortunately'. For it seems to me that the narrow view of human nature presented by Evolutionary Psychology is not only mistaken, but is also potentially bad for us. A limited view of human possibility must inevitably narrow human aspirations. And though it should perhaps also be said that aspirations can be bad as well as good, so that the openness of human possibility, of possible changes to the human developmental niche, can cut both ways, I am sufficient of an optimist to feel that opening up a better future is worth the risk of making possible one that is worse.

Afterword

Shortly after completing the original version of this essay, I heard a lecture by John Perry that made me reconsider the best way of presenting my views on the topic of free will.[15] The standard dialectic, of course, assumes that compatibilism means compatibility of freedom with determinism. Much of the debate concerns the question

---

[14] I have described my view in the past as opposed to compatibilist views of free will. Just before completing this chapter I heard John Perry's Dewey Lecture, 'Wretched Subterfuge', at the 2010 American Philosophical Association Pacific Division meeting, which convinced me that this opposition was mistaken, provided compatibilism was understood as compatibility not with determinism, but merely with naturalistic causality. Indeed, re-reading the present paragraph, and reducing these slightly portentous plans and principles to the beliefs and desires that represent them on particular occasions of action, I suspect it promotes a form of compatibilism quite consistent with that which Perry persuasively articulates. I say a little more on this in the afterword to this chapter.

[15] 'Wretched Subterfuge', the Dewey Lecture, delivered at the American Philosophical Association, Pacific Division, April 2010.

whether my doing something freely is consistent with the fact, according to determinism, that I could not have done otherwise. Perry reminds us of the slipperiness of the word 'could' and its cognates which he demonstrated on this occasion with the help of a couple of Brussels sprouts, a food he tells us he greatly dislikes. There is some perfectly clear sense in which Perry has the power to eat the sprouts: he has teeth, his jaws work properly, and he can generally feed himself without choking. But this is simply a different question from that whether, on a particular occasion when Perry refrains from eating sprouts he might, nevertheless, have eaten them. He had a great wish not to do so and no reason to do so; unless the conditions were changed in some way, for example, by the desire to prove a philosophical point, there is no possibility at all that he will eat them. The way the world works, the normal causal order, will not allow this to happen. Whether this causal order is in general deterministic or otherwise is, in this case, beside the point. There is no more chance of Perry eating this disgusting morsel than there is that I will strangle my children or dance naked down Pall Mall. All this while, however, he has the power to eat them, in the sense that he has the physiological equipment he would need to do so. In my preferred terminology, he has all the capacities required to accomplish this feat. This, for a compatibilist, is as much truth as one should expect in the statement that he could have done otherwise.

The simple realization that compatibilism should be a thesis about the compatibility of human action with the normal causal order and with no particular commitment as to whether that order is universal or patchy resolves some difficulties in my view. My view, at any rate, attempts to bring together three central ideas: first, the Humean insight that free action should, in some sense, be caused by the agent; second the Kantian thought, if liberally interpreted, that it is the fitting of action into some kind of systematic pattern that distinguishes the truly free agent from one who merely has the ability to respond to the whim of the moment; and third the ontological picture of the human agent as an entity enabled to pursue complex goals or engage in patterns of action over time by the acquisition of a uniquely rich range of capacities. This ontological picture has some affinity with agent causation theories, in that I do see the agent as an initiator of causal sequences. But of course my central point is that in this regard the human agent differs only in degree from other kinds of cause; that is the radial indeterminism I propose that goes far beyond merely the actions of humans. But given that this is what I take the normal causal order to consist of, once the compatibilist insight is separated from the determinism that I reject in the way that Perry explains, it becomes entirely congenial to my position.

As a matter of fact I think there are probably cases in which one could have done otherwise in both of the senses that Perry distinguishes. While it may be that there is no way on Earth that Perry could have eaten the sprouts given the lack of any reason to do so, it might be that when I choose an apple over an orange I could perfectly well have chosen the apple. Not just in the sense that I have the physiological equipment to pick up the apple and transfer its parts into my gut, but in the sense that given my indifference between these two fruits I might have chosen to eat either. This, though, is a different aspect of my indeterminism. Even if humans are such effective foci of causal capacity that actions for reasons always exclude the possibility of acting otherwise in the sense of rational determination, my wider view of the context in which these capacities develop is radically indeterministic. So I think compatibilist indeterminism is an ideal label for bringing together the three strands of my account of human freedom.

Bibliography

Abreu, F., J. L. Martins, T. S. Silveira, C. N. Keim, H. G. Barros, et al. (2007). '"Candidatus Magnetoglobus multicellularis", a multicellular, magnetotactic prokaryote from a hypersaline environment'. *International Journal of Systematic and Evolutionary Microbiology* 57: 1318–22.

Adams, K. L. and J. F. Wendel (2005). 'Polyploidy and genome evolution in plants'. *Current Opinion in Plant Biology* 8: 135–41.

Adl, S. M., A. G. B. Simpson, M. A. Farmer, R. A. Andersen, O. R. Anderson, et al. (2005). 'The new higher level classification of eukaryotes with emphasis on the taxonomy of protists'. *Journal of Eukaryote Microbiology* 52: 399–451.

Adler, J. (1969). 'Chemoreceptors in bacteria'. *Science* 166: 1588–97.

Aguilar, C., H. Vlamakis, R. Losick, and R. Kolter (2007). 'Thinking about *Bacillus subtilis* as a multicellular organism'. *Current Opinion in Microbiology* 10: 638–43.

Ahlquist, P. (2006). 'Parallels among positive-strand RNA viruses, reverse-transcribing viruses and double-stranded RNA viruses'. *Nature Reviews Microbiology* 4: 371–82.

Allen, E. E. and J. F Banfield (2005). 'Community genomics in microbial ecology and evolution'. *Nature Reviews Microbiology*, 3: 489–98.

Allen, T. D., D. R. Moore, X. Wang, V. Casu, R. May, et al. (2008). 'Dichotomous metabolism of *Enterococcus faecalis* induced by haematin starvation modulates colonic gene expression'. *Journal of Medical Microbiology* 57: 1193–204.

Allers T. and M. Mevarech (2005). 'Archaeal genetics: The third way'. *Nature Reviews Genetics* 6: 58–73.

Amann, R. I., W. Ludwig, and K.-H. Schleifer (1995). 'Phylogenetic identification and in situ detection of individual microbial cells without cultivation'. *Microbial Reviews* 59: 143–69.

Ameison, J. C. (2002). 'On the origin, evolution, and nature of programmed cell death: A timeline of four billion years'. *Cell Death and Differentiation* 9: 367–93.

Amend, J. P. and E. L. Shock (2001). 'Energetics of overall metabolic reactions of thermophilic and hyperthemophilic Archaea and Bacteria'. *FEMS Microbiology Reviews* 25: 175–243.

Andersson, J. O. (2000). 'Evolutionary genomics: Is *Buchnera* a bacterium or an organelle?' *Current Biology* 10: R866–R868.

Andrews, J. H. (1998). 'Bacteria as modular organisms'. *Annual Reviews of Microbiology* 52: 105–26.

Aravind, L., V. Anantharaman, and L. M. Iyer (2003). 'Evolutionary connections between bacterial and eukaryotic signaling systems: A genomic perspective'. *Current Opinions in Microbiology* 6: 490–7.

Archer, J. J. A. (2006). 'Cross-cultural differences in physical aggression between partners: A social-role analysis'. *Personality and Social Psychology Review* 10: 133–53.

Archibald, J. M. (2007). 'Nucleomorph genomes: Structure, function origin and evolution'. *BioEssays* 29: 392–402.

Atlas, R. M. and R. Bartha (1998). *Microbial Ecology: Fundamentals and Applications*, 4th edn. Menlo Park, CA: Benjamin/Cummings.

Austin, J. L. (1961). 'A Plea for Excuses'. In *Philosophical Papers*. Oxford: Clarendon Press.

Avery O. T., C. M. MacLeod, and M. McCarty (1944). 'Studies on the chemical nature of the substance inducing transformation of pneumococcal types: Induction of transformation by a desoxyribonucleic acid fraction isolated from *Pneumococcus* Type III'. *Journal of Experimental Medicine* 79: 137–58.

Baas Becking, L. G. M. (1931). *Gaia of leven en aarde*. The Hague: Nijhoff.

Bäckhed, F., H. Ding, T. Wang, L. V. Hooper, G. Y. Koh, et al. (2004). 'The gut microbiota as an environmental factor that regulates fat storage'. *Proceedings of the National Academy of Sciences* 101: 15718–23.

—— R. E. Ley, J. L. Sonnenburg, D. A. Peterson, and J. I. Gordon, J. I. (2005). 'Host–bacterial mutualism in the human intestine'. *Science* 307: 1915–20.

Baker, M. D., P. M. Wolanin, and J. B. Stock (2005). 'Signal transduction in bacterial chemo-taxis'. *BioEssays* 28: 9–22.

Bamford, D. H. (2003). 'Do viruses form lineages across different domains of life?' *Research in Microbiology* 154: 231–6.

—— J. M. Grimes, and D. I. Stuart (2005). 'What does structure tell us about virus evolution?' *Current Opinion in Structural Biology* 15: 655–63.

Bamshad, M. J., S. Wooding, W. S. Watkins, C. T. Ostler, M. A. Batzer, et al. (2003). 'Human population genetic structure and inference of group membership'. *American Journal of Human Genetics* 72: 578–89.

Bannert, N. and R. Kurth (2004). 'Retroelements and the human genome: New perspectives on an old relation'. *Proceedings of the National Academy of Sciences USA* 101 (Suppl. 2): 14572–9.

Barea, J.-M., M. J. Pozo, R. Azcón, and C. Azcón-Aguilar (2005). 'Microbial cooperation in the rhizosphere'. *Journal of Experimental Botany* 56: 1761–78.

Barkow, J., L. Cosmides, and J. Tooby (eds.) (1992). *The Adapted Mind*. New York: Oxford University Press.

Barnes, B. (1974). *Scientific Knowledge and Sociological Theory*. London and Boston: Routledge & Kegan Paul.

—— and D. Bloor (1982). 'Relativism, rationality, and the sociology of knowledge'. In M. Hollis and S. Lukes (eds.), *Rationality and Relativism*. Cambridge, MA: MIT Press, pp. 21–47.

—— and J. Dupré (2008). *Genomes and What to Make of Them*. Chicago: University of Chicago Press.

Barton, B. M., G. P. Harding, and A. J. Zuccarelli (1995). 'A general method for detecting and sizing large plasmids'. *Analytical Biochemistry* 226: 235–40.

Bassler, B. L. (2002). 'Small talk: Cell-to-cell communication in bacteria'. *Cell* 109: 421–4.

Bates, J. M., E. Mittge, J. Kuhlman, K. N. Baden, S. E. Cheesman, et al. (2006). 'Distinct signals for the microbiota promote different aspects of zebrafish gut differentiation'. *Developmental Biology* 297: 374–86.

Baumann, P. (2005). 'Biology of bacteriocyte-associated endosymbionts of plant sap-sucking insects'. *Annual Review of Microbiology* 59: 155–89.

Beadle, G. and E. Tatum (1941). 'Genetic control of biochemical reactions in Neurospora'. *Proceedings of the National Academy of Science USA* 27: 499–506.

Bechtel, W. (2006). *Discovering Cell Mechanisms*. New York: Cambridge University Press.

Bedau, M. (2003). 'Downward causation and autonomy in weak emergence'. *Principia* 6: 5–50.

Beiko, R. G., T. J. Harlow, and M. A. Ragan (2005). 'Highways of gene sharing in prokaryotes'. *Proceedings of the National Academy of Science USA* 102: 14332–7.

Béjà, O. (2004). 'To BAC or not to BAC: Marine ecogenomics'. *Current Opinion in Biotechnology* 15: 187–90.

—— L. Aravind, E. V. Koonin, M. T. Suzuki, A. Hadd, et al. (2000). 'Bacterial rhodopsin: Evidence for a new type of phototrophy in the sea'. *Science*, 289: 1902–6.

—— M. T. Suzuki, E. V. Koonin, L. Aravind, A. Hadd, et al. (2000). 'Construction and analysis of bacterial artificial chromosome libraries from a marine microbial assemblage'. *Environmental Microbiology* 2: 516–29.

—— E. N. Spudich, J. L. Spudich, M. Leclerc, and E. F. DeLong (2001). 'Proteorhodopsin phototrophy in the ocean'. *Nature* 411: 786–9.

Bell, S. D. and S. P. Jackson (1998). 'Transcription and translation in archaea: A mosaic of eukaryal and bacterial features'. *Trends in Microbiology* 6: 222–8.

Ben-Jacob, E., I. Cohen, I. Golding, D. L. Gutnick, M. Tcherpakov, et al. (2000). 'Bacterial cooperative organization under antibiotic stress'. *Physica A* 282: 247–82.

Benn, P. A. and A. R. Chapman (2009). 'Practical and ethical considerations of noninvasive prenatal diagnosis'. *JAMA* 301: 2154–6.

Berg, R. D. (1996). 'The indigenous gastrointestinal microflora'. *Trends in Microbiology* 4: 430–5.

Bertalanffy, L. von (1950). 'An outline of general system theory'. *British Journal for the Philosophy of Science* 1: 139–64.

Beurton, P. (2000). 'A unified view of the gene, or how to overcome reduction'. In P. Beurton, R. Falk, and H.-J. Rheinberger (eds.), *The Concept of the Gene in Development and Evolution*. Cambridge: Cambridge University Press, pp. 286–314.

—— R. Falk, and H.-J. Rheinberger (eds.) (2000). *The Concept of the Gene in Development and Evolution*. Cambridge: Cambridge University Press.

Bhardwaj, U., Y.-H. Zhang, Z. Rangwala, and E. R. B. McCabe (2006). 'Completely self-contained cell culture system: From storage to use'. *Molecular Genetics and Metabolism* 89: 168–73.

Bhattacharya, D. and J. M. Archibald (2006). 'The difference between organelles and endosymbionts: Response to Theissen and Martin'. *Current Biology* 16: R1017–R1018.

Biagini, G. A. and C. Bernard (2000). 'Primitive anaerobic protozoa: A false concept'. *Microbiology* 146: 1019–20.

Bingle, L. E. H. and C. M. Thomas (2001). 'Regulatory circuits for plasmid survival'. *Current Opinion in Microbiology* 4: 194–200.

Blackmore, S. (1999). *The Meme Machine*. Oxford and New York: Oxford University Press.

Bloor, D. (1976). *Knowledge and Social Imagery*. Chicago: University of Chicago Press.

Bodyt, A., P. Mackiewicz, and J. W. Stiller (2007). 'The intracellular cyanobacteria of *Paulinella chromatophora*: Endosymbionts or organelles?' *Trends in Microbiology* 15: 295–6.

Boevink, P. and K. J. Oparka (2005). 'Virus–host interactions during movement processes'. *Plant Physiology* 138: 1815–21.

Bollati, V. and A. Baccarelli (2010). 'Environmental epigenetics'. *Heredity* 105: 105–12.

Bolnick, D. A. (2008). 'Individual ancestry inference and the reification of race as a biological phenomenon'. In B. A. Koenig, S. S.-J. Lee, and S. S. Richardson (eds.), *Revisiting Race in a Genomic Age*. New Brunswick, NJ: Rutgers University Press, pp. 70–85.

Bonner, J. T. (1998). 'The origins of multicellularity'. *Integrative Biology* 1: 27–36.

Boogerd, F. C., F. J. Bruggeman, J. H. S. Hofmeyr, and H. V. Westerhoff (eds.) (2007). *Systems Biology: Philosophical Foundations*. Amsterdam: Elsevier.

Boucher, Y., C. L. Nesbo and W. F. Doolittle (2001). 'Microbial genomes: Dealing with diversity'. *Current Opinion in Microbiology* 4: 285–9.

—— C. J. Douady, R. T. Papke, D. A. Walsh, M. E. Boudreau, et al. (2003). 'Lateral gene transfer and the origins of prokaryotic groups'. *Annual Review of Genetics* 37: 283–328.

Bousset, L. and R. Melki (2002). 'Similar and divergent features in mammalian and yeast prions'. *Microbes and Infection* 4: 461–9.

Boyd, R. and P. J. Richerson (1985). *Culture and the Evolutionary Process*. Chicago: University of Chicago Press.

Brandon, R. N. (1999). 'The units of selection revisited: The modules of selection'. *Biology and Philosophy* 14: 167–80.

—— and R. M. Burian (eds.) (1984). *Genes, Organisms, Populations: Controversies over the Unit of Selection*. Cambridge, MA: MIT Press.

Braun, S. S. von and E. Schleiff (2007). 'Movement of endosymbiotic organelles'. *Current Protein and Peptide Science* 8: 426–38.

Brehm-Stecher, B. F. and E. A. Johnson (2004). 'Single-cell microbiology: Tools, technologies and applications'. *Microbiology and Molecular Biology Reviews* 68: 538–59.

Breitbart, M. and F. Rohwer (2005). 'Here a virus, there a virus, everywhere the same virus?'. *Trends in Microbiology* 13: 278–84.

—— P. Salamon, B. Andresen, J. M. Mahaffy, A. M. Segall, D. Mead, et al. (2002). 'Genomic analysis of uncultured marine viral communities'. *Proceedings of the National Academy of Sciences* 99: 14250–5.

—— I. Hewson, B. Felts, J. M. Mahaffy, J. Nulton, P. Salamon, et al. (2003). 'Metagenomic analyses of an uncultured viral community from human feces'. *Journal of Bacteriology* 185: 6220–3.

—— B. Felts, S. Kelley, J. M. Mahaffy, J. Nulton, P. Salamon, et al. (2004). 'Diversity and population structure of a nearshore marine-sediment viral community'. *Proceedings of the Royal Society London* 271B: 565–74.

Brock, T. D. (1966). *Principles of Microbial Ecology*. Englewood Cliffs, NJ: Prentice-Hall.

—— (1987). 'The study of microorganisms in situ: Progress and problems'. *Symposium of the Society for General Microbiology* 41: 1–17.

—— (1990). *The Emergence of Bacterial Genetics*. Cold Spring Harbor, NY: Cold Spring Harbor Laboratory Press.

Bromham, L. (2002). 'The human zoo: Endogenous retroviruses in the human genome'. *Trends in Ecology and Evolution* 17: 91–7.

Brown, J. H. (1932). 'The biological approach to bacteriology'. *Journal of Bacteriology* 23: 1–10.

Brown, J. R. (2001). 'Genomic and phylogenetic perspectives on the evolution of prokaryotes'. *Systematic Biology* 50: 497–512.

Brown, S. P. and R. A. Johnstone (2001). 'Cooperation in the dark: Signalling and collective action in quorum-sensing bacteria'. *Proceedings of the Royal Society of London B* 268: 961–5.

Brune, A. and M. Friedrich (2000). 'Microecology of the termite gut: Structure and function on a microscale'. *Current Opinions in Microbiology* 3: 263–9.

Bryant, C. (ed.) (1991). *Metazoan Life without Oxygen*. London: Chapman & Hall.

Bryant, D. and V. Moulton (2004). 'Neighbor-Net: An agglomerative method for the construction of phylogenetic networks'. *Molecular Biology and Evolution* 21: 255–65.

Buckley, M. R. (2004a). *The Global Genome Question: Microbes as the Key to Evolution and Ecology*. Washington, DC: American Academy of Microbiology.

—— (2004b). *Systems Biology: Beyond Microbial Genomics*. Washington, DC: American Academy of Microbiology.

Bull, A. T. and J. H. Slater (1982a). 'Historical perspectives on mixed cultures and microbial communities'. In A. T. Bull and J. H. Slater (eds.), *Microbial Interactions and Communities*. London: Academic Press, pp. 1–12.

—— —— (1982b). 'Microbial interactions and community structure'. In A. T. Bull and J. H. Slater (eds.), *Microbial Interactions and Communities*. London: Academic Press, pp. 13–44.

Buller, D. (2005). *Adapting Minds: Evolutionary Psychology and the Persistent Quest for Human Nature*. Cambridge, MA: MIT Press.

Bult, C. J., O. White, G. J. Olsen, et al. (1996). 'Complete genome sequence of the methanogenic archaeon, *Methanococcus jannaschii*'. *Science* 273: 1058–72.

Burnet, F. M. (1945). *Virus as Organism: Evolutionary and Ecological Aspects of Some Human Virus Diseases*. Cambridge, MA: Harvard University Press.

Bushman, F. (2002). *Lateral DNA Transfer: Mechanisms and Consequences*. Cold Spring Harbor, NY: Cold Spring Harbor Laboratory Press.

Buss, D. (1994). *The Evolution of Desire*. New York: Basic Books.

—— (1999). *Evolutionary Psychology: The New Science of the Mind*. New York: Doubleday.

Buss, L. W. (1987). *The Evolution of Individuality*. Princeton: Princeton University Press.

Caldwell, D. E. and J. W. Costerton (1996). 'Are bacterial biofilms constrained to Darwin's concept of evolution through natural selection?' *Microbiología SEM* 12: 347–58.

—— E. Atuku, D. C. Wilkie, K. P. Wivcharuk, S. Karthikeyan, et al. (1997). 'Germ theory vs. community theory in understanding and controlling the proliferation of biofilms'. *Advances in Dental Research* 11: 4–13.

Campbell, A. (2001). 'The origins and evolution of viruses'. *Trends in Microbiology* 9: 61.

Campbell, D. (1974). '"Downward causation" in hierarchically organized biological systems'. In F. Ayala and T. Dobzhansky (eds.), *Studies in the Philosophy of Biology*. London: Macmillan, pp. 179–86.

Carlile, M. J. (1980). 'From prokaryote to eukaryote: Gains and losses'. In G. W. Gooday, D. Lloyd, and A. P. J. Trinci (eds.), *The Eukaryotic Microbial Cell*. Cambridge: Cambridge University Press, pp. 1–40.

Carroll, S. B. (2001). 'Chance and necessity: The evolution of morphological complexity and diversity'. *Nature* 409: 1102–9.

—— (2005). *Endless Forms Most Beautiful: The New Science of Evo Devo and the Making of the Animal Kingdom*. New York: W. W. Norton.

Cartwright, N. (1983). *How the Laws of Physics Lie*. Oxford: Oxford University Press.

Casadesús, J. and R. D'Ari (2002). 'Memory in bacteria and phage'. *BioEssays* 24: 512–18.

Cases, I. and V. de Lorenzo (2005). 'Genetically modified organisms for the environment: Stories of success and failure and what we have learned from them'. *International Microbiology* 8: 213–22.

Casjens, S. (2003). 'Prophages and bacterial genomics: What have we learned so far?' *Molecular Microbiology* 49: 277–300.

Cavalier-Smith, T. (2000). 'Membrane heredity and early chloroplast evolution'. *Trends in Plant Science* 5: 174–82.

—— and J. J. Lee (1985). 'Protozoa as hosts for endosymbioses and the conversion of symbionts into organelles'. *Journal of Protozoology* 32: 376–9.

Cavalli-Sforza, L. L and M. W. Feldman (1981). *Cultural Transmission and Evolution: A Quantitative Approach.* Princeton: Princeton University Press.

Cello, J., A. V. Paul, and E. Wimme (2002). 'Chemical synthesis of poliovirus cDNA: Generation of infectious virus in the absence of natural template'. *Science* 297: 1016–18.

Centerwall, W. R. and K. Benirschke (1973). 'Male tortoiseshell and calico (T – C) cats'. *Journal of Heredity* 64: 272–8.

Chalmers, D. (1996). *The Conscious Mind: In Search of a Fundamental Theory.* New York: Oxford University Press.

Champagne, F. A. and M. J. Meaney (2006). 'Stress during gestation alters postpartum maternal care and the development of the offspring in a rodent model'. *Biological Psychiatry* 59: 1227–35.

—— I. C. Weaver, J. Diorio, S. Dymer, M. Szyf, et al. (2006). 'Maternal care associated with methylation of the estrogen receptor-alphalb promoter and estrogen receptor-alpha expression in the medial preoptic area of female offspring'. *Endocrinology* 147: 2909–15.

Charlat, S., G. D. D. Hurst and H. Merçot (2003). 'Evolutionary consequences of Wolbachia infections'. *Trends in Genetics* 19: 217–23.

Charlebois, R. L. and W. F. Doolittle (2004). 'Computing prokaryotic gene ubiquity: Rescuing the core from extinction'. *Genome Research* 14: 2469–77.

—— R. G. Beiko, and M. A. Ragan (2003). 'Microbial phylogenomics: Branching out'. *Nature* 421: 217.

Chen, K. and L. Pachter (2005). 'Bioinformatics of whole-genome shotgun sequencing of microbial communities'. *PLoS Computational Biology* 1: 106–12.

Chernoff, Y. O. (2001). 'Mutation processes at the protein level: Is Lamarck back?' *Mutation Research* 488: 39–64.

—— A. P. Galkin, E. Lewitin, T. A. Chernova, G. P. Newnam, and S. M. Belenkly (2000). 'Evolutionary conservation of prion-forming abilities of the yeast Sup35 protein'. *Molecular Microbiology* 35: 865–76.

Cho, J.-C. and J. M. Tiedje (2000). 'Biogeography and degree of endemicity of fluorescent Pseudomonas strains in soil'. *Applied and Environmental Microbiology* 66: 5448–56.

Cho, J. H., H. Jönsson, K. Campbell, P. Melke, J. W. Williams, et al. (2007). 'Self-organization in high-density bacterial colonies: Efficient crowd control'. *PLoS Biology* 5(11): e302.

Chong, S. and E. Whitelaw (2004). 'Epigenetic germline inheritance'. *Current Opinion in Genetics & Development* 14: 692–6.

Chyba, C. F. and K. P. Hand (2005). 'Astrobiology: The study of the living universe'. *Annual Review of Astronomy and Astrophysics* 43: 31–74.

Claverie, J. M., H. Ogata, S. Audic, C. Abergel, K. Shure, and P. E. Fournier (2006). 'Mimivirus and the emerging concept of "giant" virus'. *Virus Research* 117: 133–44.

Cleland, C. E. and S. D. Copley (2005). 'The possibility of alternative microbial life on earth'. *International Journal of Astrobiology* 4: 165–73.

Coenye, T., D. Gevers, Y. W. de Peer, P. Vandamme, and J. Swings (2005). 'Towards a prokaryotic genomic taxonomy'. *FEMS Microbiology Reviews* 29: 147–67.

Cohan, F. M. (2002). 'What are bacterial species?' *Annual Review of Microbiology* 56: 457–87.

Collings, D. A., J. D. I. Harper, J. Marc, R. L. Overall and R. T. Mullen (2002). 'Life in the fast lane: Actin-based motility of plant peroxisomes'. *Canadian Journal of Botany* 80: 430–41.

Collins, J. P. (2003). 'What can we learn from community genetics?' *Ecology* 84: 574–7.

Colwell, R. R. (1997). 'Microbial diversity: The importance of exploration and conservation'. *Journal of Industrial Microbiology and Biotechnology* 18: 302–7.

Committee on Metagenomics: Challenges and Functional Applications, National Research Council (2007). *The New Science of Metagenomics: Revealing the Secrets of Our Microbial Planet.* Washington, DC: National Academies Press.

Conway Morris, S. (1998). 'The evolution of diversity in ancient ecosystems: A review'. *Philosophical Transactions of the Royal Society of London B* 353: 327–45.

—— (2003). 'The Cambrian "explosion" of metazoans and molecular biology: Would Darwin be satisfied?' *International Journal of Developmental Biology* 47: 505–15.

Corliss, J. O. (1999). 'Biodiversity, classification, and numbers of species of protists'. In P. H. Raven (ed.), *Nature and Human Society: The Quest for a Sustainable World.* Washington, DC: National Academies Press, pp. 130–55.

Cosmides, L. and J. Tooby (1987). 'From evolution to behaviour: Evolutionary psychology as the missing link'. In J. Dupré (ed.), *The Latest on the Best: Essays on Evolution and Optimality.* Cambridge, MA: MIT Press, pp. 277–307.

Costerton, B. (2004). 'Microbial ecology comes of age and joins the general ecology community'. *Proceedings of the National Academy of Sciences* 101: 16983–4.

Costerton, J. W., Z. Lewandowski, D. E. Caldwell, D. R. Korber, and H. M. Lappin-Scott (1995). 'Microbial biofilms'. *Annual Review of Microbiology* 49: 711–45.

Courtois, S., C. M. Cappellano, M. Ball, F.-X. Francou, P. Normand, et al. (2003). 'Recombinant environmental libraries provide access to microbial diversity for drug discovery from natural products'. *Applied and Environmental Microbiology* 69: 49–55.

Covic, M., E. Karaca, and D. C. Lie (2010). 'Epigenetic regulation of neurogenesis in the adult hippocampus'. *Heredity* 105: 122–34.

Cowan, D., Q. Meyer, W. Stafford, S. Muyanga, R. Cameron, and P. Wittwer (2005). 'Metagenomic gene discovery: Past, present and future'. *Trends in Biotechnology* 23: 321–9.

Craver, C. F. (2005). 'Beyond reductionism: Mechanisms, multifield integration, and the unity of science'. *Studies in History and Philosophy of Biological and Biomedical Sciences* 36: 373–96.

—— and W. Bechtel (2007). 'Top-down causation without top-down causes'. *Biology and Philosophy* 22: 547–63.

Crespi, B. J. (2001). 'The evolution of social behaviour in microorganisms'. *Trends in Ecology and Evolution* 16: 178–83.

Crick, F. (1958). 'On protein synthesis'. *Symposia of the Society for Experimental Biology* 12: 139–63.

Croal, L. R., J. A. Gralnick, D. Malasarn, and D. K. Newman (2004). 'The genetics of geochemistry'. *Annual Review of Genetics* 38: 175–202.

Cutler, D. W. and L. M. Crump (1935). *Problems in Soil Microbiology.* London: Longmans, Green & Co.

Cutler, S. and D. Ehrhardt (2000). 'Dead cells don't dance: Insights from live-cell imaging in plants'. *Current Opinion in Plant Biology* 3: 532–7.

Daims, H., S. Lücker, and M. Wagner (2006). 'daime, a novel image analysis program for microbial ecology and biofilm research'. *Environmental Microbiology* 8: 200–13.

Daniel, R. (2004). 'The soil metagenome: A rich resource for the discovery of novel natural products'. *Current Opinion in Biotechnology* 15: 199–204.

Darby, A. C., N. H. Cho, H. H. Fuxelius, J. Westberg, and S. G. E. Andersson (2007). 'Intracellular pathogens go extreme'. *Trends in Genetics* 23: 511–20.

Darwin, C. (1881). *The Formation of Vegetable Mould, through the Action of Worms, with Observations on their Habits*. New York: D. Appleton.

Davey, M. E. and G. A. O'Toole (2000). 'Microbial biofilms: From ecology to molecular genetics'. *Microbiology and Molecular Biology Reviews* 64: 847–67.

Davidson, D. (1984). *Inquiries into Truth and Interpretation*. Oxford: Oxford University Press.

Davies, D. G. (2000). 'Physiological events in biofilm formation'. In D. G. Allison, P. Gilbert, H. M. Lappin-Scott, and M. Wilson (eds.), *Community Structure and Cooperation in Biofilms*. Cambridge: Cambridge University Press, pp. 37–52.

Dawkins, R. (1976). *The Selfish Gene*. Oxford: Oxford University Press.

—— (1982). *The Extended Phenotype*. Oxford: Oxford University Press.

—— (1996). *The Blind Watchmaker*. New York: W. W. Norton.

—— (2008). 'Life: A gene-centric view. Craig Venter and Richard Dawkins: A conversation in Munich' (Transcript). *Edge* 235. (accessed 6 February 2010) http://www.edge.org/documents/dawkins_venter_index.html.

de Lorenzo, V. (2005). 'Problems with metagenomic screening'. *Nature Biotechnology* 23: 1045.

del Solar, G. and M. Espinosa (2000). 'Plasmid copy number control: An ever-growing story'. *Molecular Microbiology* 37: 492–500.

——R. Giraldo, M. J. Ruiz-Echevarría, M. Espinosa, and R. Díaz-Orejas (1998). 'Replication and control of circular bacterial plasmids'. *Microbiology and Molecular Biology Reviews* 62: 434–64.

DeLong, E. F. (2002a). Microbial population genomics and ecology. *Current Opinion in Microbiology* 5: 520–4.

—— (2002b). 'Towards microbial systems science: Integrating microbial perspectives from genomes to biomes'. *Environmental Microbiology* 4: 9–10.

—— (2004a). 'Microbial population genomics and ecology: The road ahead'. *Environmental Microbiology* 6: 875–8.

—— (2004b). 'Reconstructing the wild types'. *Nature* 428: 25–6.

—— (2005). 'Microbial community genomics in the ocean'. *Nature Reviews Microbiology* 3: 459–69.

—— and N. R. Pace (2001). 'Environmental diversity of Bacteria and Archaea'. *Systematic Biology* 50: 470–8.

—— C. M. Preston, T. Mincer, V. Rich, S. J. Hallam, et al. (2006). 'Community genomics among stratified microbial assemblages in the ocean's interior'. *Science* 311: 496–503.

Dennett, D. (1995). *Darwin's Dangerous Idea: Evolution and the Meanings of Life*. New York: Simon & Schuster.

Dennis, P., E. A. Edwards, S. N. Liss, and R. Fulthorpe (2003). 'Monitoring gene expression in mixed microbial communities by using DNA microarrays'. *Applied and Environmental Microbiology* 69: 769–78.

Derkatch, I. L., M. E. Bradley, J. Y. Hong and S. W. Liebman (2001). 'Prions affect the appearance of other prions: The story of [PIN+]'. *Cell* 108: 171–82.

Devitt, M. (2008). 'Resurrecting biological essentialism'. *Philosophy of Science* 75: 344–82.

Dial, S., A. Kezouh, A. Dascal, A. Barkun, and S. Suissa (2008). 'Patterns of antibiotic use and risk of hospital admission because of *Clostridium difficile* infection'. *Canadian Medical Association Journal* 179: 767–72.

Diaz-Torres, M. L., R. McNab, D. A. Spratt, A. Villedieu, N. Hunt, et al. (2003). 'Novel tetracycline resistance determinant from the oral metagenome'. *Antimicrobial Agents and Chemotherapy* 47: 1430–2.

Dijkshoorn L., B. M. Ursing, and J. B. Ursing (2000). 'Strain, clone and species: Comments on three basic concepts of bacteriology'. *Journal of Medical Microbiology* 49: 397–401.

Dixon, B. (1994). *Power Unseen: How Microbes Rule the World*. Oxford: Freeman.

Doney, S. C., M. R. Abbott, J. J. Cullen, J. D. M. Karl, and L. Rothstein (2004). 'From genes to ecosystems: The ocean's new frontier'. *Frontiers in Ecology and Environment* 2: 457–66.

Doolittle, W. F. (1999). 'Phylogenetic classification and the universal tree'. *Science* 284: 2124–8.

—— (2002). 'Diversity squared'. *Environmental Microbiology* 4: 10–12.

—— (2003). 'Lateral gene transfer and the origins of prokaryotic groups'. *Annual Review of Genetics* 37: 283–328.

—— (2005). 'If the tree of life fell, would we recognize the sound?' In J. Sapp (ed.), *Microbial Phylogeny and Evolution: Concepts and Controversies*. Oxford: Oxford University Press, pp. 119–33.

—— and C. Sapienza (1980). 'Selfish genes, the phenotype paradigm and genome evolution'. *Nature* 284: 601–3.

—— Y. Boucher, C. L. Nesbo, et al. (2003). 'How big is the iceberg of which organellar genes are but the tip?' *Philosophical Transactions of the Royal Society of London B* 358: 39–58.

Douglas, A. E. and J. A. Raven (2003). 'Genomes at the interface between bacteria and organelles'. *Philosophical Transactions of the Royal Society London B* 358: 5–18.

Drews, G. (2000). 'The roots of microbiology and the influence of Ferdinand Cohn on microbiology of the 19th century'. *FEMS Microbiology Reviews* 24: 225–49.

Dunning-Hotopp, J. C. D., M. E. Clark, D. C. S. G. Oliveira, J. M. Foster, P. Fischer, et al. (2007). 'Widespread lateral gene transfer from intracellular bacteria to multicellular eukaryotes'. *Science* 317: 1753–6.

Dunny, G. M. and B. A. B. Leonard (1997). 'Cell–cell communication in gram-positive bacteria'. *Annual Reviews in Microbiology* 51: 527–64.

—— and S. C. Winans (1999). 'Bacterial life: Neither lonely nor boring'. In G. M. Dunny and S. C. Winans (eds.), *Cell–Cell Signalling in Bacteria*. Washington, DC: ASM, pp. 1–5.

Dupré, J. (ed.) (1987). *The Latest on the Best: Essays on Evolution and Optimality*. Cambridge, MA: MIT Press.

—— (1993). *The Disorder of Things: Metaphysical Foundations of the Disunity of Science*. Cambridge, MA: Harvard University Press.

—— (2000). 'What the theory of evolution can't tell us'. *Critical Quarterly* 42: 18–34.

—— (2001). *Human Nature and the Limits of Science*. Oxford: Oxford University Press.

—— (2002). *Humans and Other Animals*. Oxford: Oxford University Press.

—— (2003). *Darwin's Legacy: What Evolution Means Today*. Oxford: Oxford University Press.

—— (2005). 'Are there genes?' In A. O'Hear (ed.), *Philosophy, Biology and Life* (Royal Institute of Philosophy Supplements). Cambridge: Cambridge University Press, pp. 193–210.

—— (2007). 'Is biology reducible to the laws of physics?, Review of Alexander Rosenberg, *Darwinian Reductionism: Or, How to Stop Worrying and Love Molecular Biology*. *American Scientist* 95: 274–6.

—— (2010). 'The human genome, human evolution, and gender'. *Constellations* 17: 540–8.

Dworkin, M. (1985). *Developmental Biology of the Bacteria*. Reading, MA: Benjamin/Cummings.

Dworkin, M. (1996). 'Recent advances in the social and developmental biology of the Myxobacteria'. *Microbiology Reviews* 60: 70–102.

—— (1997). 'Multiculturism versus the single microbe'. In J. A. Shapiro and M. Dworkin (eds.),
Bacteria as Multicellular Organism. New York: Oxford University Press, pp. 3–13.

Dyall, S. D., M. T. Brown, and P. J. Johnson. (2004). 'Ancient invasions: From endosymbionts
to organelles'. *Science* 304: 253–7.

Dyer, B. D. (1989). 'Symbiosis and organismal boundaries'. *American Zoologist* 29: 1085–95.

Dykhuizen, D. E. (1998). 'Santa Rosalia revisited: Why are there so many species of bacteria?'
Antonie van Leeuwenhoek 73: 25–33.

Eberhard, W. G. (1980). 'Evolutionary consequences of intracellular organelle competition'.
Quarterly Review of Biology 55: 231–49.

—— (1990). 'Evolution in bacterial plasmids and levels of selection'. *Quarterly Review of Biology*
65: 3–22.

Eckberg, P. B., P. W. Lepp, and D. A. Relman (2003). 'Archaea and their potential role in
human disease'. *Infection and Immunity* 71: 591–6.

Edwards, A. W. F. (2003). 'Human genetic diversity: Lewontin's fallacy'. *BioEssays* 25: 798–801.

Edwards, O. M. (1971). 'Masculinized Turner's syndrome XY–XO mosaicism'. *Proceedings of the
Royal Society of Medicine* 64: 300–1.

Edwards, R. A. and F. Rohwer (2005). 'Viral metagenomics'. *Nature Reviews Microbiology* 3: 504–10.

Ehler, L. J. (2000). 'Gene transfers in biofilms'. In D. G. Allison, P. Gilbert, H. M. Lappin-Scott,
and M. Wilson (eds.), *Community Structure and Co-operation in Biofilms*. Cambridge: Cambridge
University Press, pp. 215–56.

Ehrlich, P. R. and E. O. Wilson (1991). 'Biodiversity studies: Science and policy'. *Science* 253:
758–62.

Eisenbach, M. (2005). 'Bacterial chemotaxis'. *Encylopedia of Life Sciences*. doi: 10.1038/npg.
els.0003952.

Ellis, B. (2001). *Scientific Essentialism*. Cambridge: Cambridge University Press.

Ellis, B. J. (1992). 'The evolution of sexual attraction: Evaluative mechanisms in women'. In
J. Barkow, L. Cosmides, and J. Tooby (eds.), *The Adapted Mind*. New York: Oxford
University Press, pp. 267–88.

Embley, T. M. and W. Martin (2006). 'Eukaryotic evolution, changes and challenges'. *Nature*
440: 623–30.

Engelberg, H. and R. Hazan (2003). 'Cannibals defy starvation and avoid sporulation'. *Science*
301: 467–8.

Ereshefsky, M. (1991). *The Units of Evolution: Essays on the Nature of Species*. Cambridge, MA:
MIT Press.

Eyers, L., I. George, L. Schuler, B. Stenuit, S. N. Agathos, and S. El Fantroussi (2004).
'Environmental genomics: Exploring the unmined richness of microbes to degrade xenobio-
tics'. *Applied Microbiology and Biotechnology* 66: 123–30.

Faguy, D. M. and K. F. Jarrell (1999). 'A twisted tale: The origin and evolution of motility and
chemotaxis in prokaryotes'. *Microbiology* 145: 279–81.

Falke, J. J., R. B. Bass, S. L. Butler, S. A. Chervitz and M. A. Danielson (1997). 'The two-
component signaling pathway of bacterial chemotaxis: A molecular view of signal transduction by receptors, kinases and adaptation enzymes'. *Annual Review of Cell and Developmental
Biology* 13: 457–512.

Falkowski, P. G. and C. de Vargas (2004). 'Shotgun sequencing in the sea: A blast from the past?' *Science* 304: 58–60.

Federle, M. J. and B. L. Bassler (2003). 'Interspecies communication in bacteria'. *Journal of Clinical Investigation* 112: 1291–9.

Feil, E. J. and B. G. Spratt (2001). 'Recombination and the population structures of bacterial pathogens'. *Annual Review of Microbiology* 55: 561–90.

Fenchel, T. (1996). 'Eukaryotic life: Anaerobic physiology'. In D. M. Roberts, P. Sharp, G. Alderson, and M. A. Collins (eds.), *Evolution of Microbial Life*. Cambridge: Cambridge University Press, pp. 185–203.

Fenn, K. and M. Blaxter (2006). '*Wolbachia* genomes: Revealing the biology of parasitism and mutualism'. *Trends in Parasitology* 22: 60–5.

Feyerabend, P. (1975). *Against Method*. London: New Left Books.

Figge, R. M. and J. W. Gober (2003). 'Cell shape, division and development: The 2002 American Society for Microbiology (ASM) conference on prokaryotic development'. *Molecular Microbiology* 47: 1475–83.

Finlay, B. J. and K. J. Clarke (1999). 'Ubiquitous dispersal of microbial species'. *Nature* 400: 828.

—— S. C. Maberly, and J. I. Cooper (1997). 'Microbial diversity and ecosystem function'. *Oikos* 80: 209–13.

Fleischmann, R. D., M. D. Adams, O. White, R. A. Clayton, E. F. Kirkness, et al. (1995). 'Whole-genome random sequencing and assembly of *Haemophilus influenzae* Rd'. *Science* 269: 496–512.

Foerstner, K. U., C. von Mering, and P. Bork (2006). 'Comparative analyses of environmental sequences: Potential and challenges'. *Philosophical Transactions of the Royal Society* 361B: 519–23.

Fogle, T. (2000). 'The dissolution of protein coding genes in molecular biology'. In P. Beurton, R. Falk, and H.-J. Rheinberger (eds.), *The Concept of the Gene in Development and Evolution*. Cambridge: Cambridge University Press, pp. 3–25.

Folkman, J. and A. Moscona (1978). 'Role of cell shape in growth control'. *Nature* 273: 345–9.

Forterre, P. (2006). 'The origin of viruses and their possible roles in major evolutionary transitions'. *Virus Research* 117: 5–16.

Foucault, M. (1977). *Discipline and Punish: The Birth of the Prison*, trans. A. Sheridan. New York: Vintage Books.

Fox, G. E., E. Stackebrandt, R. B. Hespell, J. Gitson, J. Maniloff, et al. (1980). 'The phylogeny of prokaryotes'. *Science* 209: 457–63.

Francis, C. A., J. M. Beman, and M. M. M. Kuypers (2007). 'New processes and players in the nitrogen cycle: The microbial ecology of anaerobic and archaeal ammonia oxidation'. *International Society for Microbial Ecology Journal* 1: 19–27.

Frankfurt, H. G. (1988). *The Importance of What We Care About*. Cambridge: Cambridge University Press.

Fraser, C. M., J. D. Gocayne, O. White, M. D. Adams, R. A. Clayton, et al. (1995). 'The minimal gene complement of *Mycoplasma genitalium*'. *Science* 270: 397–403.

Freeman, D. (1983). *Margaret Mead and Samoa: The Making and Unmaking of an Anthropological Myth*. Cambridge, MA: Harvard University Press.

Friedman, M. (1953). 'The methodology of positive economics'. In *Essays in Positive Economics*. Chicago: University of Chicago Press, pp. 3–43.

Fullwiley, D. (2008). 'The molecularization of race: U.S. health institutions, pharmacogenetics practice, and public science after the genome'. In B. A. Koenig, S. S.-J. Lee, and S. S. Richardson (eds.), *Revisiting Race in a Genomic Age*. New Brunswick, NJ: Rutgers University Press, pp. 149–71.

Gagnier, R. and J. Dupré (1995). 'On work and idleness'. *Feminist Economics* 1: 1–14.

Galison, P. (1987). *How Experiments End*. Chicago: University of Chicago Press.

Galperin, M. Y. (2004). 'Metagenomics: From acid mine to shining sea'. *Environmental Microbiology* 6: 543–5.

—— (2005). 'Life is not defined just in base pairs'. *Environmental Microbiology* 7: 149–52.

Gans, J., M. Wolinsky, and J. Dunbar (2005). 'Computational improvements reveal great bacterial diversity and high metal toxicity in soil'. *Science* 309: 1387–90.

Gánti, T. (1997). 'Biogenesis itself'. *Journal of Theoretical Biology* 187: 583–93.

Genereux, D. P. and J. M. Logsdon, Jr. (2003). 'Much ado about bacteria-to-vertebrate lateral gene transfer'. *Trends in Genetics* 19: 191–5.

Gerdes, K., P. B. Rasmussen, and S. Molin (1986). 'Unique type of plasmid maintenance function: Postsegregational killing of plasmid-free cells'. *Proceedings of the National Academy of Sciences USA* 83: 3116–20.

Gevers, D., F. M. Cohan, J. G. Lawrence, B. G. Spratt, T. Coenye, et al. (2005). 'Re-evaluating prokaryote species'. *Nature Reviews Microbiology* 3: 733–9.

Ghigo, J. M. (2001). 'Natural conjugative plasmids induce bacterial biofilm development'. *Nature* 412: 442–5.

Ghiselin, M. (1974). 'A radical solution to the species problem'. *Systematic Zoology* 23: 536–44.

Gibbs, A. J. and M. J. Gibbs (2006). 'A broader definition of "the virus species"'. *Archives of Virology* 151: 1419–22.

Gilbert, S. F. (2003). 'Evo-devo, devo-evo, and devgen-popgen'. *Biology and Philosophy* 18: 347–52.

Gill, S. R., M. Pop, R. T. DeBoy, P. B. Eckburg, P. J. Turnbaugh, et al. (2006). 'Metagenomic analysis of the human distal gut microbiome'. *Science* 312: 1355–9.

Glass, J. I., N. Assad-Garcia, N. Alperovich, S. Yooseph, M. R. Lewis, et al. (2006). 'Essential genes of a minimal bacterium'. *Proceedings of the National Academy of Sciences USA* 103: 425–30.

Godfrey-Smith, P. (2000). 'On the theoretical role of "genetic coding"'. *Philosophy of Science* 67: 26–44.

Gogarten, J. P. and J. P. Townsend (2005). 'Horizontal gene transfer, genome innovation and evolution'. *Nature Reviews Microbiology* 3: 679–87.

—— W. F. Doolittle, and J. G. Lawrence (2002). 'Prokaryotic evolution in the light of gene transfer'. *Molecular Biology and Evolution* 19: 2226–38.

Goodnight, C. J. and L. Stevens (1997). 'Experimental studies of group selection: What do they tell us about group selection in nature?' *American Naturalist* 150 (Supplement): S59–S79.

Gordon, J. I., R. E. Ley, R. Wilson, E. Mardis, J. Xu, et al. (2005). 'Extending our view of self: The Human Gut Microbiome Initiative (HGMI)'. http://www.genome.gov/Pages/Research/Sequencing/SeqProposals/HGMIseq.pdf (accessed 24 June 2007).

Gouin, E., C. Egile, P. Dehoux, V. Villiers, J. Adams, et al. (2004). 'The RickA protein of *Rickettsia conorii* activates the Arp2/3 complex'. *Nature* 427: 457–61.

Gould, S. J. (1994). 'The evolution of life on earth'. *Scientific American* 271: 84–91.

Graves, J. L. (2005). 'What we know and what we don't know: Human genetic variation and the social construction of race'. Posted 25 April 2005, http://raceandgenomics.ssrc.org/Graves/pf/

Gray, K. M. (1997). 'Intercellular communication and group behaviour in bacteria'. *Trends in Microbiology* 5: 184–8.

Grebe, T. W. and J. Stock (1998). 'Bacterial chemotaxis: The five sensors of a bacterium'. *Current Biology* 8: R154–R157.

Green, R. E., J. Krause, A. W. Briggs, T. Maricic, U. Stenzel, et al. (2010). 'A draft sequence of the Neandertal genome'. *Science* 328: 710–22.

Gregory, T. R. (2001). 'Coincidence, coevolution, or causation? DNA content, cell size, and the Cvalue enigma'. *Biological Reviews* 76: 65–101.

Griffin, A. S., S. A. West, and A. Buckling (2004). 'Cooperation and competition in pathogenic bacteria'. *Nature* 430: 1024–7.

Griffiths, D. J. (2001). 'Endogenous retroviruses in the human genome sequence'. *Genome Biology* 2: reviews 1017.1–1017.5.

Griffiths, P. E. (2001). 'Genetic information: A metaphor in search of a theory'. *Philosophy of Science* 68: 394–412.

Griffiths, P. E. and R. D. Gray (1994). 'Developmental systems and evolutionary explanation'. *Journal of Philosophy* 91: 277–304.

Grote, M. (2008). 'Hybridizing bacteria, crossing methods, cross-checking arguments: The transition from episomes to plasmids (1961–1969)'. *History and Philosophy of the Life Sciences* 30: 407–30.

Grzymski, J. J., B. J. Carter, E. F. DeLong, R. A. Feldman, A. Ghadiri and A. E. Murray (2006). 'Comparative genomics of DNA fragments from six Antarctic marine planktonic bacteria'. *Applied and Environmental Microbiology* 72: 1532–41.

Guala, F. (2000). 'The logic of normative falsification: Rationality and experiments in decision theory'. *Journal of Economic Methodology* 7: 59–93.

Hacking, I. (1983). *Representing and Intervening: Introductory Topics in the Philosophy of Natural Science.* Cambridge: Cambridge University Press.

—— (1995). *Rewriting the Soul: Multiple Personality and the Sciences of Memory.* Princeton: Princeton University Press.

—— (1999). *The Social Construction of What?* Cambridge, MA: Harvard University Press.

Hallam, S. J., N. Putnam, C. M. Preston, J. C. Detter, D. Rokhsar, et al. (2004). 'Reverse methanogenesis: Testing the hypothesis with environmental genomics'. *Science* 305: 1457–62.

—— T. J. Mincer, C. Schleper, C. M. Preston, K. Roberts, et al. (2006). Pathways of carbon assimilation and ammonia oxidation suggested by environmental genomic analyses of marine Crenarchaeota. *PLoS Biology* 4(4): e95.

Hambly, E. and C. A. Suttle (2005). 'The viriosphere, diversity and genetic exchange within phage communities'. *Current Opinion in Microbiology* 8: 444–50.

Hamilton, G. (2006). 'Virology: The gene weavers'. *Nature* 441: 683–5.

Handelsman, J. (2004). 'Metagenomics: Application of genomics to uncultured microorganisms'. *Microbiology and Molecular Biology Reviews* 68: 669–85.

—— M. R. Rondon, S. F. Brady, J. Clardy, and R. M. Goodman (1998). 'Molecular biological access to the chemistry of unknown soil microbes: A new frontier for natural products'. *Chemistry and Biology* 5: R245–R249.

—— M. Liles, D. Mann, C. Riesenfeld, and R. M. Goodman (2002). 'Cloning the metagenome: Culture-independent access to the diversity and functions of the uncultivated microbial world'. *Methods in Microbiology* 33: 241–55.

Harper, J. L. (1977). *Population Biology of Plants*. London: Academic Press.

Harris, R. N., R. M. Brucker, J. B. Walke, M. H. Becker, C. R. Schwantes, et al. (2009). 'Skin microbes on frogs prevent morbidity and mortality caused by a lethal skin fungus'. *The ISME Journal* 3: 818–24.

Hart, M. H., R. J. Reader, and J. N. Klironomos (2003). 'Plant coexistence mediated by arbuscular mycorrhizal fungi'. *Trends in Ecology & Evolution* 18: 418–23.

Haslanger, S. (2008). 'A social constructionist analysis of race'. In B. A. Koenig, S. S.-J. Lee, and S. S. Richardson (eds.), *Revisiting Race in a Genomic Age*. New Brunswick, NJ: Rutgers University Press, pp. 56–69.

Hausner, M. and S. Wuertz (1999). 'High rates of conjugation in bacterial biofilms as determined by quantitative in situ analysis'. *Applied and Environmental Microbiology* 65: 3710–13.

Hehemann, J., G. Correc, T. Barbeyron, W. Helbert, M. Czjzuk, and G. Michel (2010). 'Transfer of carbohydrate-active enzymes from marine bacteria to Japanese gut microbiota'. *Nature* 464: 908–12.

Heijmans, B. T., E. W. Tobi, A. D. Stein, H. Putter, G. J. Blauw, et al. (2008). 'Persistent epigenetic differences associated with prenatal exposure to famine in humans'. *Proceedings of the National Academy of Sciences, USA* 105: 17046–9.

Helvoort, Tv. (1992). 'Bacteriological and physiological research styles in the early controversy on the nature of the bacteriophage phenomenon'. *Medical History* 36: 243–70.

Hendrix, R. W. (2002). 'Bacteriophages: Evolution of the majority'. *Theoretical Population Biology* 61: 471–80.

—— M. C. M. Smith, R. N. Burns, M. E. Ford, and G. F. Hatfull (1999). 'Evolutionary relationships among diverse bacteriophages and prophages: All the world's a phage'. *Proceedings of the National Academy of Sciences USA* 96: 2192–7.

—— J. G. Lawrence, G. F. Hatfull, and S. Casjens (2000). 'The origins and ongoing evolution of viruses'. *Trends in Microbiology* 8: 504–8.

Henke J. M. and B. L. Bassler (2004). 'Bacterial social engagements'. *Trends in Cell Biology* 14: 648–56.

Hennig, W. (1966). *Phylogenetic Systematics*. Urbana: University of Illinois Press.

Hill, M. J., P. Goddard, and R. E. O. Williams (1971). 'Gut bacteria and aetiology of cancer of the breast'. *The Lancet* 298: 472–3.

Hirano, S. S. and C. D. Upper (2000). 'Bacteria in the leaf ecosystem with emphasis on Pseudomonas syringae: A pathogen, ice nucleus, and epiphyte'. *Microbiology and Molecular Biology Reviews* 64: 624–53.

Holden, C. (2005). 'Life in the air'. *Science* 307: 1558.

Holmes, A. J., M. R. Gillings, B. S. Nield, B. C. Mabbutt, K. M. H. Nevalainen, et al. (2003). 'The gene cassette metagenome is a basic resource for bacterial genome evolution'. *Environmental Microbiology* 5: 383–94.

Hood, L. and D. Galas (2003). 'The digital code of DNA'. *Nature* 421: 444–8.

Hooper, L. V. and J. I. Gordon (2001). 'Commensal host–bacterial relationships in the gut'. *Science* 292: 1115–18.

—— L. Bry, P. G. Falk, and J. I. Gordon (1998). 'Host–microbial symbiosis in the mammalian intestine: Exploring an internal ecosystem'. *BioEssays* 20: 336–43.

—— M. H. Wong, A. Thelin, L. Hansson, P. Falk and J. I. Gordon (2001). 'Molecular analysis of commensal host–microbial relationships in the intestine'. *Science* 291: 881–4.

Horikoshi, K. and W. D. Grant (eds.) (1998). *Extremophiles: Microbial Life in Extreme Environments*. New York: Wiley-Liss.

Huber, H., M. J. Hohn, R. Rachel, T. Fuchs, V. C. Wimmer, and K. O. Stetter (2002). 'A new phylum of Archaea represented by a nanosized hyperthermophilic symbiont'. *Nature* 417: 63–7.

Hugenholtz, P., B. M. Goebels, and N. R. Pace (1998). 'The impact of culture-independent studies on the emerging phylogenetic view of biodiversity'. *Journal of Bacteriology* 180: 4765–74.

Hughes, W. O. H., B. P. Oldroyd, M. Beekman, and F. L. W. Ratnieks (2008). 'Ancestral monogamy shows kin selection is key to the evolution of eusociality'. *Science* 320: 1213–16.

Hull, D. L. (1965). 'The effect of essentialism on taxonomy: 2000 years of stasis'. *British Journal for the Philosophy of Science* 15: 314–26; 16: 1–18.

—— (1974). *Philosophy of Biological Science*. Englewood Cliffs, NJ: Prentice-Hall.

—— (1976). 'Are species really individuals?' *Systematic Zoology* 25: 174–91.

—— (1980). 'Individuality and selection'. *Annual Review of Ecology and Systematics* 11: 311–32.

—— (1987a). 'The ideal species concept – and why we can't get it'. In M. F. Claridge, H. A. Dawah, and M. R. Wilson (eds.), *Species: The Units of Biodiversity*. London: Chapman & Hall, pp. 357–80.

—— (1987b). 'Genealogical actors in ecological roles'. *Biology and Philosophy* 2: 168–83.

—— (1988). *Science as a Process*. Chicago: University of Chicago Press.

Hunding, A., F. Kepes, D. Lancet, A. Minsky, V. Morris, et al. (2006). 'Compositional complementarity and prebiotic ecology in the origin of life'. *BioEssays* 28: 399–412.

Hunter-Cevera, J., D. Karl, and M. Buckley (2005). *Marine Microbial Diversity: The Key to Earth's Habitability*. Washington, DC: American Academy of Microbiology.

Hurst, C. (2000). 'An introduction to viral taxonomy and the proposal of Akamara, a potential domain for the genomic acellular agents'. In C. Hurst (ed.), *Viral Ecology*. San Diego: Academic/Elsevier, pp. 41–62.

Huson, D. H. (1998). 'Splits Tree: Analyzing and visualizing evolutionary data'. *Bioinformatics* 14: 68–73.

Ikemura, T. (1981). 'Correlation between the abundance of *Escherichia coli* transfer RNAs and the occurrence of the respective codons in its protein genes: A proposal for a synonymous codon choice that is optimal for the *E. coli* translational system'. *Journal of Molecular Biology* 151: 389–409.

Iturbe-Ormaetxe, I. and S. L. O'Neill (2007). 'Wolbachia–host interactions: Connecting phenotype to genotype'. *Current Opinion in Microbiology* 10: 221–4.

Iyer, L. M., L. Aravind, S. L. Coon, D. C. Klein and E. V. Koonin (2004). 'Evolution of cell–cell signaling in animals: Did late horizontal gene transfer from bacteria have a role?' *Trends in Genetics* 20: 292–9.

Jablonka, E. and M. J. Lamb (1995). *Epigenetic Inheritance and Evolution: The Lamarckian Dimension*. New York: Oxford University Press.

—— —— (2005). *Evolution in Four Dimensions: Genetic, Epigenetic, Behavioral, and Symbolic Variation in the History of Life*. Cambridge, MA: MIT Press.

Jacob, F. (1977). 'Evolution and tinkering'. *Science* 196: 1161–6.

—— (2001). 'Complexity and tinkering'. *Annals of the New York Academy of Sciences* 929: 71–3.

Jannasch, H. W. and G. E. Jones (1959). 'Bacterial populations in sea water as determined by different methods of enumeration'. *Limnology and Oceanography* 4: 128–39.

Jefferson, K. K. (2004). 'What drives bacteria to produce a biofilm?' *FEMS Microbiology Letters* 236: 163–73.

Jeffrey, C. (1971). 'Thallophytes and kingdoms: A critique'. *Kew Bulletin* 25: 291–9.

Jeffrey, C. J. (1999). 'Moonlighting proteins'. *Trends in Biochemical Sciences* 24: 8–11.

Johannsen, W. (1909). *Elemente der exakten Erblichkeitslehre*. Jena: Gustav Fischer.

Johnston, A. W. B., Y. Li, and L. Ogilivie (2005). 'Metagenomic marine nitrogen fixation: Feast or famine?' *Trends in Microbiology* 13: 416–20.

Jones, D. A. (2009). 'What does the British public think about human–animal hybrid embryos?' *Journal of Medical Ethics* 35: 168–70.

Joseph, S. J., P. Hugenholtz, P. Sangwan, C. A. Osborne and P. H. Janssen (2003). 'Laboratory cultivation of widespread and previously uncultured bacteria'. *Applied and Environmental Microbiology* 69: 7210–15.

Joyce, A. R. and B. Ø. Palsson (2006). 'The model organism as a system: Integrating "omics" data sets'. *Nature Reviews Molecular Cell Biology* 7: 198–210.

Kado, C. I. (1998). 'Origin and evolution of plasmids'. *Antonie van Leeuwenhoek* 73: 117–26.

Kaeberlein, T., K. Lewis, and S. S. Epstein (2002). 'Isolating "uncultivable" microorganisms in pure culture in a simulated natural environment'. *Science* 296: 1127–9.

Kaiser, D. (2001). 'Building a multicellular organism'. *Annual Review of Genetics* 35: 103–23.

Kämpfer, P. and R. Rosselló-Mora (2004). 'The species concept for prokaryotic microorganisms: An obstacle for describing diversity?' *Poiesis and Praxis* 3: 62–72.

Kan, J., T. E. Hanson, J. M. Ginter, K. Wang, and F. Chen (2005). 'Metaproteomic research of Chesapeake Bay microbial communities'. *Saline Systems* 1(7). doi: 10.1186/1746-1448-1-7.

Kaplan, J. and M. Pigliucci (2003). 'On the concept of biological race and its applicability to humans'. *Philosophy of Science* (Supplement) 70: 1161–72.

Karam, J. D. (2005). 'Bacteriophages: The viruses for all seasons of molecular biology'. *Virology Journal* 2: 19.

Kasting, J. F. and J. L. Siefert (2002). 'Life and the evolution of earth's atmosphere'. *Science* 296: 1066–8.

Keeling, P. J. and N. M. Fast (2002). 'Microsporidia: Biology and evolution of highly reduced intracellular parasites'. *Annual Review of Microbiology* 56: 93–116.

Keim, C. N., F. Abreu, U. Lins, H. Lins de Barros, and M. Farina (2004a). 'Cell organization and ultrastructure of a magnetotactic multicellular organism'. *Journal of Structural Biology* 145: 254–62.

—— J. L. Martins, F. Abreu, A. Soares Rosado, H. Lins de Barros, et al. (2004b). 'Multicellular life cycle of magnetotactic prokaryotes'. *FEMS Microbiology Letters* 240: 203–8.

—— M. Farina, and U. Lins (2007). '*Magnetoglobus*, magnetic aggregates in anaerobic environments'. *ASM News* 2: 437–45.

Keller, E. F. (2010). 'It is possible to reduce biological explanations to explanations in chemistry and/or physics?' In R. Arp and F. J. Ayala (eds.), *Contemporary Debates in Philosophy of Biology*. Hoboken, NJ: John Wiley.

Kerfeld, C. A., M. R. Sawaya, S. Tanaka, C. V. Nyugen, M. Phillips, et al. (2005). 'Protein structures forming the shell of primitive bacterial organelles'. *Science* 309: 936–8.

Kerr, R. A. (2005). 'The story of O_2'. *Science* 308: 1730–2.

Kirschner, M. and J. Gerhart (1998). 'Evolvability'. *Proceedings of the National Academy of Sciences USA* 95: 8420–7.

Kitano, H. and K. Oda (2006a). 'Robustness trade-offs and host–microbial symbiosis in the immune system'. *Molecular Systems Biology* 2: 2006.0022.

—— —— (2006b). 'Self-extending symbiosis: A mechanism for increasing robustness through evolution'. *Biological Theory* 1: 61–6.

Kitcher, P. (1985). *Vaulting Ambition: Sociobiology and the Quest for Human Nature*. Cambridge, MA: MIT Press.

—— (2001). *Science, Truth, and Democracy*. New York: Oxford University Press.

Klein, F., W. F. Amin Kotb, and I. Petersen (2009). 'Incidence of human papilloma virus in lung cancer'. *Lung Cancer* 65: 13–18.

Klekowski, E. J. (2003). 'Plant clonality, mutation, diplontic selection and mutational meltdown'. *Biological Journal of the Linnean Society* 79: 61–7.

Koch, A. L. (1995). 'The origin of intracellular and intercellular pathogens'. *Quarterly Review of Biology* 70: 423–37.

Koenig, B. A., S. S.-J. Lee, and S. S. Richardson (eds.) (2008). *Revisiting Race in a Genomic Age*. New Brunswick, NJ: Rutgers University Press.

Kohler, R. E., Jr. (1973). 'The enzyme theory and the origin of biochemistry'. *Isis* 64: 181–96.

Kolenbrander, P. E. (2000). 'Oral microbial communities: Biofilms, interactions, and genetic systems'. *Annual Review of Microbiology* 54: 413–37.

—— R. N. Andersen, D. S. Blehert, P. G. England, J. S. Foster and R. J. Palmer, Jr. et al. (2002). 'Communication among oral bacteria'. *Microbiology and Molecular Biology Reviews* 66: 486–505.

Komeili, A., Z. Li, D. K. Newman, and G. J. Jensen (2006). 'Magnetosomes are cell membrane invaginations organized by actin-like protein mamK'. *Science* 311: 242–5.

Konstantinidis, K. T. and J. M. Tiedje (2005). 'Genomic insights that advance the species definition for prokaryotes'. *Proceedings of the National Academy of Science USA* 102: 2567–72.

Koonin, E. V. (2005). 'Virology: Gulliver among the Lilliputians'. *Current Biology* 15: R167–R169.

—— K. S. Makarova, and L. Aravind (2001). 'Horizontal gene transfer in prokaryotes: Quantification and classification'. *Annual Review of Microbiology* 55: 709–42.

—— T. G. Senkevich, and V. V. Dolja (2006). 'The ancient virus world and evolution of cells'. *Biology Direct* 1: 29. doi: 10.1186/1745-6150.1-29.

Kornberg, A. (1989). *For the Love of Enzymes: The Odyssey of a Biochemist*. Cambridge, MA: Harvard University Press.

Koshland, D. E., Jr. (1979). 'A model regulatory system: Bacterial chemotaxis'. *Physiological Review* 59: 811–62.

—— (2002). 'The seven pillars of life'. *Science* 295: 2215–16.

Kreft, J.-U. (2004). 'Conflict of interest in biofilms'. *Biofilms* 1: 265–76.

Kripke, S. (1980). *Naming and Necessity*. Cambridge, MA: Harvard University Press.

Krohs, U. and W. Callebaut (2007). 'Data without models merging with models without data'. In F. Boogerd, F. Bruggeman, J. Hofmeyr, and H. Westerhoff (eds.), *Systems Biology: Philosophical Foundations*. Amsterdam: Elsevier, pp. 181–213.

Kroos, L. and J. R. Maddock (2003). 'Prokaryotic development: Emerging insights'. *Journal of Bacteriology* 185: 1128–46.

Krutzen M., J. Mann, M. R. Heithaus, R. C. Connor, L. Bejder, and W. B. Sherwin (2005). 'Cultural transmission of tool use in bottlenose dolphins'. *Proceedings of the National Academy of Sciences USA* 25: 8939–43.

Kucharski, R., J. Maleszka, S. Foret, and R. Maleszka (2008). 'Nutritional control of reproductive status in honeybees via DNA methylation'. Science 319: 1827–30.

Kuhn, T. S. ([1962] 1976). *The Structure of Scientific Revolutions*. Chicago: University of Chicago Press.

La Scola, B., C. Desnues, I. Pagnier, C. Robert, L. Barrassi, et al. (2008). 'The virophage as a unique parasite of the giant mimivirus'. *Nature* 455: 100–4.

Lai, Y., A. Di Nardo, T. Nakatsuji, A. Leichtle, Y. Yang, et al. (2009). 'Commensal bacteria regulate Toll-like receptor 3–dependent inflammation after skin injury'. *Nature Medicine* 15: 1377–82.

Lakatos, I. (1980). *The Methodology of Scientific Research Programmes. Philosophical Papers: Volume 1*, ed. J. Worrall and G. Currie. Cambridge: Cambridge University Press.

Lan, R. and P. R. Reeves (2000). 'Intraspecies variation in bacterial genomes: The need for a species genome concept'. *Trends in Microbiology* 8: 396–401.

—— —— (2001). 'When does a clone deserve a name? A perspective on bacterial species based on population genetics'. *Trends in Microbiology* 9: 419–24.

Lange, M., P. Westermann, and B. K. Ahring (2005). 'Archaea in protozoa and metazoa'. *Applied Microbiology and Biotechnology* 66: 465–74.

Lartigue, C., J. I. Glass, N. Alperovich, R. Pieper, P. P. Parmar, et al. (2007). 'Genome transplantation in bacteria: Changing one species to another'. *Science* 317: 632–8.

Latour, B. (1987). *Science in Action*. Cambridge, MA: Harvard University Press.

Lawrence, J. G. (2002). 'Gene transfer in bacteria: Speciation without species'. *Theoretical Population Biology* 61: 449–60.

—— and H. Hendrickson (2003). 'Lateral gene transfer: When will adolescence end?' *Molecular Microbiology* 50: 739–49.

—— —— (2005). 'Genome evolution in bacteria: Order beneath chaos'. *Current Opinion in Microbiology* 8: 1–7.

—— G. F. Hatfull, and R. W. Hendrix (2002). 'Imbroglios of viral taxonomy: Genetic exchange and failings of phenetic approaches'. *Journal of Bacteriology* 184: 4891–905.

Leadbetter, J. R. (2003). 'Cultivation of recalcitrant microbes: Cells are alive, well and revealing their secrets in the 21st century laboratory'. *Current Opinion in Microbiology* 6: 274–81.

Lederberg, J. (1952). 'Cell genetics and hereditary symbiosis'. *Physiological Reviews* 32: 403–30.

—— (1998). 'Personal perspective: Plasmid (1952–1997)'. *Plasmid* 39: 1–9.

—— and E. L. Tatum (1946). 'Novel genotypes in mixed cultures of biochemical mutants of bacteria'. *Cold Spring Harbor Symposia on Quantitative Biology* 11: 113–14.

Lee, K. (2004). 'There is biodiversity and biodiversity: Implications for environmental philosophers'. In M. Oksanen and J. Pietarinen (eds.), *Philosophy and Biodiversity*. New York: Cambridge University Press, pp. 152–71.

Lee, M. S. and D. A. Morrison (1999). 'Identification of a new regulator in *Streptococcus pneumoniae* linking quorum sensing to competence for genetic transformation'. *Journal of Bacteriology* 181: 5004–16.

Levin, B. R. and C. T. Bergstrom (2000). 'Bacteria are different: Observations, interpretations, speculations, and opinions about the mechanisms of adaptive evolution in prokaryotes'. *Proceedings of the National Academy of Science USA* 97: 6951–85.

Lewis, C. S. (1945). *That Hideous Strength*. New York: Scribner.

Lewis, K. (2000). 'Programmed death in bacteria'. *Microbiology and Molecular Biology Reviews* 64: 503–14.

Lewontin, R. C. (1972). 'The apportionment of human diversity'. *Evolutionary Biology* 6: 381–98.

—— (2005). 'Confusions about race'. Posted, 20 April 2005, http://raceandgenomics.ssrc.org/Lewontin/pf/

—— S. Rose, and L. J. Kamin (1984). *Not In Our Genes: Biology, Ideology and Human Nature.* New York: Random House.

Ley, R. E., F. Bäckhed, P. Turnbaugh, C. Lozupone, R. Knight, and J. I. Gordon (2005). 'Obesity alters gut microbial ecology'. *Proceedings of the National Academy of Sciences* 102: 11070–5.

—— D. A. Peterson, and J. I. Gordon (2006). 'Ecological and evolutionary forces shaping microbial diversity in the human intestine'. *Cell* 124: 837–48.

Lezon, T. R., J. R. Banavar, and A. Maritan (2006). 'The origami of life'. *Journal of Physics and Condensed Matter* 18: 847–88.

Liebman, S. W. and I. L. Derkatch (1999). 'The yeast [PSI+] prion: Making sense of nonsense'. *Journal of Biological Chemistry* 274: 1181–4.

Liles, M. R., B. F. Manske, S. B. Bintrim, J. Handelsman, and R. M. Goodman (2003). 'A census of rRNA genes and linked genomic sequences within a soil metagenomic library'. *Applied and Environmental Microbiology* 69: 2684–91.

Lindell, D., J. D. Jaffe, Z. I. Johnson, G. M. Church, and S. W. Chisholm (2005). 'Photosynthesis genes in marine viruses yield proteins during host infection'. *Nature* 438: 86–9.

Lloyd, D. (2004). '"Anaerobic protists": Some misconceptions and confusions'. *Microbiology* 150: 1115–16.

Lloyd, E. A. (1989). 'A structural approach to defining units of selection'. *Philosophy of Science* 56: 395–418.

—— (1994). *The Structure and Confirmation of Evolutionary Theory.* Princeton: Princeton University Press.

—— (2000). 'Groups on groups: Some dynamics and possible resolution of the units of selection debates in evolutionary biology'. *Biology and Philosophy* 15: 389–401.

Lo, Y. M. D. (2000). 'Fetal DNA in maternal plasma: Biology and diagnostic applications'. *Clinical Chemistry* 46: 1903–6.

Longino, H. (1990). *Science as Social Knowledge: Values and Objectivity in Scientific Inquiry.* Princeton: Princeton University Press.

—— (2002). *The Fate of Knowledge.* Princeton: Princeton University Press.

Looijen, R. C. (1998). *Holism and Reductionism in Biology and Ecology: The Mutual Dependence of Higher and Lower Level Research Programmes.* Kluwer: Dordrecht.

Loreau, M., S. Naeem, P. Inchausti, J. Bengtsson, J. P. Grime, et al. (2001). 'Biodiversity and ecosystem functioning: Current knowledge and future challenges'. *Science* 294: 804–8.

Lorenz, P. and J. Eck (2005). 'Metagenomics and industrial application'. *Nature Reviews Microbiology* 3: 510–16.

Lovley, D. R. (2003). 'Cleaning up with genomics: Applying molecular biology to bioremediation'. *Nature Reviews Microbiology* 1: 35–44.

Lowe, M. and F. A. Barr (2007). 'Inheritance and biogenesis of organelles in the secretory pathway'. *Nature Reviews Molecular Cell Biology* 8: 429–39.

Luisi, P. L. (1998). 'About various definitions of life'. *Origins of Life and Evolution of the Biosphere* 28: 613–22.

Lupi, O., P. Dadalti, E. Cruz, and P. R. Sanberg (2006). 'Are prions related to the emergence of early life?' *Medical Hypotheses* 67: 1027–33.

—— P. Dadalti, and C. Goodheart (2007). 'Did the first virus self-assemble from self-replicating prion proteins and RNA?' *Medical Hypotheses* 69: 724–30.

Luria, S. E. (1947). 'Recent advances in bacterial genetics'. *Journal of Bacteriology* 11: 1–40.

—— and M. Delbrück (1943). 'Mutations of bacteria from virus sensitivity to virus resistance'. *Genetics* 28: 491–511.

—— J. E. Darnell, D. Baltimore, and A. Campbell (eds.) (1978). *General Virology*, 3rd edn. New York: John Wiley.

Lyon, P. (2007). 'From quorum to cooperation: Lessons from bacterial sociality for evolutionary theory'. *Studies in History and Philosophy of Biological and Biomedical Sciences* 38: 820–33.

Maamar, H., A. Raj, and D. Dubnau (2007). 'Noise in gene expression determines cell fate in *Bacillus subtilis*'. *Science* 317: 526–9.

McFall-Ngai, M. J. (2001). 'Identifying "prime suspects": Symbioses and the evolution of multicellularity'. *Comparative Biochemistry and Physiology Part B* 129: 711–23.

—— (2002). 'Unseen forces: The influence of bacteria on animal development'. *Developmental Biology* 242: 1–14.

McGowan, P. O., A. Sasaki, A. C. D'Alessio, S. Dymor, B. Labonté, et al. (2009). 'Epigenetic regulation of the glucocorticoid receptor in human brain associates with childhood abuse'. *Nature Neuroscience* 12: 342–8.

Machamer, P. K., L. Darden, and C. F. Craver (2000). 'Thinking about mechanisms'. *Philosophy of Science* 67: 1–25.

McShea, D. W. (2004). 'A revised Darwinism'. *Biology and Philosophy* 19: 45–53.

Mackie, J. L. (1974). *The Cement of the Universe: A Study of Causation*. Oxford: Oxford University Press.

Magasanik, B. (1999). 'A midcentury watershed: The transition from microbial biochemistry to molecular biology'. *Journal of Bacteriology* 181: 357–8.

Maier, R. M., I. L. Pepper, and C. P. Gerba (2000). *Environmental Microbiology*. San Diego: Academic Press.

Mallet, F., O. Bouton, S. Proudhomme, V. Cheynet, G. Oriol, et al. (2004). 'The endogenonous retroviral locus ERVWE1 is a bona fide gene involved in hominoid placental physiology'. *Proceedings of the National Academy of Sciences USA* 101: 1731–6.

Mallet, J. (2008). 'Hybridization, ecological races, and the nature of species: Empirical evidence for the ease of speciation'. *Philosophical Transactions of the Royal Society B* 363: 2971–86.

Manchester, K. L. (2000). 'Biochemistry comes of age: A century of endeavour'. *Endeavour* 24: 22–7.

Manichanh, C., L. Rigottier-Gois, E. Bonnaud, K. Gloux, E. Pelletier, et al. (2006). 'Reduced diversity of faecal microbiota in Crohn's disease revealed by a metagenomic approach'. *Gut* 55: 205–11.

Manuelidis, L. (2004). 'A virus behind the mask of prions?' *Folia Neuropathology* 42, *Supplement B*: 10–23.

Marais, G. A. B., A. Calteau, and O. Tenaillon (2007). 'Mutation rate and genome reduction in endosymbiotic and free-living bacteria'. *Genetica* 134. doi: 10.1007/s1079-007-9226-6.

Margulis, L. (1970). *Origin of Eukaryotic Cells*. New Haven: Yale University Press.

—— (1998). *Symbiotic Planet: A New Look at Evolution*. New York: Basic Books.

—— and D. Sagan (2002). *Acquiring Genomes: A Theory of the Origins of Species*. New York: Basic Books.

Marri, P. R., W. Hao, and G. B. Golding (2007). 'The role of laterally transferred genes in adaptive evolution'. *BMC Evolutionary Biology* 7 (Supp. 1): S8.

Marsh, P. D. (2004). 'Dental plaque as a microbial biofilm'. *Caries Research* 38: 204–11.

Marshall Graves, J. A. and C. M. Disteche (2007). 'Does gene dosage really matter?' *Journal of Biology* 6: 1.

Martin, W. (2003). 'Gene transfer from organelles to the nucleus: Frequent and in big chunks'. *Proceedings of the National Academy of Sciences USA* 100: 8612–14.

—— and M. J. Russell (2003). 'On the origin of cells: A hypothesis for the evolutionary transitions from abiotic geochemistry to chemoautotrophic prokaryotes, and from prokaryotes to nucleated cells'. *Philosophical Transactions of the Royal Society of London B* 358: 59–85.

Martiny J. B. H., B. J. M. Bohannan, J. H. Brown, R. K. Colwell, J. A. Fuhrman, et al. (2006). 'Microbial biogeography: Putting microorganisms on the map'. *Nature Reviews Microbiology* 4: 102–12.

Masel, J. and A. Bergman (2003). 'The evolution of the evolvability of the yeast prion [PSI+]'. *Evolution* 57: 1498–512.

Maynard Smith, J. (1995). 'Do bacteria have population genetics?' In S. Baumberg, J. P. W. Young, J. R. Saunders, and E. M. H. Wellington (eds.), *Population Genetics of Bacteria, Society for General Microbiology Symposium 52*. Cambridge: Cambridge University Press, pp. 1–12.

—— (2000). 'The concept of information in biology'. *Philosophy of Science* 67: 177–94.

—— and E. Szathmáry (1995). *The Major Transitions in Evolution*. New York: W. H. Freeman.

—— N. H. Smith, M. O'Rourke, and B. G. Spratt (1993). 'How clonal are bacteria?' *Proceedings of the National Academy of Sciences USA* 90: 4384–8.

—— E. J. Feil, and N. H. Smith (2000). 'Population structure and evolutionary dynamics of pathogenic bacteria'. *BioEssays* 22: 1115–22.

Mayr, E. (1961). 'Cause and effect in biology'. *Science* 131: 1501–6.

—— (1963). *Animal Species and Evolution*, Cambridge, MA: Harvard University Press.

—— (1998). 'Two empires or three?' *Proceedings of the National Academy of Sciences USA* 95: 9720–3.

Maze, I., H. E. Covington III, D. M. Dietz, Q. LaPlant, W. Renthal, et al. (2010). 'Essential role of the histone methyltransferase G9a in cocaine-induced plasticity'. *Science* 327: 213–16.

Mazzotti, M. (2008). *Knowledge as Social Order: Rethinking the Sociology of Barry Barnes*. Farnham, Surrey: Ashgate.

Mead, M. (1949). *Male and Female*. New York: Morrow.

Meaney, M. J., M. Szyf, and J. R. Seckl (2007). 'Epigenetic mechanisms of perinatal programming of hypothalamic-pituitary-adrenal function and health'. *Trends in Molecular Medicine* 13: 269–77.

Medini, D., C. Donati, H. Tettelin, V. Masignani, and R. Rappuoli (2005). 'The microbial pan-genome'. *Current Opinion in Genetics and Development* 15: 589–94.

Meseguer, M. A., A. Álvarez, M. T. Rejas, C. Sanchez, J. C. Perez-Diaz and F. Baquero (2003). '*Mycoplasma pneumoniae*: A reduced-genome intracellular bacterial pathogen'. *Infection, Genetics and Evolution* 3: 47–55.

Méthot, P.-O. (2011). 'Research traditions and evolutionary explanations in medicine'. *Journal of Theoretical Medicine and Bioethics* 32: 75–90.

Michael, C. A., M. R. Gillings, A. J. Holmes, L. Hughes, N. R. Andrew, et al. (2004). 'Mobile gene cassettes: A fundamental resource for evolution'. *American Naturalist* 164: 1–12.

Michod, R. E. (1997a). 'Cooperation and conflict in the evolution of individuality. I. Multilevel selection of the organism'. *American Naturalist* 149: 607–45.

—— (1997b). 'Evolution of the individual'. *American Naturalist* 150: S5–S21.

Miller, G. (2010). 'The seductive allure of behavioral epigenetics'. *Science* 329: 24–7.

Miller, M. B. and B. L. Bassler (2001). 'Quorum sensing in bacteria'. *Annual Review of Microbiology* 55: 165–99.

Milne, R. (2010). 'Drawing bright lines: Food and the futures of biopharming'. In S. Parry and J. Dupré (eds.), *Nature after the Genome*. Sociological Review Monograph Series. Malden, MA: Blackwell Publishing, pp. 133–51.

Mindell, D. P. and L. P. Villarreal (2003). 'Don't forget about viruses'. *Science* 302: 1677.

—— J. S. Rest, and L. P. Villarreal (2003). 'Viruses and the tree of life'. In J. Cracraft and M. Donoghue (eds.), *Assembling the Tree of Life*. New York: Oxford University Press, pp. 107–18.

Mold, J. E., J. Michaëlsson, T. D. Burt, M. O. Muench, K. P. Beckerman, et al. (2008). 'Maternal alloantigens promote the development of tolerogenic fetal regulatory T cells in utero'. *Science* 322: 1562–5.

Molin, S. and T. Tolker-Nielsen (2003). 'Gene transfer occurs with enhanced efficiency in biofilms and induces stabilisation of the biofilm structure'. *Current Opinion in Biotechnology* 14: 255–61.

Moran, N. A. (2006). 'Symbiosis'. *Current Biology* 16: R866–R871.

—— and P. H. Degnan (2006). 'Functional genomics of *Buchnera* and the ecology of aphid hosts'. *Molecular Ecology* 15: 1251–61.

Moreira, D. and P. López-García (2005). 'Comment on "The 1.2-megabase genome sequence of Mimivirus"'. *Science* 308: 1114a.

—— —— (2009). 'Ten reasons to exclude viruses from the tree of life'. *Nature Reviews Microbiology* 7: 306–11.

Morgan, D., K. A. Grant, H. D. Gage, R. H. Mach, J. R. Kaplan, et al. (2002). 'Social dominance in monkeys: Dopamine D2 receptors and cocaine self-administration'. *Nature Neuroscience* 5: 169–74.

Morgan, D. K. and E. Whitelaw (2008). 'The case for transgenerational epigenetic inheritance in humans'. *Mammalian Genome* 19: 394–7.

Morgan, T. H., A. Sturtevant, H. J. Müller, and C. Bridges (1915). *The Mechanism of Mendelian Heredity*. New York: Holt.

Morton, T. A., S. A. Haslam, T. Postmes, and M. K. Ryan (2006). 'We value what values us: The appeal of identity-affirming science'. *Political Psychology* 27: 823–38

Moss, L. (2003). *What Genes Can't Do*. Cambridge, MA: MIT Press.

Müller, H. J. (1927). 'Artificial transmutation of the gene'. *Science* 46: 84–7.

—— (1966). 'The gene material as the initiator and the organizing basis of life'. *American Naturalist* 100: 493–517.

Munro, S. (2004). 'Organelle identity and the organization of membrane traffic'. *Nature Cell Biology* 6: 469–72.

Myers, G., I. Paulsen, and C. Fraser (2006). 'The role of mobile DNA in the evolution of prokaryotic genomes'. In L. H. Caporale (ed.), *The Implicit Genome*. Oxford: Oxford University Press, pp. 121–37.

Nanney, D. (1999). 'When is a rose? The kinds of Tetrahymena'. In R. A. Wilson (ed.), *Species: New Interdisciplinary Essays*. Cambridge, MA: MIT Press, pp. 97–118.

Nee, S. (2004). 'More than meets the eye'. *Nature* 429: 804–5.

—— (2005). 'The great chain of being'. *Nature* 435: 429.

Nelson, K. E. (2003). 'The future of microbial genomics'. *Environmental Microbiology* 5: 1223–5.

Newman, D. K. and J. F. Banfield (2002). 'Geomicrobiology: How molecular scale interactions underpin biogeochemical systems'. *Science* 296: 1071–7.

Nicholson, J. K., E. Holmes, J. C. Lindon, and I. D. Wilson (2004). 'The challenges of modelling mammalian biocomplexity'. *Nature Biotechnology* 22: 1268–74.

—— E. Holmes, and I. D. Wilson (2005). 'Gut microorganisms, mammalian metabolism and personalized health care'. *Nature Reviews Microbiology* 3: 431–8.

Niftrik, L. A. V., J. A. Fuerst, J. S. S. Damsté, J. G. Kuenen, M. S. M. Jetten, and M. Strous (2004). 'The ammoxosome: An introcytoplasmic compartment in anammox bacteria'. *FEMS Microbiology Letters* 233: 7–13.

Nisbet, E. G. and N. H. Sleep (2001). 'The habitat and nature of early life'. *Nature* 409: 1083–91.

Noonan, J. P., M. Hofreiter, D. Smith, J. R. Priest, N. Rohland, et al. (2005). 'Genomic sequencing of Pleistocene cave bears'. *Science* 309: 597–9.

Norris, V., A. Hunding, F. Kepes, D. Lancet, A. Minsky, et al. (2007). 'Question 7: The first units of life were not cells'. *Origins of Life and Evolution of the Biosphere* 37: 429–32.

Nunnari, J. and P. Walter (1996). 'Regulation of organelle biogenesis'. *Cell* 84: 389–94.

Nussbaum, M. and A. Sen (eds.) (1993). *The Quality of Life*. Oxford: Oxford University Press.

O'Donnell, A. G., M. Goodfellow, and D. L. Hawksworth (1994). 'Theoretical and practical aspects of the quantification of biodiversity among microorganisms'. *Philosophical Transactions of the Royal Society of London B* 345: 65–73.

O'Hear, A. (ed.) (2005). *Philosophy, Biology and Life* (Royal Institute of Philosophy Supplements). Cambridge: Cambridge University Press.

O'Malley, M. A. and Y. Boucher (2005). 'Paradigm change in evolutionary microbiology'. *Studies in the History and Philosophy of Biological and Biomedical Sciences* 36: 183–208.

—— and J. Dupré (2005). 'Fundamental issues in systems biology'. *BioEssays* 12: 1270–6.

O'Toole, G., H. B. Kaplan, and R. Kolter (2000). 'Biofilm formation as microbial development'. *Annual Review of Microbiology* 54: 49–79.

Ochman, H., J. G. Lawrence, and E. A. Groisman (2000). 'Lateral gene transfer and the nature of bacterial innovation'. *Nature* 405: 299–304.

Odling-Smee, F. J., K. N. Laland, and M. W. Feldman (2003). *Niche Construction: The Neglected Process in Evolution*. Princeton: Princeton University Press.

Okasha, S. (2003). 'Recent work on the levels of selection problem'. *Human Nature Review* 3: 349–56.

—— (2004). 'Multi-level selection and the major transitions in evolution'. *Philosophy of Science* 72: 1013–25.

Oksanen, M. and J. Pietarinen (2004). *Philosophy and Biodiversity*. New York: Cambridge University Press.

Olsen, G. J., D. J. Lane, S. J. Giovannoni, and N. R. Pace (1986). 'Microbial ecology and evolution: A ribosomal RNA approach'. *Annual Review of Microbiology* 40: 337–65.

—— C. R. Woese, and R. Overbeek (1994). 'The winds of (evolutionary) change: Breathing new life into microbiology'. *Journal of Bacteriology* 176: 1–6.

Ordovas, J. M. and V. Mooser (2006). 'Metagenomics: The role of the microbiome in cardiovascular diseases'. *Current Opinion in Lipidology* 17: 157–61.

Orgel, L. E. and F. H. C. Crick (1980). 'Selfish DNA: The ultimate parasite'. *Nature* 284: 604–7.

Osteryoung, K. W. and J. Nunnari (2003). 'The division of endosymbiotic organelles'. *Science* 302: 1698–704.

Oyama, S. (1985). *The Ontogeny of Information: Developmental Systems and Evolution.* Cambridge: Cambridge University Press.

—— P. E. Griffiths, and R. D. Gray (eds.) (2001). *Cycles of Contingency: Developmental Systems and Evolution.* Cambridge, MA: MIT Press.

Pääbo, S. (2001). 'The human genome and our view of ourselves'. *Science* 291: 1219–20.

Pace, N. R. (1997). 'A molecular view of microbial diversity and the biosphere'. *Science* 27: 734–40.

—— D. A. Stahl, G. J. Olsen, and D. J. Lane (1985). 'Analyzing natural microbial populations by rRNA sequences'. *ASM News* 51: 4–12.

Page, R. A. and M. J. Ryan (2006). 'Social transmission of novel foraging behavior in bats: Frog calls and their referents'. *Current Biology* 16: 1201–5.

Pál, C. (2001). 'Yeast prions and evolvability'. *Trends in Genetics* 17: 167–9.

—— B. Papp, and M. J. Lercher (2005). 'Adaptive evolution of bacterial metabolic networks by horizontal gene transfer'. *Nature Genetics* 37: 1372–5.

Palsson, B. (2000). 'The challenges of in silico biology'. *Nature Biotechnology* 18: 1147–50.

Pályi, G., C. Zucchi, and L. Caglioti (eds.) (2002). *Fundamentals of Life.* Paris: Elsevier.

Palys, T., L. K. Nakamura, and F. M. Cohan (1997). 'Discovery and classification of ecological diversity in the bacterial world: The role of DNA sequence data'. *International Journal of Systematic Bacteriology* 47: 1145–56.

Papke, R. T. and D. M. Ward (2004). 'The importance of physical isolation to microbial diversification'. *FEMS Microbiology Ecology* 48: 293–303.

Park, S., P. M. Wolanin, E. A. Yuzbashyan, P. Silberzan, J. B. Stock, and R. H. Austin (2003). 'Motion to form a quorum'. *Science* 301: 188.

Parker, V. T. (2004). 'The community of an individual: Implications for the community concept'. *Oikos* 104: 27–34.

Parkin, E. T., N. T. Watt, I. Hussain, E. A. Eckman, C. B. Eckman, et al. (2007). 'Cellular prion protein regulates β-secretase cleavage of the Alzheimer's amyloid precursor protein'. *Proceedings of the National Academy of Sciences USA* 104: 11062–7.

Parry, S. (2010). 'Interspecies entities and the politics of nature'. In S. Parry and J. Dupré (eds.), *Nature after the Genome.* Sociological Review Monograph Series. Malden, MA: Blackwell Publishing, pp. 113–29.

—— and J. Dupré (eds.) (2010). *Nature after the Genome.* Sociological Review Monograph Series. Malden, MA: Blackwell Publishing.

Parsek, M. R. and C. Fuqua (2004). 'Biofilms 2003: Emerging themes and challenges in studies of surface-associated microbial life'. *Journal of Bacteriology* 186: 4427–40.

Pater, W. (1873). *Studies in the History of the Renaissance.* London: Macmillan and Co.

Paterson, H. E. H. (1985). 'The Recognition Concept of Species'. In E. Vrba (ed.), *Species and Speciation,* Transvaal Museum Monograph No. 4. Pretoria: Transvaal Museum.

Pauling, L. and E. Zuckerkandl (1963). 'Chemical paleogenetics: Molecular "restoration studies" of extinct forms of life'. *Acta Chemica Scandinavica* 17: S9–S16.

Paulsson, J. (2002). 'Multileveled selection on plasmid replication'. *Genetics* 161: 1373–84.

Paxson, H. (2008). 'Post-Pasteurian cultures: The microbiopolitics of raw-milk cheese in the United States'. *Cultural Anthropology* 23: 15–47.

Pazos, F., A. Valencia, and V. de Lorenzo (2003). 'The organization of the microbial biodegradation network from a systems biology perspective'. *EMBO Reports* 4: 994–9.

Pearson, H. (2002). 'Human genetics: Dual identities'. *Nature* 417: 10–11.

—— (2008). '"Virophage" suggests viruses are alive'. *Nature* 454: 677.

Pembrey, M. E. (2010). 'Research into the epigenetic impact of assisted conception'. *Bionews*, 18 January 2010. http://www.bionews.org.uk/page_53453.asp.

—— L. O. Bygren, G. Kaati, S. Edvinsson, K. Northstone, M. Sjöström, et al. (2006). 'Sex-specific, male-line transgenerational responses in humans'. *European Journal of Human Genetics* 14: 159–66.

Penn, M. and M. Dworkin (1976). 'Robert Koch and two visions of microbiology'. *Bacteriological Reviews* 40: 276–83.

Pennisi, E. (2005). 'Reading ancient DNA the community way'. *Science* 308: 1401.

—— (2007). 'Replacement genome gives microbe new identity'. *Science* 316: 1827.

Pérez-Brocal, V., R. Gil, S. Ramos, A. Lamelas, M. Postigo, et al. (2006). 'A small microbial genome: The end of a long symbiotic relationship?' *Science* 314: 312–13.

Perlin, M. H. (2002). 'The subcellular entities a.k.a. plasmids'. In U. N. Streips and R. E. Yasbin (eds.), *Modern Microbial Genetics* (2nd edn.). New York: Wiley, pp. 507–60.

Perotto, S. and P. Bonfante (1997). 'Bacterial associations with mycorrhizal fungi: Close and distant friends in the rhizosphere'. *Trends in Microbiology* 5: 496–501.

Peterson, S. N., C. K. Sung, R. Cline, B. V. Desai, E. C. Snesrud, et al. (2004). 'Identification of competence pheromone responsive genes in *Streptococcus pneumoniae* by use of DNA microarrays'. *Molecular Microbiology* 51: 1051–70.

Pigliucci, M. and J. Kaplan (2006). *Making Sense of Evolution: The Conceptual Foundations of Evolutionary Biology*. Chicago: University of Chicago Press.

Pineda-Krch, M. and T. Fagerström (1999). 'On the potential for evolutionary change in meristematic cell lineages through interorganismal selection'. *Journal of Evolutionary Biology* 12: 681–8.

Pinker, S. (1997). *How the Mind Works*. New York: W. W. Norton.

—— (2002). *The Blank Slate: The Modern Denial of Human Nature*. London: Penguin/Allen Lane.

Poinar, H. N., C. Schwarz, J. Qi, B. Shapiro, R. D. E. MacPhee, et al. (2006). 'Metagenomics to paleogenomics: Large-scale sequencing of mammoth DNA'. *Science* 311: 392–4.

Popa, R. (2004). *Between Necessity and Probability: Search for the Definition and Origin of Life*. Berlin: Springer-Verlag.

Poretsky, R. S., N. Bano, A. Buchan, et al. (2005). 'Analysis of microbial gene transcripts in environmental samples'. *Applied and Environmental Microbiology* 71: 4121–6.

Postgate, J. R. (1976). 'Death in macrobes and microbes'. In T. R. G. Gray and J. R. Postgate (eds.), *The Survival of Vegetative Microbes*. Cambridge: Cambridge University Press, pp. 1–18.

Powell, A. and J. Dupré (2009). 'From molecules to systems: The importance of looking both ways'. *Studies in History and Philosophy of Biological and Biomedical Sciences* 40: 54–64.

Powell, M. J., N. J. Sutton, C. E. Del Castillo, and A. T. Timperman (2005). 'Marine proteomics: Generation of sequence tags for dissolved proteins in seawater using tandem mass spectrometry'. *Marine Chemistry* 95: 183–98.

Pradeu, T. (2012). *The Limits of the Self: Immunology and Biological Identity*. Oxford: Oxford University Press.

Price, P. B. (2000). 'A habitat for psychrophiles in deep Antarctic ice'. *Proceedings of the National Academy of Science USA* 97: 1247–51.

Prusiner, S. B. (1998). 'Prions'. *Proceedings of the National Academy of Science USA* 95: 13363–83.

Puck, T. T. (1972). *The Mammalian Cell as a Microorganism: Genetic and Biochemical Studies in Vitro.* San Francisco: Holden-Day.

Queller, D. C. (2004). 'Kinship is relative'. *Nature* 430: 975–6.

Ram, R. J., N. C. VerBerkmoes, M. P. Thelen, G. W. Tyson, M. Shah, et al. (2005). 'Community proteomics of a natural microbial biofilm'. *Science* 308: 1915–20.

Ramasarma, T. (1999). 'Is it fair to describe a protein recruited for many cellular chores as "moonlighting" and "promiscuous"?' *Current Science* 77: 1401–5.

Raoult, D. (2005). 'The journey from *Rickettsia* to Mimivirus'. *ASM News* 71: 278–84.

—— S. Audic, C. Robert, C. Abergel, P. Renesto, et al. (2004). 'The 1.2-megabase genome sequence of Mimivirus'. *Science* 306: 1344–50.

Raser, J. M. and E. K. O'Shea (2005). 'Noise in gene expression: Origins, consequences, and control'. *Science* 309: 2010–13.

Rawls, J. F., B. S. Samuel, and J. I. Gordon (2004). 'Gnotobiotic zebrafish reveal evolutionarily conserved responses to the gut microbiota'. *Proceedings of the National Academy of Sciences USA* 101: 4596–601.

Rayner, A. D. M. (1997). *Degrees of Freedom: Living in Dynamic Boundaries.* London: Imperial College Press.

Reanney, D. C., W. P. Roberts, and W. J. Kelly (1982). 'Genetic interactions among microbial communities'. In A. T. Bull and J. H. Slater (eds.), *Microbial Interactions and Communities.* London: Academic Press, pp. 287–322.

Redfield, R. J. (2002). 'Is quorum sensing a side effect of diffusion sensing?' *Trends in Microbiology* 10: 365–70.

Relman, D. A. and S. Falkow (2001). 'The meaning and impact of the human genome sequence for microbiology'. *Trends in Microbiology* 9: 206–8.

Rheinberger, H.-J. (2000). 'Gene concepts: Fragments from the perspective of molecular biology'. In P. Beurton, R. Falk, and H.-J. Rheinberger (eds.), *The Concept of the Gene in Development and Evolution.* Cambridge: Cambridge University Press, pp. 219–39.

Rice, K. C. and K. W. Bayles (2003). 'Death's toolbox: Examining the molecular components of bacterial programmed cell death'. *Molecular Microbiology* 50: 729–38.

Richerson, P. J. and R. Boyd (2005). *Not By Genes Alone: How Culture Transformed Human Evolution.* Chicago: Chicago University Press.

Riesenfeld, C. S., P. D. Schloss, and J. Handelsman (2004). 'Metagenomics: Genomic analysis of microbial communities'. *Annual Reviews of Genetics* 38: 525–52.

Robert, J. S. (2004). *Embryology, Epigenesis, and Evolution: Taking Development Seriously.* Cambridge: Cambridge University Press.

Rode, B. M., W. Flader, C. Sotriffer, and A. Righi (1999). 'Are prions a relic of an early stage of peptide evolution?' *Peptides* 20: 1513–16.

Rodríguez-Ezpeleta, N. and H. Phillipe (2006). 'Plastid origin: Replaying the tape'. *Current Biology* 16: R53–R56.

Rodríguez-Valera, F. (2002). 'Approaches to prokaryotic diversity: A population genetics approach'. *Environmental Microbiology* 4: 628–33.

—— (2004). 'Environmental genomics: The big picture'. *FEMS Microbiology Letters* 231: 153–8.

Rohwer, F. (2003). 'Global phage diversity'. *Cell* 113: 141.

—— and R. Edwards (2002). 'The phage proteomic tree: A genome-based taxonomy for phages'. *Journal of Bacteriology* 184: 4529–35.

Rondon, M. R., P. R. August, A. D. Bettermann, S. F. Brady, T. H. Grossman, et al. (2000). 'Cloning the soil metagenome: A strategy for accessing the genetic and functional diversity of uncultured microorganisms'. *Applied and Environmental Microbiology* 66: 2541–7.

Roselló-Mora, R. and R. Amann (2001). 'The species concept for prokaryotes'. *FEMS Microbiology Reviews* 25: 39–67.

Rosen, R. (1970). *Dynamical Systems Theory in Biology*. New York: Wiley Interscience.

Rosenberg, A. (1994). *Instrumental Biology, or, The Disunity of Science*. Chicago: University of Chicago Press.

—— (2006). *Darwinian Reductionism: Or, How to Stop Worrying and Love Molecular Biology*. Chicago: University of Chicago Press.

Rosenberg, N. A., J. K. Pritchard, J. L. Weber, H. M. Cann, K. K. Kidd, et al. (2002). 'Genetic structure of human populations'. Science 298: 2981–5.

Rottem, S. and Y. Naot (1998). 'Subversion and exploitation of host cells by mycoplasmas'. *Trends in Microbiology* 6: 436–40.

Roughgarden, J. (2009). *The Genial Gene: Deconstructing Darwinian Selfishness*. Berkeley: University of California Press.

Rowe-Magnus, D. A., A.-M. Guerot, and D. Mazel (2002). 'Bacterial resistance evolution by recruitment of super-integron gene cassettes'. *Molecular Microbiology* 43: 1657–69.

Ruiz-Mirazo, K., J. Pereto, and A. Moreno (2004). 'A universal definition of life: Autonomy and open-ended evolution'. *Origins of Life and Evolution of the Biosphere* 34: 323–46.

Ruskin, J. (1905/1849). *The Seven Lamps of Architecture, Vol. 1 of The Complete Works of John Ruskin*. New York: Thomas Y. Cromwell & Co.

Sanders, I. R. (2002). 'Ecology and evolution of multigenomic Arbuscular mycorrhizal fungi'. *American Naturalist* 160: S128–S141.

Sapp, J. (1987). *Beyond the Gene: Cytoplasmic Inheritance and the Struggle for Authority in Genetics*. New York: Oxford University Press.

—— (1994). *Evolution by Association: A History of Symbiosis*. New York: Oxford University Press.

—— (2003). *Genesis: The Evolution of Biology*. Oxford: Oxford University Press.

—— (2005a). *Microbial Phylogeny and Evolution: Concepts and Controversies*. Oxford: Oxford University Press.

—— (2005b). 'The prokaryote–eukaryote dichotomy: Meanings and mythology'. *Microbiology and Molecular Biology Reviews* 69: 292–305.

Sarkar, S. (2002). 'Defining "biodiversity": Assessing biodiversity'. *Monist* 85: 131–55.

Saunders, N. J., P. Boonmee, J. F. Peden, and S. A. Jarvis (2005) 'Inter-species horizontal transfer resulting in core-genome and niche-adaptive variation within *Helicobacter pylori*'. *BMC Genomics* 6: 9.

Savage, D. C. (1977). 'Microbial ecology of the gastrointestinal tract'. *Annual Review of Microbiology* 31: 107–33.

Schloss, P. D. and J. Handelsman (2003). 'Biotechnological prospects from metagenomics'. *Current Opinion in Biotechnology* 14: 303–10.

—— —— (2004). 'Status of the microbial census'. *Microbiology and Molecular Biology Reviews* 68: 686–91.

—— —— (2005). 'Metagenomics for studying unculturable organisms: Cutting the Gordian knot'. *Genome Biology* 6: 229.

Schmeisser, C., C. Stöckigt, C. Raasch, J. Wingender, K. N. Timmis, et al. (2003). 'Metagenome survery of biofilms in drinking-water networks'. *Applied and Environmental Microbiology* 69: 7298–309.

Schoolnik, G. K. (2001). 'The accelerating convergence of genomics and microbiology'. *Genome Biology* 2: 4009.1–4009.2.

Schuch, R. and V. A. Fischetti (2009). 'The secret life of the anthrax agent *Bacillus anthracis*: Bacteriophage-mediated ecological adaptations'. *PLoS One* 4: e6532.

Schulman, R. and E. Winfree (2007). 'How crystals that sense and respond to their environments could evolve'. *Natural Computing* 7. doi: 10.1007/s11047-007-9046-8.

Schulz, H. N. and B. B. Jørgensen (2001). 'Big bacteria'. *Annual Review of Microbiology* 55: 105–37.

Schulze, W. X., G. Gleixner, K. Kaiser, G. Guggenberger, M. Mann, and E.-D. Schulze (2005). 'A proteomic fingerprint of dissolved organic carbon and of soil particles'. *Oecologia* 142: 335–43.

Sebat, J. L., F. S. Colwell, and R. L. Crawford (2003). 'Metagenomic profiling: Microarray analysis of an environmental genomic library'. *Applied and Environmental Microbiology* 69: 4927–34.

Segré, D. and D. Lancet (2000). 'Composing life'. *EMBO Reports* 1: 217–22.

Seufferheld, M., M. C. F. Viera, F. A. Fuiz, C. O. Rodrigues, S. N. Moreno, and R. Docampo (2003). 'Identification of organelles in bacteria similar to acidocalcisomes of unicellular eukaryotes'. *Journal of Biological Chemistry* 278: 29971–8.

Shapiro, J. A. (1997). 'Multicellularity: The rule, not the exception'. In J. A. Shapiro and M. Dworkin (eds.), *Bacteria as Multicellular Organisms*. New York: Oxford University Press, pp. 14–49.

—— (1998). 'Thinking about bacterial populations as multicellular organisms'. *Annual Review of Microbiology* 52: 81–104.

—— and M. Dworkin (eds.) (1997). *Bacteria as Multicellular Organisms*. New York: Oxford University Press.

Shimkets, L. J. (1999). 'Intercellular signalling during fruiting-body development of Myxococcus xanthus'. *Annual Review of Microbiology* 53: 525–49.

—— and Y. V. Brun (2000). 'Prokaryotic development: Strategies to enhance survival'. In Y. V. Brun and L. J. Shimkets (eds.), *Prokaryotic Development*. Washington, DC: ASM Press, pp. 1–7.

Shiner, E. K., K. P. Rumbaugh, and S. C. Williams (2005). 'Interkingdom signaling: Deciphering the language of acyl homoserine lactones'. *FEMS Microbiology Reviews* 29: 935–47.

Shorter, J. and S. Lindquist (2005). 'Prions as adaptive conduits of memory and inheritance'. *Nature Reviews Genetics* 6: 435–50.

Simpson A. G. B. and A. J. Roger (2004). 'The real "kingdoms" of eukaryotes'. *Current Biology* 14: R693–R696.

Slater J. H. and A. T. Bull (1978). 'Interactions between microbial populations'. In A. T. Bull and P. M. Meadow (eds.), *Companion to Microbiology: Selected Topics for Further Study*. London: Longman, pp. 181–206.

Slater, P. J. B. (1986). 'The cultural transmission of bird song'. *Trends in Ecology & Evolution* 1: 94–7.

Smith, A. ([1776] 1994). *The Wealth of Nations*, ed. E. Cannan. New York: The Modern Library.

Smith, H. O., C. A. Hutchison III, C. Pfannkoch, and J. C. Venter (2003). 'Generating a synthetic genome by whole genome assembly: ΦX174 bacteriophage from synthetic nucleotides'. *Proceedings of the National Academy of Sciences USA* 100: 15440–5.

Smith, M. L., J. N. Bruhn, and J. B. Anderson (1992). 'The fungus *Armillaria bulbosa* is among the largest and oldest living organisms'. *Nature* 356: 428–31.

Snow, C. P. ([1959] 1993). *The Two Cultures*. Cambridge: Cambridge University Press.

Sober, E. and D. S. Wilson (1994). 'A critical review of philosophical work on the units of selection problem'. *Philosophy of Science* 61: 534–55.

—— —— (1998). *Unto Others: The Evolution and Psychology of Unselfish Behavior*. Cambridge, MA: Harvard University Press.

Solomon, J. M. and A. D. Grossman (1996). 'Who's competent and when: Regulation of natural competence in bacteria'. *Trends in Genetics* 12: 150–5.

Solomon, M. (2001). *Social Empiricism*. Cambridge, MA: MIT Press.

Sonea, S. and L. G. Mathieu (2001). 'Evolution of the genomic systems of prokaryotes and its momentous consequences'. *International Microbiology* 4: 67–71.

Sørensen, S. J., M. Bailey, L. H. Hansen, N. Kroer, and S. Wuertz (2005). 'Studying plasmid horizontal transfer in situ: A critical review'. *Nature Reviews Microbiology* 3: 700–10.

Soto, C. and G. P. Saborio (2001). 'Prions: Disease propagation and disease therapy by conformational transmission'. *Trends in Molecular Medicine* 7: 109–14.

Stackebrandt, E., W. Frederisksen, G. M. Garrity, P. A. Grimont, P. Kämpfer, et al. (2002). 'Report of the ad hoc committee for the re-evaluation of the species definition in bacteriology'. *International Journal of Systematic and Evolutionary Microbiology* 52: 1043–7.

Stahl, D. A. and J. M. Tiedje (2002). *Microbial Ecology and Genomics: A Crossroads of Opportunity*. Washington, DC: American Academy of Microbiology.

—— D. J. Lane, G. J. Olsen, and N. R. Pace (1985). 'Characterization of a Yellowstone hot spring microbial community by 5S rRNA sequences'. *Applied and Environmental Microbiology* 49: 1379–84.

Staley, J. T. (1997). 'Biodiversity: Are microbial species threatened?' *Current Opinion in Biotechnology* 8: 340–5.

—— and J. J. Gosink (1999). 'Poles apart: Biodiversity and biogeography of sea ice bacteria'. *Annual Review of Microbiology* 53: 189–215.

—— and A. Konopka (1985). 'Measurements of in situ activities of nonphotosynthetic microorganisms in aquatic and terrestrial habitats'. *Annual Review of Microbiology* 39: 321–46.

Stanier, R. Y. and C. B. Van Niel (1941). 'The main outlines of bacterial classification'. *Journal of Bacteriology* 42: 437–66.

—— —— (1962). 'The concept of a bacterium'. *Archiv für Mikrobiologie* 42: 17–35.

—— M. Doudoroff, and E. A. Adelberg (1957). *The Microbial World*. Englewood Cliffs, NJ: Prentice-Hall.

Stanley, W. M. (1941). 'Some chemical, medical and philosophical aspects of viruses'. *Science* 93: 145–51.

—— (1957). 'On the nature of viruses, cancer, genes, and life: A declaration of dependence'. *Proceedings of the American Philosophical Society* 101: 317–24.

Starmer, C. (2000). 'Developments in non-expected utility: The hunt for a descriptive theory of choice under risk'. *Journal of Economic Literature* 38: 332–82.

Stein, J. L., T. L. Marsh, K. Y. Wu, H. Shizuya, and E. F. DeLong (1996). 'Characterization of uncultivated prokaryotes: Isolation and analysis of a 40-kilobase-pair genome fragment from a planktonic marine archaeon'. *Journal of Bacteriology* 178: 591–9.

Sterelny, K. (1999). 'Species as ecological mosaics'. In R. A. Wilson (ed.), *Species: New Interdisciplinary Essays*. Cambridge, MA: MIT Press, pp. 119–38.

—— (2004). 'Symbiosis, evolvability and modularity'. In G. Schlosser and G. P. Wagner (eds.), *Modularity in Evolution and Development*. Chicago: University of Chicago Press, pp. 490–518.

—— and P. E. Griffiths (1999). *Sex and Death: An Introduction to the Philosophy of Biology*. Chicago: University of Chicago Press.

Stewart, P. S. and J. W. Costerton (2001). 'Antibiotic resistance of bacteria in biofilms'. *The Lancet* 358: 135–38.

Stokes, H. W., A. J. Holmes, B. S. Nield, M. P. Holley, K. M. Nevalainen, et al. (2001). 'Gene cassette PCR: Sequence-independent recovery of entire genes from environmental DNA'. *Applied and Environmental Microbiology* 67: 5240–6.

Stoodley, P., K. Sauer, D. G. Davies, and J. W. Costerton (2002). 'Biofilms as complex differentiated communities'. *Annual Review of Microbiology* 56: 187–209.

Stotz, K. and P. E. Griffiths (2004). 'Genes: Philosophical analyses put to the test'. *History and Philosophy of the Life Sciences* 26: 5–28.

—— P. E. Griffiths, and R. Knight (2004). 'How scientists conceptualize genes: An empirical study'. *Studies in History & Philosophy of Biological and Biomedical Sciences* 35: 647–73.

Stouthamer, R., J. A. J. Breeuwer, and G. D. D. Hurst (1999). '*Wolbachia pipientis*: Microbial manipulator of arthropod reproduction'. *Annual Review of Microbiology* 53: 71–102.

Streit, W. R. and R. A. Schmitz (2004). 'Metagenomics: The key to uncultured microbes'. *Current Opinion in Microbiology* 7: 492–8.

Stroud, B. (1996). 'The charm of naturalism', *Proceedings and Addresses of the American Philosophical Association* 70(2): 43–55.

Sturtevant, A. (2001). *A History of Genetics*. Cold Spring Harbor, NY: Cold Spring Harbor Press.

Summers, W. C. (1991). 'From culture as organism to organism as cell: Historical origins of bacterial genetics'. *Journal of the History of Biology* 24: 171–90.

Suttle, C. A. (2005). 'Viruses in the sea'. *Nature* 437: 356–644.

Szathmáry, E. (2006). 'The origin of replicators and reproducers'. *Philosophical Transactions of the Royal Society B* 361: 1761–76.

Tamas, I., L. M. Klasson, J. P. Sandström, and S. G. E. Andersson (2001). 'Mutualists and parasites: How to paint yourself into a (metabolic) corner'. *FEBS Letters* 498: 135–9.

Teixeira L., A. Ferreira, and M. Ashburner (2008). 'The bacterial symbiont *Wolbachia* induces resistance to RNA viral infections in *Drosophila melanogaster*'. *PLoS Biology* 6(12). doi:10.1371/journal.pbio.1000002

Thiessen, U. and W. Martin (2006). 'The difference between organelles and endosymbionts'. *Current Biology* 16: R1016–R1018.

Thomas, C. M. (2000). 'Paradigms of plasmid organization'. *Molecular Microbiology* 37: 485–91.

—— (2006). 'Transcription regulatory circuits in bacterial plasmids'. *Biochemical Society Transactions* 34: 1072–4.

—— and K. M. Nielsen (2005). 'Mechanisms of, and barriers to, horizontal gene transfer between bacteria'. *Nature Reviews Microbiology* 3: 711–21.

Thong, H.-Y., S.-H. Jee, C.-C. Sun, and R. E. Boissy (2003). 'The patterns of melanosome distribution in keratinocytes of human skin as one determining factor of skin colour'. *British Journal of Dermatology* 149: 498–505.

Thornhill, R. and C. T. Palmer (2000). *A Natural History of Rape: Biological Bases of Sexual Coercion*. Cambridge, MA: MIT Press.

—— and N. W. Thornhill (1992). 'The evolutionary psychology of men's coercive sexuality'. *Behavioral and Brain Sciences* 15: 363–421.

Timmis, J. N., M. A. Ayliffe, C. Y. Huang, and W. Martin (2004). 'Endosymbiotic gene transfer: Organelle genomes forge eukaryotic chromosomes'. *Nature Reviews Genetics* 5: 123–35.

Tooby, J. and L. Cosmides (1992). 'The psychological foundations of culture'. In J. Barkow, L. Cosmides, and J. Tooby (eds.), *The Adapted Mind*. New York: Oxford University Press, pp. 19–136.

Torsvik, V. and L. Øvreås (2002). 'Microbial diversity and function in soil: From genes to ecosystems'. *Current Opinion in Molecular Biology* 5: 240–5.

Travis, C. B. (ed.) (2003). *Evolution, Gender, and Rape*. Cambridge, MA: MIT Press.

Travisano, M. and G. J. Velicer (2004). 'Strategies of microbial cheater control'. *Trends in Microbiology* 12: 72–8.

Traweek, S. (1988). *Beamtimes and Lifetimes: The World of High Energy Physicists*. Cambridge, MA: Harvard University Press.

Treusch, A. H., S. Leininger, S. Schuster, A. Kletzin, H. P. Klenk, and C. Schleper (2005). 'Novel genes for nitrate reduction and Amo-related proteins indicate a role of uncultivated mesophilic crenarchaeota in nitrogen cycling'. *Environmental Microbiology* 7: 1985–95.

Tringe, S. G. and E. M. Rubin (2005). 'Metagenomics: DNA sequencing of environmental samples'. *Nature Reviews Genetics* 6: 805–14.

—— C. von Mering, A. Kobayashi, A. A. Salamov, K. Chen, et al. (2005). 'Comparative metagenomics of microbial communities'. *Science* 308: 554–7.

Trivers, R. (1972). 'Parental investment and sexual selection'. In B. Campbell (ed.), *Sexual Selection and the Descent of Man*. New York: Aldine de Gruyter, pp. 136–79.

True, H. L. and S. L. Lindquist (2000). 'A yeast prion provides a mechanism for genetic variation and phenotypic diversity'. *Nature* 407: 477–83.

Tumpey, T. M., C. F. Basler, P. V. Aguilar, H. Zeng, A. Solorzano, et al. (2005). 'Characterization of the reconstructed 1918 Spanish influenza pandemic virus'. *Science* 310: 77–80.

Twine, R. (2010). 'Genomic natures read through posthumanisms'. In S. Parry and J. Dupré (eds.), *Nature after the Genome*. Sociological Review Monograph Series. Malden, MA: Blackwell Publishing, pp. 175–95.

Tyson, G. W. and J. F. Banfield (2005). 'Cultivating the uncultivated: A community genomics perspective'. *Trends in Microbiology* 13: 411–15.

—— J. Chapman, P. Hugenholz, E. E. Allen, R. J. Ran, et al. (2004). 'Community structure and metabolism through reconstruction of microbial genomes from the environment'. *Nature* 428: 37–43.

Umesaki, Y. and H. Setoyama (2000). 'Structure of the intestinal flora responsible for development of the gut immune system in a rodent model'. *Microbes and Infection* 2: 1343–51.

Underwood, A. J. (1996). 'What is a community?' In D. M. Raup and D. Jablonski (eds.), *Patterns and Processes in the History of Life*. Berlin: Springer-Verlag, pp. 351–67.

Uptain, S. M. and S. Lindquist (2002). 'Prions as protein-based genetic elements'. *Annual Review of Microbiology* 56: 703–41.

Valdivia, R. H. and J. Heitman (2007). 'Endosymbiosis: The evil within'. *Current Biology* 17: R408–R410.

van Haastert, P. J. M. and P. N. Devreotes (2004). 'Chemotaxis: Signalling the way forward'. *Nature Reviews Molecular Cell Biology* 5: 626–34.

van Regenmortel, M. H. V. (2007). 'Virus species and virus identification: Past and current controversies'. *Infection, Genetics and Evolution* 7: 133–44.

van Valen, L. (1976). 'Ecological species, multispecies, oaks'. *Taxon* 25: 233–9.

Vandamme, P., B. Pot, M. Gillis, P. de Vos, K. Kersters and J. Swings (1996). 'Polyphasic taxonomy: A consensus approach to bacterial systematics'. *Microbiological Reviews* 60: 407–38.

Velicer, G. J. (2003). 'Social strife in the microbial world'. *Trends in Microbiology* 11: 330–7.

Venkataraman, S., S. P. Reddy, J. Loo, N. Idamakanti, P. L. Hallenbeck, and V. S. Reddy (2008). 'Structure of Seneca Valley Virus-001: An oncolytic picornavirus representing a new genus'. *Structure* 16: 1555–61.

Venter, J. C., K. Remington, J. F. Heidelberg, A. L. Halpern, D. Rusch, et al. (2004). 'Environmental genome shotgun sequencing of the Sargasso Sea'. *Science* 304: 66–74.

Villarreal, L. P. (2004a). 'Are viruses alive?' *Scientific American* 291: 100–5.

—— (2004b). 'Can Viruses Make Us Human?' *Proceedings of the American Philosophical Society* 148: 296–323.

Vines, G. (1998). 'Hidden inheritance'. *New Scientist* 2162: 27–30.

Visick, K. L. and C. Fuqua (2005). 'Decoding microbial chatter: Cell–cell communication in bacteria'. *Journal of Bacteriology* 187: 5507–19.

Voget, S., C. Leggewie, A. Uesbeck, C. Raasch, K.-E. Jaeger, and W. R. Streit (2003). 'Prospecting for novel biocatalysts in a soil metagenome'. *Applied and Environmental Microbiology* 69: 6235–42.

Wadhams, G. H. and J. P. Armitage (2004). 'Making sense of it all: Bacterial chemotaxis'. *Nature Reviews Molecular Cell Biology* 5: 1024–37.

Waggoner, B. (2001). 'Eukaryotes and multicells: Origins'. *Encyclopedia of Life Sciences* (http://www.els.net).

Wainwright, M. (2003). 'An alternative view of the early history of microbiology'. *Advances in Applied Microbiology* 52: 333–55.

Walsh, D. A. and W. F. Doolittle (2005). 'The real "domains" of life'. *Current Biology* 15: R237–R240.

Walsh, J. B. (1992). 'Intracellular selection, conversion bias, and the expected substitution rates of organelle genes'. *Genetics* 130: 939–46.

Walters, S. M. (1961). 'The shaping of angiosperm taxonomy'. *New Phytologist* 60: 74–84.

Ward, B. B. (2002). 'How many species of prokaryotes are there?' *Proceedings of the National Academy of Science USA* 99: 10234–6.

Ward, D. M. (1998). 'A natural species concept for prokaryotes'. *Current Opinion in Microbiology* 1: 271–7.

Ward, N. (2006). 'New directions and interactions in metagenomics'. *FEMS Microbial Ecology* 55: 331–8.

—— and C. M. Fraser (2005). 'How genomics has affected the concept of microbiology'. *Current Opinion in Microbiology* 8: 564–71.

Warren, G. and W. Wickner (1996). 'Organelle inheritance'. *Cell* 84: 395–400.

Waters, K. (1990). 'Why the anti-reductionist consensus won't survive: The case of classical Mendelian genetics'. *Proceedings of the Biennial Meeting of the Philosophy of Science Association* 1990(1): 125–39.

—— (1994). 'Genes made molecular.' *Philosophy of Science* 61: 163–85.

Watnick, P. and R. Kolter (2000). 'Biofilm, city of microbes'. *Journal of Bacteriology* 182: 2675–9.

Weaver, I. C., N. Cervoni, F. A. Champagne, A. C. D'Alessio, S. Sharma, et al. (2004). 'Epigenetic programming by maternal behavior'. *Nature Neuroscience* 7: 847–54.

Webb, J. S., M. Givskov, and S. Kjelleberg (2003). 'Bacterial biofilms: Prokaryotic adventures in multicellularity'. *Current Opinion in Microbiology* 6: 578–85.

Weber, B. H. (2007). 'Emergence of life'. *Zygon* 42: 837–56.

Webre, D. J., P. M. Wolanin, and J. B. Stock (2003). 'Bacterial chemotaxis'. *Current Biology* 13: R47–R49.

Webster, N. S. and L. L. Blackall (2009). 'What do we really know about sponge-microbial symbioses?' *The ISME Journal* 3: 1–3.

Weeks, A. R., K. T. Reynolds, and A. A. Hoffman (2002). '*Wolbachia* dynamics and host effects: What has (and has not) been demonstrated?' *Trends in Ecology and Evolution* 17: 257–62.

Wegrzyn, G. (2005). 'What does "plasmid biology" currently mean? Summary of the Plasmid Biology 2004 Meeting'. *Plasmid* 53: 14–22.

Weinbauer, M. G. (2004). 'Ecology of prokaryotic viruses'. *FEMS Microbiology Reviews* 28: 127–81.

—— and F. Rassoulzadegan (2004). 'Are viruses driving microbial diversification and diversity?' *Environmental Microbiology* 6: 1–11.

Weissmann, C. (2004). 'The state of the prion'. *Nature Reviews Microbiology* 2: 861–71.

—— M. Enari, P.-C. Klöhn, D. Rossi, and E. Flechsig (2002). 'Transmission of prions'. *Journal of Infectious Diseases* 186 (Suppl. 2): S157–S165.

Wellington, E. M. H., A. Berry, and M. Krsek (2003). 'Resolving functional diversity in relation to microbial community structure in soil: Exploiting genomics and stable isotope probing'. *Current Opinion in Microbiology* 6: 295–301.

Wernegreen, J. J. (2002). 'Genome evolution in bacterial endosymbionts of insects'. *Nature Reviews Genetics* 3: 850–61.

—— (2004). 'Endosymbiosis: Lessons in conflict resolution'. *PLoS Biology* 2: e68.

Werren, J. H. (1997). 'Biology of *Wolbachia*'. *Annual Review of Entomology* 42: 587–609.

Wertz J. E., C. Goldstone, D. M. Gordon, and M. A. Riley (2003). 'A molecular phylogeny of enteric bacteria and implications for a bacterial species concept'. *Journal of Evolutionary Biology* 16: 1236–48.

West Eberhard, M. J. (2003). *Developmental Plasticity and Evolution*. New York: Oxford University Press.

Whitaker, R. J., D. W. Grogan, and J. W. Taylor (2003). 'Barriers isolate endemic populations of hyperthermophilic archaea'. *Science* 301: 976–8.

Whitham, T. G., W. P. Young, G. D. Martinsen, C. A. Gehring, J. A. Schweitzer, et al. (2003). 'Community and ecosystem genetics: A consequence of the extended phenotype'. *Ecology* 84: 559–73.

Whitman, W. B., D. C. Coleman, and W. J. Wiebe (1998). 'Prokaryotes: The unseen majority'. *Proceedings of the National Academy of Science USA* 95: 6578–83.

Whitworth, T. L., R. D. Dawson, H. Magalon, and E. Baudry (2007). 'DNA barcoding cannot reliably identify species of the blowfly genus *Protocalliphora (Diptera: Calliphoridae)*'. *Proceedings of the Royal Society B* 274: 1731–9.

Wickner, R. B., H. K. Edskes, E. D. Ross, M. M. Pierce, U. Baxa, et al. (2004). 'Prion genetics: New rules for a new kind of gene'. *Annual Review of Genetics* 38: 681–707.

—— H. K. Edskes, F. Shewmaker, and T. Nakayashiki (2007). 'Prions of fungi: Inherited structures and biological roles'. *Nature Reviews Microbiology* 5: 611–18.

Wilhelm, S. W. and C. A. Suttle (1999). 'Viruses and nutrient cycles in the sea'. *BioScience* 49: 781–8.

Wilkins, J. S. (2003). 'How to be a chaste species pluralist-realist: The origin of species modes and the synapomorphic species concept'. *Biology and Philosophy* 18: 621–38.

Wilmes, P. and P. L. Bond (2006). 'Metaproteomics: Studying functional gene expression in microbial ecosystems'. *Trends in Microbiology* 14: 92–7.

Wilson, D. S. (1997). 'Altruism and organism: Disentangling the themes of multilevel selection theory'. *American Naturalist* 150 (Supplement): S122–S134.

Wilson, E. O. (1975). *Sociobiology: The New Synthesis*. Cambridge, MA: Harvard University Press.

Wilson, J. A. (2000). 'Ontological butchery: Organism concepts and biological generalizations'. *Philosophy of Science* 67 (Proceedings): S301–S311.

Wilson, R. A. (1999). *Species: New Interdisciplinary Essays*. Cambridge, MA.: MIT Press.

Wimmer, E. (2006). 'The test-tube synthesis of a chemical called poliovirus'. *EMBO Reports* 7 (Special Issue): S3–S9.

Wimpenny, J. (2000). 'An overview of biofilms as functional communities'. In D. G. Allison, P. Gilbert, H. M. Lappin-Scott, and M. Wilson (eds.), *Community Structure and Cooperation in Biofilms*. Cambridge: Cambridge University Press, pp. 1–24.

Winsor, M. P. (2003). 'Non-essentialist methods in pre-Darwinian taxonomy'. *Biology and Philosophy* 18: 387–400.

Wittgenstein, L. ([1953] 2001). *Philosophical Investigations*. Oxford: Blackwell.

Woese, C. R. (1987). 'Bacterial evolution'. *Microbiology Reviews* 51: 221–71.

—— (2005). 'Evolving biological organization'. In J. Sapp (ed.), *Microbial Phylogeny and Evolution: Concepts and Controversies*. Oxford: Oxford University Press, pp. 99–117.

—— and G. G. Fox (1977). 'Phylogenetic structure of the prokaryotic domain: The primary kingdoms'. *Proceedings of the National Academy of Science USA* 11: 5088–90.

—— O. Kandler, and M. L. Wheelis (1990). 'Towards a natural system of organisms: Proposal for the domains Archaea, Bacteria, and Eucarya'. *Proceedings of the National Academy of Science USA* 87: 4576–9.

—— G. J. Olsen, M. Ibba, and D. Söll (2000). 'Aminoacyl-tRNA synthetases, the genetic code, and the evolutionary process'. *Microbiology and Molecular Biology Reviews* 64: 202–36.

Woodward, J. (2003). *Making Things Happen: A Theory of Causal Explanation*. New York: Oxford University Press.

Wright, R. (1994). *The Moral Animal: Evolutionary Psychology and Everyday Life*. New York: Pantheon.

Wu, D., S. C. Daugherty, S. E. Van Aken, G. H. Pai, K. L. Watkins, et al. (2006). 'Metabolic complementarity and genomics of the dual bacterial symbiosis of sharpshooters'. *PLoS Biology* 4: e188.

Wu, L., D. K. Thompson, G. Li, R. A. Hurt, J. M. Tiedje, and J. Zhou (2001). 'Development and evaluation of functional gene arrays for detection of selected genes in the environment'. *Applied and Environmental Microbiology* 67: 5780–90.

Xu, J. (2006). 'Molecular ecology in the age of genomics and metagenomics: Concepts, tools, and recent advances'. *Molecular Ecology* 15: 1713–31.

—— and J. I. Gordon (2003). 'Honor thy symbionts'. *Proceedings of the National Academy of Science USA* 100: 10452–9.

Young, J. M. (2001). 'Implications of alternative classifications and horizontal gene transfer for bacterial taxonomy'. *International Journal of Systematic and Evolutionary Microbiology* 51: 945–53.

Yu, N., M. S. Kruskall, J. J. Yunis, J. H. Knoll, L. Uhl, et al. (2002). 'Disputed maternity leading to identification of tetragametic chimerism'. *New England Journal of Medicine* 346: 1545–52.

Zhang, T., L. C. Buoen, B. E. Seguin, G. R. Ruth, and A. F. Weber (1994). 'Diagnosis of freemartinism in cattle: The need for clinical and cytogenic evaluation'. *Journal of the American Veterinary Medicine Association* 204: 1672–75.

Zhou, J. (2003). 'Microarrays for bacterial detection and microbial community analysis'. *Current Opinion in Microbiology* 6: 288–94.

Zhuravlev, Y. N. and V. A. Avetisov (2006). 'The definition of life in the context of its origins'. *Biogeosciences* 3: 281–91.

Zoetendel, E. G., E. E. Vaughan, and W. M. de Vos (2006). 'A microbial world within us'. *Molecular Microbiology* 59: 1639–50.

Zuckerkandl, E. and L. Pauling (1965). 'Molecules as documents of evolutionary history'. *Journal of Theoretical Biology* 8: 357–66.

Index

sequencing, DNA (*cont.*)
  whole-genome 169
sex *see* sexual reproduction
sexism 245 n.3
sexual assault 86; *see also* rape; consent, sexual
sexual differences, human 245 n.3, 258,
    273, 278
sexual reproduction 76
sexual selection 59
sexuality 250, 258
simplicity (epistemic virtue) 26
simulation 131, 197
single-celled organisms, social 164; *see also*
    unicellular organisms
single nucleotide polymorphisms *see* SNPs
skin
  diseases 234
    fungal infections of amphibian 234 n.4
    microbial communities living on 233, 234
  colour (in relation to race) 268; *see also* race
slime moulds 164; *see also Dictyostelium*
Smith, Adam 65
Snow, C.P. 37, 39
SNPs 270–271; *see also* genetic markers
Sober, Elliott 154
sociability 277
social change (and developmental
    obsolescence) 259
social construction of science 40; *see also* social
    constructivism
social constructivism 40–54
  danger of 54
  symmetry principle 42
social epistemology 43
social norms (and criminality) 46
social sciences
  whether scientific 25
  and human nature 275–294
sociality
  emergence of 226; *see also* collaboration;
    cooperation
  human 277
societies (and human nature) 275–294
sociobiology 275–276
sociology of scientific knowledge 42, 245
somatic cell mutations 109 n.6
souls 21
space, occupation of 22
spatial boundedness (as criterial of life) 223;
    see also life, criteria of
speciation
  insect 151; *see also Wolbachia*
  microbial 78
species 73
  American Society of Microbiology definition
    of 183
  asexual 74

Biological Species Concept 74, 75, 76, 182
classifications of 47–50;
  see also classification; kinds
concepts, monistic 184
fates contingent on human beliefs 78–79
as individuals 47, 78
individuation of 97–98, 182;
  see also boundaries
operational measures 183–184
number of prokaryotic 185
interactions 78
specificity, host *see* host specificity
splicing (RNA) 51, 80
  alternative 138, 251, 264
sponges 233
SSK *see* sociology of scientific knowledge
stability, dynamic 202
standard of living 62–63
'Standard Social Sciences Model' 285
Sterelny, Kim 166 n.17
Stone Age 59, 99, 248; *see also* Pleistocene
  human diet in 246
  origins of human psychology 257
Stotz, Karola 139
stress response 256
  in plants 256
strong emergence 131–132; *see also* emergence;
    emergent capacities; emergent properties
Stroud, Barry 21
structural hierarchy 69; *see also* reduction;
    reductionism
Structure of Scientific Revolutions, The 38
Sturtevant, Alfred 50
subatomic particles, discovery of 27
supernatural powers 37
supernaturalism 21
  pessimism of 23
superorganism 165, 175, 194
supervenience 37, 142
symbiome 165 n.15, 222
symbionts, reduced extracellular 218–220
symbiosis 86, 124–126, 152, 233–235
  and immunology 240
  and vertebrate development 197–198
  evolutionary significance of 223
  in human body 198; *see also* gut, human
symmetry principle (Edinburgh School) 42
  compatible with partial realism 54
syncytium (of placental mammals) 149
synthetic biology 206;
  minimal cell research 219
  synthetic viral genomes 216
systems
  biological *see* biological systems
  causal powers of 92; *see also* downward
    causation; causal influence; causal
    interaction